Advances in Wood Composites

Advances in Wood Composites

Special Issue Editor

Antonios N. Papadopoulos

MDPI • Basel • Beijing • Wuhan • Barcelona • Belgrade • Manchester • Tokyo • Cluj • Tianjin

Special Issue Editor
Antonios N. Papadopoulos
International Hellenic University
Greece

Editorial Office
MDPI
St. Alban-Anlage 66
4052 Basel, Switzerland

This is a reprint of articles from the Special Issue published online in the open access journal *Polymers* (ISSN 2073-4360) (available at: https://www.mdpi.com/journal/polymers/special_issues/wood_Compos).

For citation purposes, cite each article independently as indicated on the article page online and as indicated below:

LastName, A.A.; LastName, B.B.; LastName, C.C. Article Title. *Journal Name* **Year**, *Article Number*, Page Range.

ISBN 978-3-03928-584-6 (Pbk)
ISBN 978-3-03928-585-3 (PDF)

Contents

About the Special Issue Editor

Antonios N. Papadopoulos is a specialist in Wood Science, Chemistry and Technology. He is Associate Professor and Head of the Department of Forestry and Natural Environment at the International Hellenic University. He is an M.Sc and Ph.D holder, both from University of North Wales, Bangor. His M.Sc thesis focused on wood composites and his Ph.D thesis on chemical and thermal modification of wood. His main areas of research interests includes: chemical and thermal modification of wood, nanotechnology and wood science, composites, wood based panels and adhesives. He has published more than 200 peer reviewed papers, books, and book chapters. He is on various Editorial Board of journals in the fields of wood science and technology.

Editorial

Advances in Wood Composites

Antonios N. Papadopoulos

Laboratory of Wood Chemistry and Technology, Department of Forestry and Natural Environment, International Hellenic University, GR-661 00 Drama, Greece; antpap@for.ihu.gr

Received: 26 December 2019; Accepted: 26 December 2019; Published: 30 December 2019

Wood composites are manufactured from a variety of materials. They usually contain the same woods that are used in lumber, but they are combined to make them stronger and more durable. Wood has long been used as a construction material and revered for its strength and natural aesthetics. However, with forests being chopped down across the globe to meet our insatiable demands, it's time to look towards an alternative solution and wood composite may be the answer. Wood composites include a range of different derivative wood products, all of which are created by binding the strands, fibers, or boards of wood together. It's also known as manmade wood, manufactured board or engineered wood, as well as wood-plastic composite (WPC) when using wood fibers and thermoplastics. Similar composite products can also be made from vegetable fibers using lignin-containing materials such as hemp stalks, sugar cane residue, and rye and wheat straw, with chemical additives enabling the integration of polymer and wood flour while helping facilitate optimal processing conditions. They are fixed using adhesives and are engineered to certain specifications, resulting in a material that can have diverse applications. The best part about wood composites is that they can be created using wood waste materials and smaller trees, reducing the need to fell old-growth forests. Composite wood products can be used in a variety of different ways, including both home and industrial construction, and are often used to replace steel for joists and beams in building projects. Their most widespread use, however, is in outdoor deck flooring, but they are also popular for railings, fencing, benches, window and door frames, cladding, and landscaping work. While composite wood can be used in most applications traditionally using solid wood, it is also a popular material for making flat-pack furniture due to its low manufacturing costs and light weight properties.

One of the main advantages of wood composite is that because it is man-made, it can be designed for specific qualities or performance requirements. It can be made into different thicknesses, grades, sizes and exposure durabilities, as well as manufactured to take advantage of the natural strength characteristics of wood (and sometimes results in a greater structural strength and stability than regular wood). As a result, it can be used in a diverse array of applications, from industrial scale to small home projects, and enable more design options without sacrificing structural requirements. Wood composites have also some disadvantages. Some of these, does require more primary energy for its manufacture when compared to solid lumber. Some particle and fiber-based composite woods are also not suitable for outdoor use as they can absorb water and be more prone to humidity-induced warping than solid woods. Another concern regarding wood composites is the adhesives used in their design with some resins releasing toxic formaldehyde in the finished product (particularly those made with urea-formaldehyde bonded products which is one of the cheapest and most common adhesives). The plastic materials often used in the creation of wood composites also have a higher fire hazard when compared to solid wood products, due to their higher chemical heat content and melting properties. With the widespread use of wood composites in the modern world—from panel products to engineered lumber—there is a need to understand their strengths and weaknesses with respect to weathering and decay.

Wood composites have shown very good performance, and substantial service lives when correctly specified for the exposure risks present. Selection of an appropriate product for the job should be accompanied by decisions about the appropriate protection, whether this is by design, by preservative treatment or by wood modification techniques. In this Special Issue, Advances in Wood Composites presents recent progress in enhancing and refining the performance and properties of wood composites by chemical and thermal modification and the application of smart nanomaterials, which have made them a particular area of interest for researchers. In addition, it reviews some important aspects in the field of wood composites, with particular focus on their materials, applications, and engineering and scientific advances, including solutions inspired biomimetrically by the structure of wood and wood composites. This Special Issue, with a collection of 13 original contributions, provides selected examples of recent Advances in Wood Composites.

The main drawbacks of wood, namely susceptibility to biodegradability by microorganisms and a dimensional instability when subjected to a varied moisture content, are mainly due to the cell wall main polymers and, in particular, due to their high abundance of hydroxyl groups (OH) [1–3]. This is solved, so far, either by using imported tropical woods or by using conventional biocides which are usually based in the use of toxic chemicals. In addition, the disposal of the treated wood has caused many restrictions which have to be taken into consideration upon the utilization of conventional chemical treatments. However, both these options are nowadays under political and consumer pressure and alternatives have become an imperative need [2]. This need led to attention to non-biocide treatments, such as the chemical and thermal modification of wood, and inorganic modification by sol-gel technologies.

Acetylation and in situ polymerization are two typical chemical modifications that are used to improve the dimensional stability of lignocellulosic materials, like bamboo. The combination of chemical modification of vinyl acetate (VA) acetylation and methyl methacrylate (MMA) in situ polymerization of bamboo was investigated [4]. Performances of the treated bamboo were evaluated in terms of dimensional stability, wettability, thermal stability, chemical structure, and dynamic mechanical properties. The results revealed that the dimensional stability of bamboo after the combined treatment of VA and MMA was remarkably improved because of the decrease in hydrophilic hydroxyl groups. VA and MMA were mainly grafted onto the surface of the cell walls or in the bamboo cell lumen. From TG analysis, an increase in the extent of pretreatment via chemical modifications resulted in the peak temperatures of the maximum weight loss gradually moving toward the side of higher pyrolysis temperature.

Thermal modification is an ecological and low-cost pretreatment method and it is normally performed at between 160 to 260 °C in a vacuum, nitrogen, air, or oil environments [5,6]. A recent study investigated the relevance between the evolution of chemical structure and the physical and mechanical properties during wood thermal modification processes [7]. Moreover, the volatility of compounds (VOCs) was analyzed using a thermogravimetric analyzer coupled with Fourier transform infrared spectrometry (TGA-FTIR) and a pyrolizer coupled with gas chromatography/mass spectrometer (Py–GC/MS). The designed temperatures for heat treatment were 160, 180, and 200 °C and the durations were 3, 6, and 9 h. The results indicated that the dimensional stability improved markedly due to the reduction of hydrophilic hydroxyl groups. Based on a TGA-FTIR analysis, the small molecular gaseous components were composed of H_2O, CH_4, CO_2, and CO, where H_2O was the dominant component with the highest absorbance intensity. Based on the Py–GC/MS analysis, the VOCs were shown to be mainly composed of acids, aldehydes, ketones, phenols, furans, alcohols, sugars, and esters, where acids were the dominant compounds.

Among the sol-gel technologies, the sol-gel derived wood-inorganic composite (WIC) approaches have received a lot of attention over the last few years [8–10]. These WICs have been proven to be effective in improving the flame retardancy, thermal stability, UV stability, and fungal resistance compared to wood. In an interesting study, methyltrimethoxysilane (MTMOS), methyltriethoxysilane (MTEOS), tetraethoxysilane (TEOS), and titanium (IV) isopropoxide (TTIP) were used as precursor sols to prepare wood-inorganic composites [11]. (WICs) by a sol-gel process, and subsequently, the long-term creep behavior of these composites was estimated by application of the stepped isostress method (SSM). The results revealed that the flexural modulus of wood and WICs were in the range of 9.8–10.5 GPa, and there were no significant differences among them. However, the flexural strength of the WICs was stronger than that of wood. Additionally, based on the SSM processes, smooth master curves were obtained from different SSM testing parameters, and they fit well with the experimental data. These results demonstrated that the SSM was a useful approach to evaluate the long-term creep behavior of wood and WICs. Furthermore, the WICs exhibited better performance on the creep resistance than that of wood, except for the WICMTEOS. The reduction of time-dependent modulus for the WIC prepared from MTMOS was 26% at 50 years, which is the least among all WICs tested. These findings clearly indicate that treatment with suitable metal alkoxides could improve the creep resistance of wood.

Another option to overcome the main drawbacks of wood, namely susceptibility to biodegradability by microorganisms and a dimensional instability when subjected to a varied moisture content, perhaps a more attractive one, is to investigate the potentials that nanotechnology may offer. It is well documented in literature, that the cell wall of wood presents a significant porosity; this porosity is of molecular scale [12]. The main reason of employing nanotechnology in the so called wood science and technology, is the unique characteristic of nano-based materials to penetrate deeply on the wood substrates with an effective way, which in turns results in the alteration altering of its surface chemistry. This subsequently causes an improvement in wood properties [13,14]. Any potential change in the wood properties due to treatment with nanomaterials, is based in the higher interfacial area which is developed due to treatment. This is happened because the amount of particles is significantly reduced to nanoscale. The nanomaterials improve the properties of wood as raw material and alter its original feature to a limited extent. Currently, nano-based materials may be effectively applied to wood through the following distinctive ways: (a) through impregnation of nano-based materials, (b) as a polymeric nanocarriers through impregnation of nano-based materials, and (c) as coatings [14–17].

Bayani et al. [18] investigated the physical and mechanical properties of thermally modified beech wood impregnated with silver nano-suspension and examined their relationship with the crystallinity of cellulose. Specimens were impregnated with a 400 ppm nanosilver suspension. Heat treatment took place in a laboratory oven at three temperatures, namely 145, 165, and 185 °C. Physical properties and mechanical properties of treated wood demonstrated statistically insignificant fluctuations at low temperatures compared to control specimens. On the other hand, an increase of temperature to 185 °C had a significant effect on all properties. As a consequence of the thermal modification, part of amorphous cellulose was changed to crystalline cellulose. At low temperatures an increased crystallinity caused some of the properties to be improved. Change of amorphous cellulose to crystalline cellulose, as well as cross-linking in lignin, partially ameliorated the negative effects of thermal degradation at higher temperatures and therefore, compression parallel to grain and modulus of elasticity did not decrease significantly. Previous studies reported increased thermal conductivity in solid wood and wood-based composites with the addition of nano-metals and nano-minerals [19,20]. The increased thermal conductivity ultimately caused higher degradation in the main wood components and therefore, mechanical properties and pull-off strength decreased in specimens thermally modified at temperature higher than 165 °C. It was concluded that impregnating specimens with silver nano-suspension prior to thermal modification enhanced the effects of thermal modification as a result of improved thermal conductivity [18].

Wollastonite (a silicate mineral, $CaSiO_3$) makes bonds with cell-wall polymers of wood. Formation of these bonds results in significant improvement in physical and mechanical properties in composite panels treated with nanowollastonite suspension [21]. Moreover, high thermal conductivity coefficient of wollastonite, along with its non-combustible mineral nature, was reported to ultimately cause improvement in fire properties [22]. Hassani et al. [23] examined the effect of fortification level of nanowollastonite on UF resin and its effect on the mechanical and physical properties of oriented strand lumber (OSL). Two resin contents were applied, namely 8% and 10%. Nanowollastonite was mixed with the resin at two levels of 10% and 20%. It was found that that the fortification of UF resin with 10% nanowollastonite can be considered as an optimum level. When nanowollastonite content was higher (that is, 20%), higher volume of UF resin was absorbed by wollastonite nanofibers, being left from the process of sticking the strips together. The mechanism involved in the fortification of UF resin with nanowollastonite, which resulted in of thickness swelling values can be attributed to the following two factors: (i) nanowollastonite compounds made active bonds with the cellulose hydroxyl groups, putting them out of reach for the water molecules to make bonds with these, and (ii) high thermal conductivity coefficient of wollastonite improved the transfer of heat to different layers of OSL mat, facilitating better and more complete resin curing. Since nanowollastonite contributes in making bonds between the wood strips and consequently improves physical and mechanical properties, its use can be safely recommended in the OSL production process to improve physical and mechanical properties of the panel.

The adhesives used in the manufacture of wood composites have arguably the most influence on the composites properties. The adhesives influence all aspects of the composites, from their mechanical properties and their ability to perform in wet conditions to their effects on the environment (both localized and the wider environment). Composites industry uses almost solely petrol-based adhesives, such as urea-formaldehyde and melamine-urea-formaldehyde. There are two main drivers to replace these synthetic systems with formaldehyde-free bio-based alternatives; lowered formaldehyde limits and the need for sustainability. Bio-based adhesives can be fully formaldehyde-free and have the additional benefit of increased sustainability. However, to reach 100% bio-based formulations is challenging, and the focus has been on increasing the content of bio-based material with a stepwise approach [24].

The most researched biomaterial for wood panel adhesives is lignin. One of the main uses of lignin residues is lignin-phenol-formaldehyde resins, where lignin is used to partially replace phenol. The phenol replacement amounts are typically below 50%, as the addition of lignin lowers the reactivity of the resin, leading to increased reaction times [25]. The utilization of proper crosslinker is an important factor for developing a lignin-based adhesive that can meet the reaction speed and adhesion strength requirements of the wood panel industry. In a recent study [26], two crosslinkers were tested for ammonium lignosulfonate (ALS)—bio-based furfuryl alcohol (FOH) and synthetic polymeric 4,4′-diphenylmethane diisocyanate (pMDI). The derived ALS adhesives were used for gluing 2-layered veneer samples and particleboards. Differential scanning calorimetry showed a reduction of curing temperature and heat for the samples with crosslinkers. Light microscopy showed that the FOH crosslinked samples had thicker bondlines and higher penetration, which occurred mainly through vessels. Tensile shear strength values of two-layered veneer samples glued with crosslinked ALS adhesives were at the same level as the melamine reinforced urea-formaldehyde (UmF) reference. For particleboards, the FOH crosslinked samples showed a significant decrease in these mechanical properties: internal bond (IB), modulus of elasticity (MOE), modulus of rupture (MOR), and thickness swelling. For pMDI crosslinked samples, these properties increased compared to the UmF. Although the FOH crosslinked ALS samples can be classified as non-added-formaldehyde adhesives, their emissions were higher than what can be expected to be sourced from the particles.

In order to reduce the cost of plywood and save edible resources (wheat flour), a cheap and resourceful clay, sepiolite, was used to modify urea formaldehyde (UF) resin [27]. The performances of filler-filled UF resins were characterized by measuring the thermal behavior, cross section, and functional groups. Results showed that cured UF resin with SEP (sepiolite) formed a toughened fracture surface, and the wet shear strength of the resultant plywood was maximum improved by 31.4%. The tunnel structure of SEP was beneficial to the releasing of formaldehyde, as a result, the formaldehyde emission of the plywood bonded by UF resin with SEP declined by 43.7% compared to that without SEP. This study provided a new idea to reduce the formaldehyde emission, i.e., accelerating formaldehyde release before the product is put into use.

Bekhta and Sedliacik [28], presented the first effort to develop and evaluate composites based on alder veneers and high-density polyethylene (HDPE) film, since it was known that thermoplastic films exhibit good potential to be used as adhesives for the production of veneer-based composites. Three types of adhesives were used: urea-formaldehyde (UF), phenol-formaldehyde (PF), and HDPE film. UF and PF adhesives were used for the comparison. The findings of this work indicate that formaldehyde-free HDPE film adhesive gave values of mechanical properties of alder plywood panels that are comparable to those obtained with traditional UF and PF adhesives, even though the adhesive dosage and pressing pressure were lower than when UF and PF adhesives were used. It was concluded that environmentally-friendly high-density polyethylene-bonded formaldehyde-free alder plywood panels have been successfully produced using thermoplastic polymers as an adhesive.

The majority of wood composites in their final use, contain additives. These may include protective coatings, coatings to improve its aesthetic appearance, preservatives for protection against fire or biological factors like fungi and insects, and even plastics to successfully create new types of products [29]. When wood is used in outdoor applications, its surface is very much exposed to many agents and its protection is an imperative need. Its protection therefore can be achieved with an efficient manner, which includes coating, this can be done with surface or bulk treatment. Coatings, which are currently applied to wood surface can be categorized as air-drying coatings, reaction-curing coatings, water-soluble coatings, stains, and oils and waxes [16,29]. A fast water-based ultraviolet light (UV) curing polyurethane-acrylate (PUA) wood coating was prepared and applied on oak (*Quercus alba* L.) at different coating amounts [30]. The coating amounts affected the coating properties after curing on oak. With the increase of coating amount, the adhesion, hardness, and gloss value of the surface increased to different extents. Meanwhile, the surface of sample became smooth gradually because the voids of the oak were filled. Thus, higher coating amount resulted in better coating properties. However, no significant increase of penetration depth was found. During curing, the hydroxyl groups of the wood reacted with the coating. The optimal parameter in this study was the coating amount of 120 g/m^2, where the adhesion reached 1 (with 0%–5% cross-cut area of flaking along the edges), with the hardness of 2H and the gloss of 92.56°.

Another interesting topic is the preparation of biomorphic porous SiC ceramics from bamboo by combining sol-gel impregnation and carbothermal reduction [31]. This study investigated the feasibility of using bamboo to prepare biomorphic porous silicon carbide (bio-SiC) ceramics through a combination of sol-gel impregnation and carbothermal reduction. The effects of sintering temperature, sintering duration, and sol–gel impregnation cycles on the crystalline phases and microstructure of bio-SiC were investigated. X-ray diffraction patterns revealed that when bamboo charcoal-SiO$_2$ composites (BcSiCs) were sintered at 1700 °C for more than 2 h, the resulting bio-SiC ceramics exhibited significant SiC diffraction peaks. In addition, when the composites were sintered at 1700 °C for 2 h, scanning electron microscopy micrographs of the resulting bio-SiC ceramic prepared using a single impregnation cycle showed the presence of SiC crystalline particles and nanowires in the cell wall and cell lumen of the carbon template, respectively. However, bio-SiC prepared using three and five repeated cycles of sol-gel impregnation exhibited a foam-like microstructure compared with that prepared using a single impregnation cycle. Similar results were observed for the bio-SiC ceramics prepared from bamboo-SiO$_2$ composites (BSiCs). Accordingly, bio-SiC ceramics can be directly and successfully prepared from BSiCs, simplifying the manufacturing process of SiC ceramics.

The fast growing wood polymer composites (WPC) sector, and the closely related natural fiber composites (NFC) sector, presents many new opportunities for utilizing wood as well as natural fibers or agricultural residues, as filler or reinforcement in polymer profiles and moldings. The field combines wood processing techniques for fiber or filler preparation with polymer science and engineering, and a range of polymer processing techniques including extrusion, injection molding, compression molding, and pultrusion. Wood plastic composites (WPCs) incorporating graphene nano-platelets (GNPs) were fabricated using hot-pressed technology to enhance thermal and mechanical behavior [32]. The influences of thermal filler content and temperature on the thermal performance of the modified WPCs were investigated. The results showed that the thermal conductivity of the composites increased significantly with the increase of GNPs fillers, but decreased with the increase of temperature. Moreover, thermogravimetric analysis demonstrated that coupling GNPs resulted in better thermal stability of the WPCs. The limiting oxygen index test also showed that addition of GNPs caused good fire retardancy in WPCs. Incorporation of GNPs also led to an improvement in mechanical properties as compared to neat WPCs. Through a series of mechanical performance tests, it could be concluded that the flexural and tensile moduli of WPCs were improved with the increase of the content of fillers.

A novel wood-plastic composite (WPC) lumber has shown potential to replace high-density polyethylene (HDPE) lumber in the construction of aquacultural geodesic spherical cage structures [33]. Six HDPE and six WPC assemblies, which are representative of typical full-size cage dimensions, were fabricated by bolting pairs of triangular panel components made with connected struts. Half of the panel assemblies had a plastic-coated steel wire mesh to simulate the actual restraint in field applications of the cages. The objective of the research was to characterize the structural performance of the panel assemblies under compressive loading. To determine the critical buckling load for the panel assemblies made from WPC and HDPE struts with and without wire mesh, Southwell's method was implemented. A two-dimensional (2D) linear finite element analysis model was developed to determine axial forces in the struts of the panel assembly for the applied load and boundary conditions. This model was used to determine strut compressive forces corresponding to the Southwell's method buckling load and the experimental failure load. It was found that the wire mesh increased the load capacity of both HDPE and WPC panel assemblies by a factor of two. The typical failure mode of the panels made from HDPE lumber struts, with and without wire mesh, was buckling of the struts, whereas the failure mode of the WPC panels, with and without wire mesh, was fracture at the notched section corresponding to the location of the bolts. The load capacity of the panel assemblies made from WPC lumber struts was three times and 2.5 times higher than the load capacity of the panel assemblies made from HDPE lumber struts with and without wire mesh, respectively.

Virgin thermoplastics, such as polypropylene (PP), polyethylene (PE), polystyrene (PS), and polyvinyl chloride (PVC), are commonly used as matrices for manufacturing WPCs [with wood flour, but using recycled plastic as a matrix was only reported in a few studies [34]. A paper in this Special Issue examined the effects of selected types of thermoplastics on the physical and mechanical properties of polymer-triticale boards [35]. The investigated thermoplastics differed in their type (polypropylene (PP), polyethylene (PE), polystyrene (PS)), form (granulate, agglomerate) and origin (native, recycled). The resulting five-ply boards contained layers made from different materials (straw or pine wood) and featured different moisture contents (2%, 25%, and 7% for the face, middle, and core layers, respectively). Thermoplastics were added only to two external layers, where they substituted 30% of straw particles. This study demonstrated that, irrespective of their type, thermoplastics added to the face layers most favorably reduced the hydrophobic properties of the boards, i.e., thickness, swelling, and V100, by nearly 20%. The bending strength and modulus of elasticity were about 10% lower in the experimental boards than in the reference ones, but still within the limits set out in standard for P7 boards according to EN 312.

As a conclusion or a general remark regarding the future of wood composites, it can be said that the ability of wood composites to be tailored to specific uses, together with their strength properties and affordability, makes them a viable solution to reducing the need for solid wood. They have been successfully applied in all forms of building, from small home projects to industrial construction work, and as technology surrounding their manufacture only advances, the future looks bright.

Conflicts of Interest: The author declares no conflict of interest.

References

1. Rowell, R.M. *Handbook of Wood Chemistry and Wood Composites*, 2nd ed.; CRC Press, Taylor and Francis Group: Boca Raton, FL, USA, 2012.
2. Hill, C.A.S. *Wood Modification—Chemical, Thermal and other Processes*; John Wiley and Sons Ltd.: West Sussex, UK, 2006.
3. Papadopoulos, A.N. Chemical modification of solid wood and wood raw materials for composites production with linear chain carboxylic acid anhydrides: A brief review. *BioResources* **2010**, *5*, 499–506.
4. Huang, S.; Jiang, Q.; Yu, B.; Nie, Y.; Ma, Z.; Ma, L. Combined chemical modification of bamboo material prepared using vinyl acetate and methyl methacrylate: Dimensional stability, chemical structure, and dynamic mechanical properties. *Polymers* **2019**, *11*, 1651. [CrossRef]
5. Hill, C.A.S.; Abdul Khalil, H.P.S. Effect of fiber treatments on mechanical properties of coir or oil palm fiberreinforced polyester composites. *J. Appl. Polym. Sci.* **2000**, *78*, 1685–1697. [CrossRef]
6. Boonstra, M.J.; Tjeerdsma, B. Chemical analysis of heat treated softwoods. *Holz Roh Werkst.* **2006**, *64*, 204–211. [CrossRef]
7. Xu, J.; Zhang, Y.; Shen, Y.; Li, C.; Wang, Y.; Ma, Z.; Sun, W. New perspective on wood thermal modification: Relevance between the evolution of chemical structure and physical-mechanical properties, and online analysis of release of VOCs. *Polymers* **2019**, *11*, 1145. [CrossRef]
8. Shabir Mahr, M.; Hübert, T.; Stephan, I.; Bücker, M.; Militz, H. Reducing copper leaching from treated wood by sol-gel derived TiO_2 and SiO_2 depositions. *Holzforschung* **2013**, *67*, 429–435. [CrossRef]
9. Wang, X.; Liu, J.; Chai, Y. Thermal, mechanical, and moisture absorption properties of wood-TiO_2 composites prepared by a sol-gel process. *Bioresources* **2012**, *7*, 893–901.
10. Pries, M.; Mai, C. Fire resistance of wood treated with a cationic silica sol. *Eur. J. Wood Prod.* **2013**, *71*, 237–244. [CrossRef]
11. Hung, K.-C.; Wu, T.-L.; Wu, J.-H. Long-term creep behavior prediction of sol-gel derived SiO_2_ and TiO_2-wood composites using the stepped isostress method. *Polymers* **2019**, *11*, 1215. [CrossRef]
12. Hill, C.A.S.; Papadopoulos, A.N. A review of methods used to determine the size of the cell wall microvoids of wood. *J. Inst. Wood Sci.* **2001**, *15*, 337–345.
13. Wegner, T.H.; Jones, P. Advancing cellulose-based nanotechnology. *Cellulose* **2005**, *13*, 115–118. [CrossRef]
14. Papadopoulos, A.N.; Bikiaris, D.N.; Mitropoulos, A.C.; Kyzas, G.Z. Nanomaterials and chemical modification technologies for enhanced wood properties: A review. *Nanomaterials* **2019**, *9*, 607. [CrossRef] [PubMed]
15. Teng, T.; Arip, M.; Sudesh, K.; Lee, H. Conventional technology and nanotechnology in wood preservation: A review. *BioResources* **2018**, *13*, 9220–9252. [CrossRef]
16. Papadopoulos, A.N.; Taghiyari, H.R. Innovative wood surface treatments based on nanotechnology. *Coatings* **2019**, *9*, 866. [CrossRef]
17. Taghiyari, H.; Esmailpour, A.; Papadopoulos, A. Paint pull-off strength and permeability in nanosilver-impregnated and heat-treated beech wood. *Coatings* **2019**, *9*, 723. [CrossRef]
18. Bayani, S.; Taghiyari, H.R.; Papadopoulos, A.N. Physical and mechanical properties of thermally-modified beech wood impregnated with silver nano-suspension and their relationship with the crystallinity of cellulose. *Polymers* **2019**, *11*, 1535. [CrossRef]
19. Taghiyari, H.R.; Schimdt, O. Nanotachnology in wood based composite panels. *Int. J. Bio-Inorg. Hybrid Nanomater.* **2014**, *3*, 65–73.
20. Taghiyari, H.R.; Enayati, A.; Gholamiyan, H. Effects of nano-silver impregnation on brittleness, physical and mechanical properties of heat-treated hardwoods. *Wood Sci. Technol.* **2012**, *47*, 467–480. [CrossRef]
21. Taghiyari, H.R.; Nouri, P. Effects of nano-wollastonite on physical and mechanical properties of medium-density fiberboard. *Maderas Cienc. Tecnol.* **2015**, *17*, 833–842. [CrossRef]
22. Esmailpour, A.; Taghiyari, H.R.; Majidi, R.; Morrell, J.J.; Mohammad-Panah, B. Nano-wollastonite to improve fire retardancy in medium-density fiberboard (MDF) made from wood fibers and camel-thorn. *Wood Mater Sci. Eng.* **2019**. [CrossRef]
23. Hassani, V.; Taghiyari, H.R.; Schmidt, O.; Maleki, S.; Papadopoulos, A.N. Mechanical and physical properties of oriented strand lumber (OSL): The effect of fortification level of nanowollastonite on UF resin. *Polymers* **2019**, *11*, 1884. [CrossRef] [PubMed]

24. Hemmilä, V.; Adamopoulos, S.; Karlsson, O.; Kumar, A. Development of sustainable bio-adhesives for engineered wood panels: A review. *RSC Adv.* **2017**, *7*, 38604–38630. [CrossRef]

25. Danielson, B.; Simonson, R. Kraft lignin in phenol formaldehyde resin. Part 1. Partial replacement of phenol by kraft lignin in phenol formaldehyde adhesives for plywood. *J. Adhes. Sci. Technol.* **1998**, *12*, 923. [CrossRef]

26. Hemmilä, V.; Adamopoulos, S.; Hosseinpourpia, R.; Ahmed, S.A. Ammonium lignosulfonate adhesives for particleboards with pMDI and furfuryl alcohol as crosslinkers. *Polymers* **2019**, *11*, 1633. [CrossRef]

27. Li, X.; Gao, Q.; Xia, C.; Li, J.; Zhou, X. Urea formaldehyde resin resultant plywood with rapid formaldehyde release modified by tunnel-structured sepiolite. *Polymers* **2019**, *11*, 1286. [CrossRef]

28. Bekhta, P.; Sedliačik, J. Environmentally-friendly high-density polyethylene-bonded plywood panels. *Polymers* **2019**, *11*, 1166. [CrossRef]

29. Sanberg, D. Additives in wood products-today and future development. In *Environmental Impacts of Traditional and Innovative Forest-based Bioproducts*; Kutnar, A., Muthu, S.S., Eds.; Springer Science and Business Media: Singapore, 2016.

30. Wang, J.; Wu, H.; Liu, R.; Long, L.; Xu, J.; Chen, M.; Qiu, H. Preparation of a fast water-based UV cured polyurethane-acrylate wood coating and the effect of coating amount on the surface properties of Oak (*Quercus alba* L.). *Polymers* **2019**, *11*, 1414. [CrossRef]

31. Hung, K.-C.; Wu, T.-L.; Xu, J.-W.; Wu, J.-H. Preparation of biomorphic porous SiC ceramics from bamboo by combining Sol-Gel impregnation and carbothermal reduction. *Polymers* **2019**, *11*, 1442. [CrossRef]

32. Zhang, X.; Zhang, J.; Wang, R. Thermal and mechanical behavior of wood plastic composites by addition of graphene nanoplatelets. *Polymers* **2019**, *11*, 1365. [CrossRef]

33. Alrubaie, M.A.A.; Gardner, D.J.; Lopez-Anido, R.A. Structural performance of HDPE and WPC lumber components used in aquacultural geodesic spherical cages. *Polymers* **2020**, *12*, 26. [CrossRef]

34. Najafi, S.K.; Tajvidi, M.; Hamidina, E. Effect of temperature, plastic type and virginity on the water uptake of sawdust/plastic composites. *Holz Roh Werkst.* **2007**, *65*, 377–382. [CrossRef]

35. Mirski, R.; Bekhta, P.; Dziurka, D. Relationships between thermoplastic type and properties of polymer-triticale boards. *Polymers* **2019**, *11*, 1750. [CrossRef] [PubMed]

 *polymers*

Article

Structural Performance of HDPE and WPC Lumber Components Used in Aquacultural Geodesic Spherical Cages

Murtada Abass A. Alrubaie [1,*], Douglas J. Gardner [2] and Roberto A. Lopez-Anido [1]

[1] Department of Civil and Environmental Engineering, Advanced Structures and Composites Center, University of Maine, Orono, ME 04469, USA; rla@maine.edu
[2] School of Forest Resources, Advanced Structures and Composites Center, University of Maine, Orono, ME 04469, USA; douglasg@maine.edu
* Correspondence: murtada.alrubaie1@maine.edu

Received: 27 November 2019; Accepted: 19 December 2019; Published: 21 December 2019

Abstract: Based on previous research, a novel wood–plastic composite (WPC) lumber has shown potential to replace high-density polyethylene (HDPE) lumber in the construction of aquacultural geodesic spherical cage structures. Six HDPE and six WPC assemblies, which are representative of typical full-size cage dimensions, were fabricated by bolting pairs of triangular panel components made with connected struts. Half of the panel assemblies had a plastic-coated steel wire mesh to simulate the actual restraint in field applications of the cages. The objective of the research was to characterize the structural performance of the panel assemblies under compressive loading. To determine the critical buckling load for the panel assemblies made from WPC and HDPE struts with and without wire mesh, Southwell's method was implemented. A two-dimensional (2D) linear finite element analysis model was developed to determine axial forces in the struts of the panel assembly for the applied load and boundary conditions. This model was used to determine strut compressive forces corresponding to the Southwell's method buckling load and the experimental failure load. It was found that the wire mesh increased the load capacity of both HDPE and WPC panel assemblies by a factor of two. The typical failure mode of the panels made from HDPE lumber struts, with and without wire mesh, was buckling of the struts, whereas the failure mode of the WPC panels, with and without wire mesh, was fracture at the notched section corresponding to the location of the bolts. The load capacity of the panel assemblies made from WPC lumber struts was three times and 2.5 times higher than the load capacity of the panel assemblies made from HDPE lumber struts with and without wire mesh, respectively.

Keywords: buckling; WPC; HDPE; Southwell's method; finite element analysis; Abaqus; aquacultural; structural analysis; wood; plastic; composite

---

1. Introduction

Aquaculture cages for fish farming are made in different ways. Unlike other types of aquaculture fish cage structures, the Aquapod Net Pen cage is a rigid-frame geodesic spherical cage structure. The cage structure is comprised of individual triangular panels made from high-density polyethylene (HDPE) lumber (strut), and these panels are fastened to each other to form the geodesic spherical shape of the cage [1]. These triangular panels are covered with wire mesh netting, which is affixed to the struts of the panels by mechanical fastening (stapling). Five panels (P1, P2, P3, P4, and P5) are the main structural components to construct the spherical shape of the Aquapod cage, as shown in Figure 1. These triangular panel components are designed to contribute to facile construction by reducing the time and manpower required to construct the cage structure. The cage structure is utilized in a fully submerged situation; except for cleaning, where it will be partially (30%) exposed to the air [1].

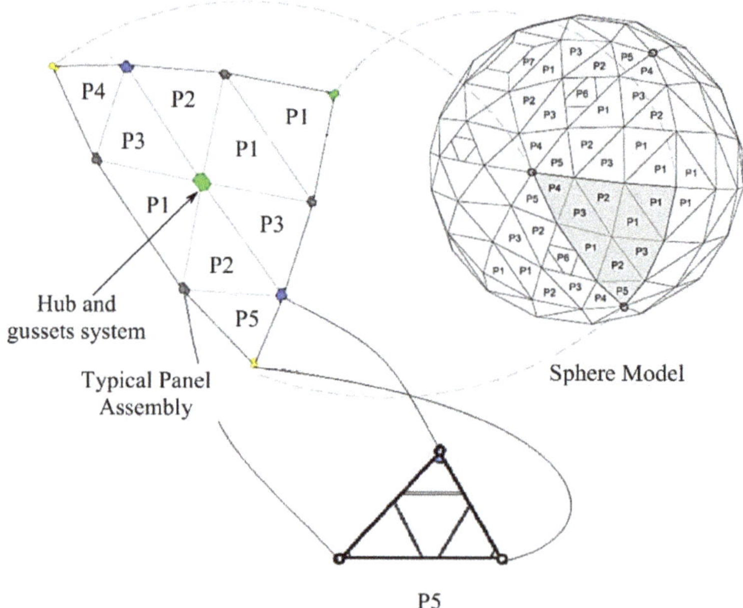

Figure 1. Details of connected struts in the panels and the connected panels to form the cage faces, types of hubs, and the types of the panels of the geodesic spherical cage structure with an approximate diameter of 21 m [1,2].

Because of the increased demand for aquaculture structures that have useful features (high volume capacity, rigid frame, and durable structure for up to 10 years) compared with other aquaculture cage structures, the geodesic spherical cage structure has been constructed in different volume capacities and diameters since 2006 [3] to its most recent product, Aquapod 4700, with a volume capacity of 4700 m^3 (dia. of approximately 21 m) [3]. The cages function under submersion without any apparent problems from the marine exposure. However, damage to cage structures was reported in the Gulf of Mexico in 2015 [4], when the structures were exposed during cleaning to destructive surface waves during a hurricane. InnovaSea Systems, Inc. decided to explore a better material option to replace the HDPE lumber (struts). An extruded wood–plastic composite (WPC) lumber made from high-strength styrenic copolymer and thermally modified wood flour appears to be a promising alternative to replace HDPE, attributable to its desirable mechanical properties compared with HDPE lumber. For instance, the elastic modulus of the WPC lumber is approximately five time the elastic modulus of the HDPE lumber. Although WPCs have been investigated for structural applications [5–12], the performance of WPCs requires evaluation to be utilized for marine applications, where the material will be exposed to the combined effect of temperature and saltwater immersion. Nevertheless, it is very difficult to evaluate the structural performance of the full-scale structure of the cage structure that is made from HDPE or WPC lumber in such combined conditions (temperature and water immersion). WPCs are similar to other thermoplastic materials that exhibit viscoelastic behavior, hence their time-dependent behavior was investigated. Previous studies have focused on the time-dependent behavior of WPCs [5,13–17]. For the WPC lumber that is considered as an alternative to HDPE lumber in the construction of the cage structure, Alrubaie et al. conducted a 180-day creep experiment to compare the time-dependent behavior of HDPE and WPC lumber under similar conditions (temperature 23 ± 2 °C and relative humidity 50% ± 5%). Furthermore, the short-term time-dependent behavior of the WPC lumber (that is considered an alternative for the HDPE in the construction of the aquacultural geodesic spherical cage structure) was investigated and modeled under the synergistic effect of elevated temperature and water immersion [18,19]. InnovaSea Systems, Inc. conducted mechanical testing at the Advanced Manufacturing Center, University of Maine, Orono in 2006 to evaluate the buckling capacity of the full-scale fastened panels (with and without netting) of Aquapod A4700 made from glass bar-reinforced HDPE lumber.

The objective of the research presented here was to experimentally investigate and characterize the buckling capacity of two connected panels made from WPC and HDPE struts, with and without metallic mesh, to compare the structural performance of WPC lumber in aquacultural panel structures.

In this study, 24 triangular panels with struts length: 965, 1003, and 1321 mm were tested in compression along the longest strut. Twelve panels were made from WPC struts and twelve panels were made from HDPE struts. Six of each of these panels were constructed with plastic-coated steel wire mesh with 38.1 mm openings and 2.8 mm thickness of the steel wire [20]. A set of two panels were connected using three steel galvanized bolts with a diameter of 12.7 mm and two steel-galvanized square washers with dimension of 51 mm to each bolt. Four types of panels were experimentally investigated in the buckling experiment: WPC panels without the steel mesh condition (WPC-panel), WPC panels with the steel mesh condition (WPC-M-panel), HDPE panels without steel mesh (HDPE-panel), and HDPE panels with steel mesh (HDPE-M-panel). Three sets were tested for each panel type [21].

2. Experimental

2.1. Materials

The commercially available HDPE lumber with cross section dimensions (b = 140 mm and h = 38.1 mm) was provided by InnovaSea Systems Inc. (Morrill, Maine, USA) and used in the manufacture of the HDPE triangular panels with and without steel wire mesh. The struts (made from WPC and HDPE lumber) of the triangular panels were connected to each other to form the panel via a triangular blocks (gussets) made from the same HDPE of the struts. The WPC lumber with cross section dimensions b = 139 mm and h = 33.5 mm was produced using a twin-screw Davis-Standard Woodtruder^{TM} (Orono, Maine, USA) in the Advanced Structures and Composites Center at the University of Maine's Orono campus and were used in the manufacture of the WPC triangular panels with and without steel wire mesh. The WPC lumber examined is based on a patent-pending formulation that combines a thermally modified wood flour that was produced at a sawmill in Uimaharju, Finland and a high-strength styrenic copolymer system in an equivalent weight ratio to each of the two constituents. Flexural and compression tests were conducted to obtain the flexural and compressive properties of the WPC and HDPE lumber. Modulus of elasticity, flexural strength, and compressive strength of both materials were the properties obtained from these two tests and reported in Table 1. The number of samples was five samples for each test (flexure and compression). The length of the samples (L) of WPC and HDPE lumber tested in flexure was 545 and 620 mm to be in agreement with the required length (L) to depth (h) ratio in ASTM D6109 to be 16. Similarly, the length (L) of the WPC and HDPE lumber samples tested in compression was 160 and 183 mm to be in agreement with the ASTM D198 [22] of having the ratio of length (L) to the radius of gyration (r) to be less than 17. Prior to the testing of the WPC and HDPE lumber, the samples were conditioned (temperature 23 ± 2 °C and relative humidity 50% ± 5%) in accordance with ASTM D618 [23] for 96 h and tested under the same conditions in a climate-controlled mechanical laboratory at the Advanced Structures and Composites Center at the University of Maine, Orono, Maine. Figures 2 and 3 show representative specimens of WPC and HDPE lumber tested in accordance with ASTM D6109 and ASTM D6108 [24] to obtain the modulus of elasticity, and the flexural and compressive strength, respectively. Unlike the WPC lumber, HDPE lumber did not exhibit failure in either the flexure or compression test. Thus, the flexural and compressive strengths were selected as the stress values that corresponded to the 3% strain in the stress–strain relationship in accordance with ASTM D6109 [24] and ASTM D6108 [25], respectively.

Table 1. Mechanical properties of wood–plastic composite (WPC) and high-density polyethylene (HDPE) lumber used as struts of the geodesic components of the Aquapod net pen geodesic spherical cage structure.

Type of Test	Obtained Properties	WPC	HDPE
Flexure (ASTM D6109)	Modulus of elasticity (E)/GPa	4.3 ± 0.3	0.9 ± 0.03
	Modulus of rupture (strength)/MPa	41.2 ± 4.5	14.1 ± 0.7
Compression (ASTM D6108)	Modulus of elasticity (E)/GPa	4.3 ± 1.0	0.8 ± 0.1
	Strength/MPa	50.8 ± 1.9	8.8 ± 0.1

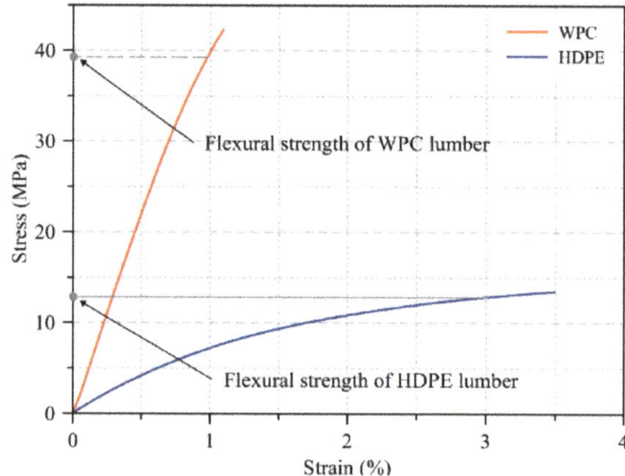

Figure 2. Typical stress versus strain relationship to obtain the elastic modulus and the flexural strength of WPC and HDPE lumber in accordance with ASTM D6109.

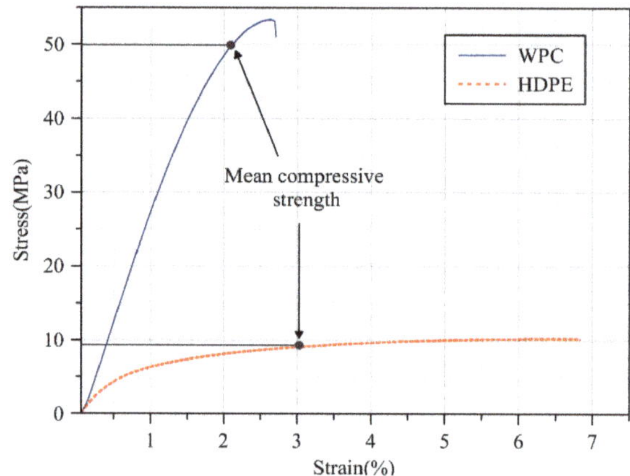

Figure 3. Typical stress versus strain relationship to obtain the elastic modulus and the compressive strength of WPC and HDPE lumber in accordance with ASTM D6108.

2.2. Equipment and Test Setup

The panel manufacture was conducted by InnovaSea Systems, Inc. The fixture to test the connected panels in compression was manufactured at the Advanced Structures and Composite Center, University of Maine, Orono, Maine. Figure 4 shows the connected triangular panels and the test fixture. An Instron (Norwood, MA, USA) test frame with a load cell capacity of 1334 kN was used. The data acquisition system (DAQ) with a written labview software was used to collect: the applied axial load (negative *Y*-direction in Figure 4), the axial displacement (of the actuator, the in-plane displacement of the struts), the in-plane displacement (negative and positive direction of *X*-axis as shown in Figure 4), and the out-of-plane displacement (negative and positive direction of *Z*-axis as shown in Figure 4). The tests were conducted in displacement control with a crosshead speed of 1 mm/min.

Figure 4. (**Left**) the test frame and the buckling test setup, (**right**) the WPC- and HDPE-connected triangular panels with and without metallic mesh.

2.3. Degrees of Freedom of the Supports System of the Triangular Panels

The two-dimensional (2D) free body diagram shows the degree of freedom at each support as shown in Figure 5 and Table 2.

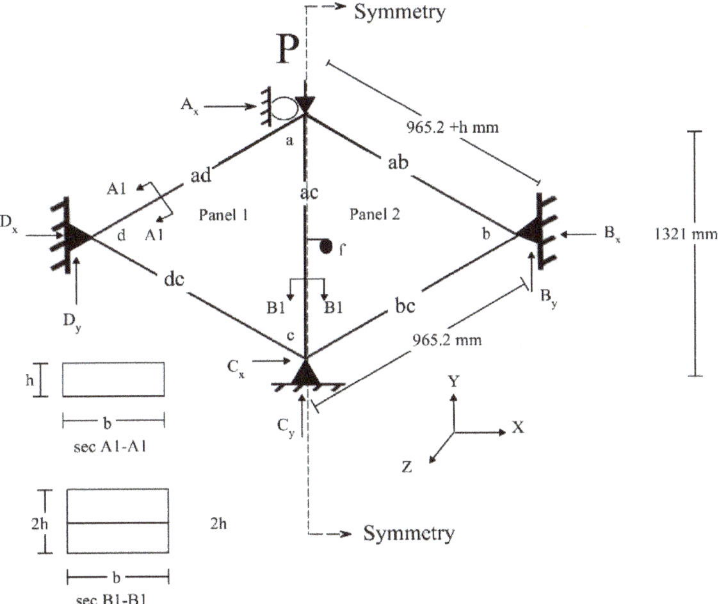

Figure 5. Schematic of the two-dimensional (2D) free body diagram of the tested connected (bolted) WPC and HDPE triangular panels with and without mesh.

Table 2. Two-dimensional degrees of freedom of the triangular panels during the buckling experiment.

Supports Boundary Conditions (Free = 0, Fixed = 1)	Supports			
	a	b	c	d
U_x	1	1	1	1
U_y	0	1	1	1
θ_z	0	0	0	0

3. Results and Discussion

The relationship between the applied buckling load and lateral deflection of the middle vertical strut ac at point f (Figure 5) for the connected components (panels) made from WPC and HDPE struts are reported in Figure 6. The buckling capacity of the panels made from WPC struts was three times the buckling capacity of the panels made from HDPE struts. The steel wire mesh contributed to the increased buckling capacity of the panels. A 2D finite element (FE) analysis model provided a useful assessment of the multiplier factor (α) that can be used in the computation of the reactions and the member of forces under different values of applied loads. Table 3 reports the values of α. Table 4 reports the average maximum buckling load of the HDPE and WPC strut-connected panels at each condition (with and without metallic mesh) and their corresponding type of failure.

Table 3. Reactions and member forces computed from the 2D finite element (FE) linear analyses obtained from applying unit load on panel 1 (Figure 5 at point a) of the geodesic spherical cage structure for four sample types: WPC with steel mesh (WPC-M)-panel, WPC-panel, HDPE with steel mesh (HDPE-M)-panel, and HDPE-panel.

Reactions and Member Forces (N)	Condition of the Panels Made from WPC Struts		Condition of the Panels Made from HDPE Struts	
	M-Panel	Panel	M-Panel	Panel
A_x	−0.4	−0.414	−0.37	−0.414
C_x	−0.004	0.001	0.01	0.001
C_y	0.66	0.64	0.7	0.64
D_x	0.404	0.413	0.38	0.413
D_y	0.34	0.36	0.32	0.36
F_{ad}	−0.54	−0.55	−0.54	−0.55
F_{ac}	−0.60	−0.64	−0.54	−0.64
F_{cd}	−0.04	0	−0.1	0

The buckling failure of the cage components (panels) made from WPC struts was the dominant type of failure at one of the regions of the galvanized bolts that connect the two panels, causing a net section failure at the region where the bolts were located, whereas, no such net section failure was noticed at the buckling failure occurred for the panels made from HDPE lumber. This is attributable to the brittle behavior of the high wood-flour content WPC lumber compared with the HDPE plastic lumber. Table 4 summarizes the types of failure of the structural components of the cage structure with and without metallic mesh. Table 5 summarizes the implementation of the multiplier load factor (α) to compute the allowable member force of strut ac based on the buckling load values obtained from Southwell's method [26]. Southwell's method can be summarized by creating a plot based on the relationship between: the ratio of the lateral displacement (deflection) (Δ) over the applied buckling load (P), and the lateral displacement (Δ). If this relationship can be described by a linear relationship, then the inverse of the slope of this line represents the critical buckling load (P_{cr}) and the buckling mode is global. This critical load does not account for imperfections or mode of interactions.

Table 4. The experimental maximum buckling load and the failure type and occurrence sequence in the structural components of the geodesic spherical cage structure.

Connected Triangular Panels		Experimental Buckling Load for Each Panel (kN)			Failure Type and Location of Occurrence in the Component	
Struts Material	Mesh Condition M-panel = with Mesh Panel = without Mesh	P1	P2	P3	M-Panel	Panel
WPC	M-panel	294	207	270	Struts Buckling failure (X-axis) and metallic mesh buckling (Z-axis)	Buckling failure (X-axis)
WPC	Panel	120	111	131		
HDPE	M-panel	85	92	74	Metallic mesh buckling (Z-axis) followed by struts buckling (X-axis)	Buckling failure (X-axis) in the struts
HDPE	panel	54	55	35		

Table 5. The buckling load of the member (strut) ac based on multiplying the multiplier value α obtained from the 2D FE linear analyses by the value of critical load obtained from Southwell's method.

Connected Panels	Experimental	Southwell's Method	Euler's Method	Allowable Buckling Load
	Max. Average Load (kN)		$\frac{4\pi^2 EI}{l^2}$	
Material and Mesh Condition	F_{ac}	$F_{ac\text{-}Southwell\text{-}critical}$ (kN)	$F_{ac\text{-}Euler\text{-}critical}$ (kN)	F_{ac} (kN)
WPC-M-panel	129	116	NA	70
WPC-panel	60	61	43	39
HDPE-M-panel	42	44	NA	24
HDPE-panel	24	27	14	17

Regarding the connected components (panels) made from WPC lumber (strut) without metallic mesh, the buckling failure tended to be abrupt after reaching the maximum applied load, as shown in Figure 6A. A similar pattern of the failure propagation was observed in the panels made from HDPE lumber without metallic mesh, the panels showed propagated deformation after reaching the maximum applied load without an abrupt failure, as shown in Figure 6B. However, panel number three (HDPE-panel-3), as shown in Figure 6B, exhibited a different load-lateral deflection curve. The panels (the middle strut ac) started deforming with the propagation of the applied load. This can be attributed to the geometry of the panels or to an eccentricity that developed while the load was imposed to the panel.

Regarding the failure behavior of the panels made from WPC with metallic mesh, the metallic mesh contributed to an increase in the buckling capacity (maximum applied load) approximately three times of the buckling capacity (maximum applied load) of the panels made without the metallic mesh. However, as regards to improving the ductility of the panels, the metallic mesh did not contribute to improving the ductility of the panels made from WPC struts. The panels showed a lateral deformation smaller than 2 mm before reaching the maximum applied load and then experiencing abrupt failure. Moreover, the failure mode (of the panels made from WPC struts with mesh) in the struts did not change from the failure mode of the panels without the metallic mesh, which is the net section failure at the connected struts attributable to the buckling in the X-axis at the strut ac (Figure 6C). Figure 7 shows the failure modes of the panels made from HDPE and WPC lumber for the four different cases.

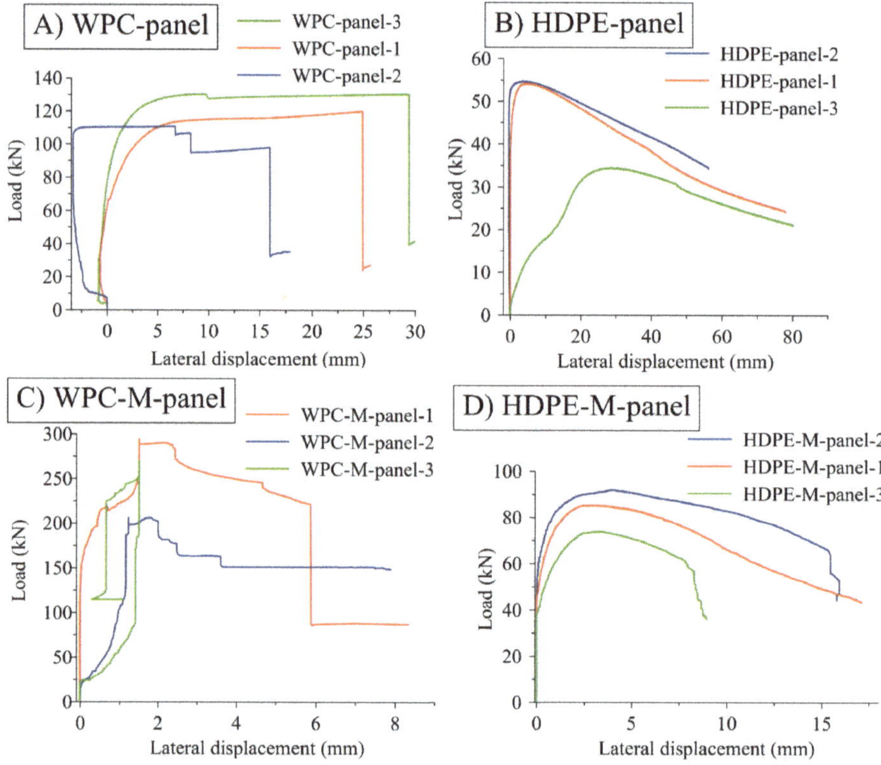

Figure 6. The relationship between the applied buckling load and lateral mid-span deflection of the vertical strut *ac* in the panels made from WPC and HDPE struts, and with and without metallic mesh.

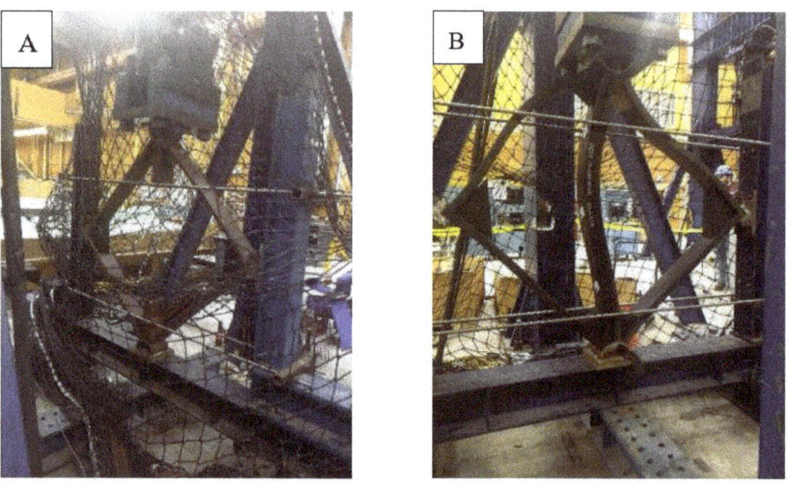

Figure 7. *Cont.*

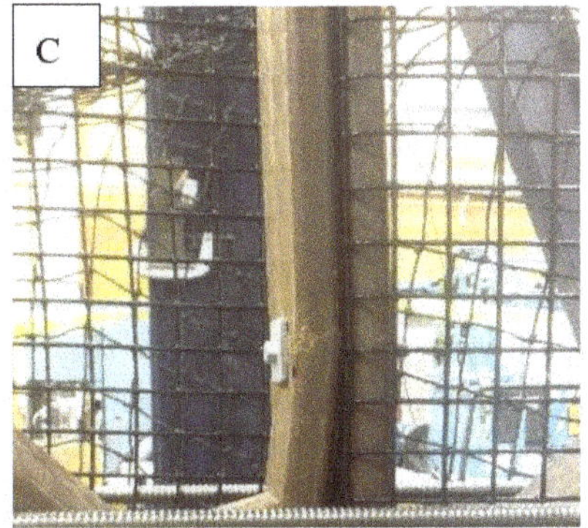

Figure 7. Failure modes of the panels made from HDPE and WPC lumber for the four different cases; (**A**) net section failure at the middle strut *ac* at the location of the bolt connection of the panels made from WPC without metallic mesh, (**B**) buckling mode failure of the strut *ac* of the panels made from HDPE without metallic mesh, (**C**) net section failure mode of panels made from WPC struts with metallic mesh, and (**D**) buckling failure mode of the panels made from HDPE struts with metallic mesh.

The metallic mesh is affixed to the WPC and HDPE struts by staples on the perimeter of the struts (on the width b of the cross section of the struts (sections A1-A1 and B1-B1) in Figure 5). The staples have shown good resistance to the applied load and in developing the buckling capacity of the panels with the metallic mesh. This can be observed with the small lateral displacement of the panels (with metallic mesh) made from WPC and HDPE struts of 1.5 and 3.4 mm, corresponding to the maximum buckling capacity, respectively. Whereas, the lateral displacements for the same type of panels without metallic mesh were 20.1 and 12.5 mm, respectively.

3.1. Structural Analysis of the Tested Structural Components (Panels) of the Geodesic Spherical Cage Structure

To compute the reaction and the section forces at the supports and the struts of the connected, respectively, 2D (two-dimensional) finite element (FE) linear elastic analyses models were conducted to the four types of the test panels (WPC-M-panel, WPC-panel, HDPE-M-panel, and HDPE-panel) using commercially available software Abaqus/CAE with the following assumptions:

1. Based on the symmetry of the connected panels, panel 1 in Figure 5 was used on the 2D FE model to compute the member forces and the support reactions.

2. The supports at points b, c, and d were assumed to be as pin supports (vertical (*Y*-axis) and horizontal (*X*-axis) movement restriction), whereas, point a was assumed to be as s roller support (horizontal (*X*-axis) movement restriction). This assumption was made based on the design of the fixture used in the experiment and the ability of the structure to have rotation at the points a, b, c, and d.

3. A slender beam element B23 (cubic beam in plane) was chosen from the available types of beam elements available in the used commercial software Abaqus and was used in the 2D linear finite element (FE) analysis of the structural components of the geodesic spherical cage structure. The selection was made based on the assumption that both the struts of the components and the metallic mesh are slender even some of the beams have a slender ratio (span (l)/ radius of gyration (r)) less than 200. This assumption eliminated the need to have the values of Poisson's ratio of the materials of the components (steel of the metallic mesh, WPC struts, and HDPE struts), i.e., the elastic moduli were the required input for the mechanical properties of the materials in the 2D FE linear analysis model.

4. Regarding the connected panels with metallic mesh, the metallic mesh was modeled as vertical and horizontal beam elements [type B23] (each wire mesh modeled as a beam) spaced 38.1 mm from each other and has a circular cross-section with a diameter of 3 mm to each beam. The geometry and

the space of the wire mesh was implemented based on the specification of the metallic wire mesh, Aquamesh®, used in the manufacture of the structural panels of the geodesic spherical cage structure.

5. The elastic moduli of the WPC and HDPE lumber used in the structural analysis were obtained from the 4-point bending test conducted on specimens with a span to depth ratio of 16 to be 4430 and 930 MPa as reported in Table 1, respectively. The elastic modulus of the wire mesh was assumed to be the elastic modulus of steel, E_{steel} = 200 GPa.

6. The 2D FE model was conducted to investigate the response of the structure in the linear region. Thus, the values of the applied load were assumed to be a unit load (1 N) to be applied to the structure. The computed member forces and reactions at the supports represented a multiplier coefficient that can be used to compute the reactions and member forces at any value of the applied load.

The reaction values at the supports were computed from the 2D FE model. Furthermore, the member forces were computed for the tested panels in the four cases, to provide an understanding to the distribution of the applied load through the struts of the panels. However, the strut *cd* for the panels made from HDPE or WPC lumber without metallic mesh had no member force. Whereas, the metallic mesh contributed into distributing the applied load among the struts; *ac*, *cd*, and *ad*. Furthermore, the value of the member force varied along the length of the strut attributable to the presence of the metallic mesh. The maximum values of member forces of the struts of the panels made from HDPE and WPC struts with metallic mesh are shown and reported in Table 3 and Figure 8, respectively.

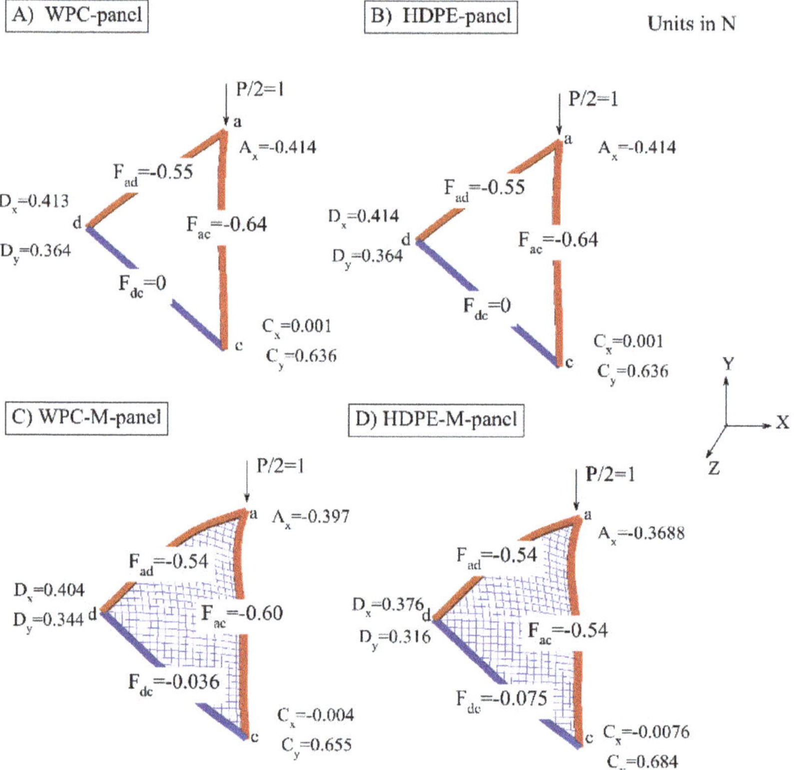

Figure 8. Reactions and member forces in N units of panel 1 of the geodesic spherical cage structure obtained from 2D FE linear analyses; (**A**) WPC-panel, (**B**) HDPE-panel, (**C**)WPC-M-panel, and (**D**) HDPE-M-panel.

3.2. Implementation of Southwell's Method to Determine the Critical Load

To investigate the critical load mode for the four cases of the panels made from WPC and HDPE struts in the cases of metallic wire mesh and without wire mesh, Southwell's method was implemented. The method was implemented on the relationships reported in Figure 9 after modifying the relationship to include the load vs later deflection only at the limit of the maximum applied load (i.e., the data points after the maximum applied load has not been considered in the application of Southwell's method). By using linear regression to obtain the slope of the equation of the line, hence, the inverse of the slope of the line represents the value of P_{cr}. Figure 9 shows the implication of Southwell's method on the four sample panels (WPC-panel, WPC-M-panel, HDPE-panel, and HDPE-M-panel) and the obtained slope of each tested set of panels. Based on the linear relationship between Δ/P versus Δ, the critical load can be determined.

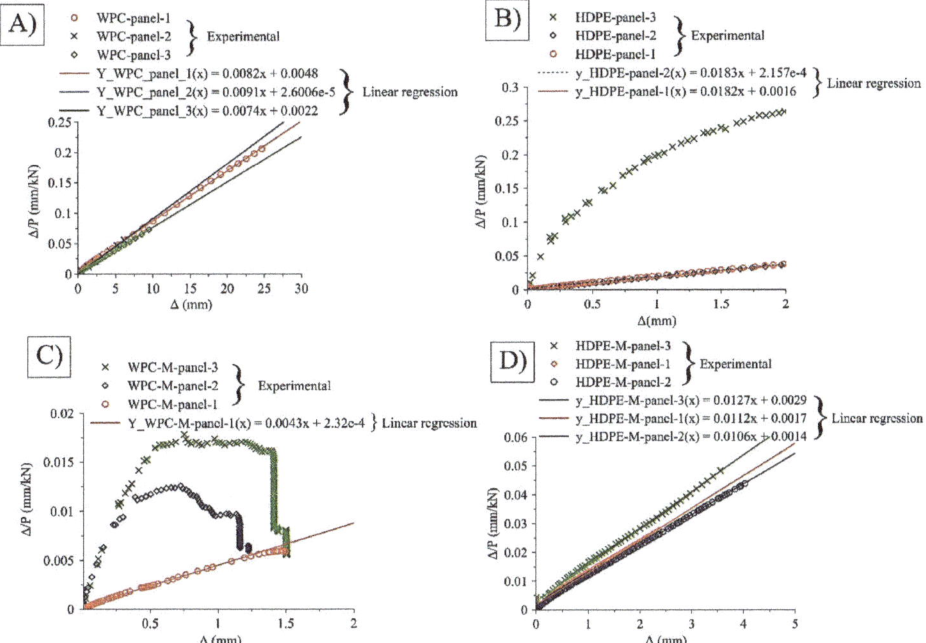

Figure 9. Application of Southwell's method to obtain the critical buckling load of the structural panels of the cage structure made from: (**A**) WPC-panel, (**B**) HDPE-panel, (**C**) WPC-M-panel, and (**D**) HDPE-M-panel.

4. Conclusions

1. The buckling behavior of the structural components of the geodesic spherical cage structure made from HDPE and WPC lumber was experimentally investigated and characterized. The buckling capacity (load) of the triangular panels made from WPC struts and with mesh was 256.81 kN, whereas the buckling capacity of the same type of panels made from HDPE struts was 83.80 kN. Furthermore, the buckling capacity of the panels made from WPC struts and without steel mesh was 120.42 kN, which was 2.5 times the buckling capacity of the same condition of panels but made from HDPE struts.

2. The metallic mesh contributed into distributing the member forces through the struts: *ac*, *ca* and *da* of the component (panel). Whereas, the panels without metallic mesh experienced strut (*cd*) without member force (Figure 8).

3. The structural analyses conducted on the triangular components (panels) of the geodesic spherical cage structure, the compression, and flexure tests have shown that the failure occurrence in the triangular components (panels) was attributable to bending in the struts (Table 1).

4. Attributable to the brittleness behavior of WPC lumber (50 wt. % wood flour) compared with the ductile behavior of HDPE lumber (100 wt. % plastic), an abrupt failure to panels made from WPC was observed in the experiments.

5. According to the linear structural analysis, the short struts in the connected panels (struts *bc* and strut *cd*) did not carry load values and this contributed to the buckling occurrence to be initiated in the longest strut (*ac*) and then at the shorter struts (*ab* and *ad*) for the panels made from WPC and HDPE lumber and without metallic mesh.

6. As a containment aquaculture structure system in open ocean environments, it is preferable to have a structural material (strut) that shows an indication prior to failure, or to defect without breakage, than an abrupt failure, so that the member can be replaced properly.

7. Attributable to the viscoelastic behavior and the brittleness behavior of the WPC in this study, it is preferable to consider using the WPC lumber in structural applications where the applied load should be at a low level compared with strength of the WPC, to avoid the abrupt failure of the structural member during the service life of the structure.

8. The finite element analyses (Figure 8) conducted in this study by applying a unit load considered a useful tool that can be used to compute the reactions and the member forces in the struts of a similar test setup subjected to different values of loading. This analyses also help to compute the reactions and the member forces in the struts for similar test setup panels but in a different scale.

9. The findings of this study that the loading capacity of the connected panels with metallic mesh is twice the loading capacity of the panels made without metallic mesh for both HDPE and WPC struts, is considered a powerful tool to minimize the computational efforts in the design and the analysis of similar structures with and without metallic mesh. Thus, the structures can be analyzed by ignoring the metallic mesh and then can be multiplied by a factor of two to consider the effect of the metallic mesh.

Author Contributions: For research articles with several authors, a short paragraph specifying their individual contributions must be provided. The following statements should be used "conceptualization, M.A.A.A. and D.J.G.; methodology, M.A.A.A.; software, M.A.A.A.; validation, M.A.A.A., D.J.G. and R.A.L.-A.; formal analysis, M.A.A.A.; investigation, M.A.A.A. and D.J.G.; resources, D.J.G.; data curation, M.A.A.A.; writing—original draft preparation, M.A.A.A.; writing—review and editing, M.A.A.A., D.J.G., and R.A.L.-A.; visualization, M.A.A.A., D.J.G., and R.A.L.-A.; supervision, D.J.G.; project administration, D.J.G.; funding acquisition, D.J.G.", please turn to the CRediT taxonomy for the term explanation. Authorship must be limited to those who have contributed substantially to the work reported. All authors have read and agreed to the published version of the manuscript.

Funding: This research received no external funding.

Acknowledgments: This project was supported by the USDA National Institute of Food and Agriculture, Hatch (or McIntire-Stennis, Animal Health, etc.) project number ME0-41809 through the Maine Agricultural and Forest Experiment Station. Maine Agricultural and Forest Experiment Station Publication Number 3680. The work described in this document was conducted at the Advanced Structures and Composites Center at the University of Maine, Orono, Maine (USA). The University of Maine research reinvestment funds (RRF) Seed Grant entitled 'Development of structural wood plastic composite timber for innovative marine application' and the United Stated Department of Agriculture (USDA)—the agricultural research service (ARS) Funding Grant Number (58-0204-6-003) have provided the financial support for this project. The wood–plastic composite is based on a patent-pending formulation that has the publication number (WO2018/142314). The thermally modified wood fiber used in this research is supplied by Stora Enso (Finland).

Conflicts of Interest: The authors declare no conflict of interest.

References

1. Vandenbroucke, K.; Metzlaff, M. Abiotic Stress Tolerant Crops: Genes, Pathways and Bottlenecks. In *Sustainable Food Production*; Springer: New York, NY, USA, 2013; pp. 1–17.
2. Page, S.H. Aquapod Systems for Sustainable Ocean Aquaculture. In *Sustainable Food Production*; Springer: New York, NY, USA, 2013; pp. 223–235.
3. InnovaSea Systems, Inc. A4700 Bridle System in Grid Mooring Cell. 2016. Available online: https://www.innovasea.com (accessed on 13 September 2017).
4. InnovaSea Systems, Inc. *Report on Structural Damage to A4800 AquaPod*; Building Design & Structural Engineering Consultants: Morrill, ME, USA, 2015.
5. Alvarez-Valencia, D.; Dagher, H.J.; Davids, W.G.; Lopez-Aodio, R.A.; Gardner, D.J. Structural performance of wood plastic composite sheet piling. *J. Mater. Civ. Eng.* **2010**, *22*, 1235–1243. [CrossRef]
6. Balma, D.A. Evaluation of Bolted Connections in Wood Plastic Composites. Master's Thesis, Washington State University, Pullman, WA, USA, December 1999.
7. Brandt, W.C.; Fridley, K.J. Load-duration behavior of wood-plastic composites. *J. Mater. Civ. Eng.* **2003**, *15*, 524–536. [CrossRef]
8. Brandt, W.C.; Fridley, K.J. Effect of load rate on flexural properties of wood-plastic composites. *Wood Fiber Sci.* **2007**, *35*, 135–147.
9. Dura, M.J. Behavior of Hybrid Wood Plastic Composite-Fiber Reinforced Polymer Structural Members for Use in Sustained Loading Applications. Master's Thesis, The University of Maine, Orono, ME, USA, 2005.
10. Haiar, K.J. Performance and Design of Prototype Wood-Plastic Composite Sections. Master's Thesis, Washington State University, Pullman, WA, USA, May 2000.
11. Slaughter, A.E. Design and Fatigue of a Structural Wood-Plastic Composite. Master's Thesis, Washington State University, Pullman, WA, USA, 2006.
12. Tamrakar, S. Effect of Strain Rate and Hygrothermal Environment in Wood Plastic Composite Sheet Piles. Electronic Theses and Dissertations. 2011. Available online: https://digitalcommons.library.umaine.edu/etd/1576 (accessed on 21 December 2019).
13. Chang, F.-C. Creep Behaviour of Wood-Plastic Composites. Ph.D. Thesis, University of British Columbia, Vancouver, BC, Canada, 2011.
14. Hamel, S.E. Modeling the Time-Dependent Flexural Response of Wood-Plastic Composite Materials. Ph.D. Thesis, University of Alaska Anchorage, Anchorage, AK, USA, January 2011.
15. Pooler, D.J.; Smith, L.V. Nonlinear viscoelastic response of a wood–plastic composite including temperature effects. *J. Thermoplast. Compos. Mater.* **2004**, *17*, 427–445. [CrossRef]
16. Sandeep, T.; Lopez-Anido, R.A.; Kiziltas, A.; Gardner, D.J. Time and temperature dependent response of a wood–polypropylene composite. *Compos. Part A Appl. Sci. Manuf.* **2011**, *42*, 834–842.
17. Alrubaie, M.A. Investigating the Time-dependent and the Mechanical Behavior of Wood Plastic Composite Lumber Made from Thermally Modified Wood in the Use of Marine Aquacultural Structures. Electronic Theses and Dissertations. 2019. Available online: https://digitalcommons.library.umaine.edu/etd/3026 (accessed on 15 March 2019).
18. Alrubaie, M.A.A.; Lopez-Anido, R.A.; Gardner, D.J.; Tajvidi, M.; Han, Y. Experimental investigation of the hygrothermal creep strain of wood–plastic composite lumber made from thermally modified wood. *J. Thermoplast. Compos. Mater.* **2019**. [CrossRef]
19. Alrubaie, M.A.A.; Lopez-Anido, R.A.; Gardner, D.J.; Tajvidi, M.; Han, Y. Modeling the hygrothermal creep behavior of wood plastic composite (WPC) lumber made from thermally modified wood. *J. Thermoplast. Compos. Mater.* **2019**. [CrossRef]
20. Caccese, V. *Structural Testing of Various Configurations for the AquaPod Net Pen*; Advanced Manufacturing Center: Orono, ME, USA, 2006; p. 40.
21. Tangent Technologies LLC. Polyforce Sturctural Recycled Plastic Lumber. [Cited 2018 November; Mechanical Propertied of the HDPE Polyforce Lumber]. 2015. Available online: http://tangentusa.com/wp-content/uploads/2016/01/PolyForce_DataSheet_01_20_16.pdf (accessed on 10 Feberuary 2019).
22. ASTM International. *Standard Test Methods of Static Tests of Lumber in Structural Sizes, D198-15*; ASTM International: West Conshohocken, PA, USA, 2015.

23. ASTM International. *Standard Practice for Conditioning Plastics for Testing, D618-13*; ASTM International: West Conshohocken, PA, USA, 2013.
24. ASTM International. *Standard Test Methods for Flexural Properties of Unreinforced and Reinforced Plastic Lumber and Related Products, D6109-13*; ASTM International: West Conshohocken, PA, USA, 2013.
25. ASTM International. *Standard Test Method for COmpressive Properties of Plastiv Lumber and Shapes, D6108-19*; ASTM International: West Conshohocken, PA, USA, 2019.
26. Barbero, E.; Tomblin, J. Euler buckling of thin-walled composite columns. *Thin Walled Struct.* **1993**, *17*, 237–258. [CrossRef]

Article

Mechanical and Physical Properties of Oriented Strand Lumber (OSL): The Effect of Fortification Level of Nanowollastonite on UF Resin

Vahid Hassani [1], Hamid R. Taghiyari [1,*], Olaf Schmidt [2], Sadegh Maleki [3] and Antonios N. Papadopoulos [4,*]

1   Wood Science and Technology Department, Faculty of Materials Engineering and New Technologies, Shahid Rajaee Teacher Training University, Tehran 14115, Iran; hassanivahid1988@gmail.com
2   Department of Wood Biology, University of Hamburg, Leuschnerstr. 91, 21031 Hamburg, Germany; olaf.Schmidt@uni-hamburg.de
3   Department of Wood Science and Technology, Faculty of Natural Resources, Tarbiat Modares University, Tehran 14115, Iran; s.maleki33@gmail.com
4   Department of Forestry and Natural Environment, International Hellenic University, Laboratory of Wood Chemistry and Technology, GR-661 00 Drama, Greece
*   Correspondence: htaghiyari@sru.ac.ir (H.R.T.); antpap@teiemt.gr (A.N.P.)

Received: 16 October 2019; Accepted: 10 November 2019; Published: 14 November 2019

Abstract: The aim of this work is to investigate the effect of the fortification level of nanowollastonite on urea-formaldehyde resin (UF) and its effect on mechanical and physical properties of oriented strand lumbers (OSL). Two resin contents are applied, namely, 8% and 10%. Nanowollastonite is mixed with the resin at two levels (10% and 20%). It is found that the fortification of UF resin with 10% nanowollastonite can be considered as an optimum level. When nanowollastonite content is higher (that is, 20%), higher volume of UF resin is left over from the process of sticking the strips together, and therefore is absorbed by wollastonite nanofibers. The mechanism involved in the fortification of UF resin with nanowollastonite, which results in an improvement of thickness swelling values, can be attributed to the following two main factors: (i) nanowollastonite compounds making active bonds with the cellulose hydroxyl groups, putting them out of reach for bonding with the water molecules and (ii) high thermal conductivity coefficient of wollastonite improving the transfer of heat to different layers of the OSL mat, facilitating better and more complete resin curing. Since nanowollastonite contributes to making bonds between the wood strips, which consequently improves physical and mechanical properties, its use can be safely recommended in the OSL production process to improve the physical and mechanical properties of the panel.

Keywords: oriented strand lumber (OSL); nanowollastonite; mechanical and physical properties; UF resin

---

1. Introduction

Oriented strand lumber (OSL) is a structural panel with consistent properties from one unit to another, which is capable of handling large loads. OSL is made by aligning long strands of wood in parallel and binding them together using adhesives, pressure, and heat. It replaces softwood timber in some residential building applications, but because it can attain dimensions not possible for a single piece of wood, it has additional applications in nonresidential construction. OSL is also used for industrial purposes such as furniture manufacturing [1].

A major difference in the performance of structural panels compared to solid wood is the greater thickness swell that occurs when the panels are exposed to relative humidity and/or direct contact with water. This is due to the higher pressure required to consolidate the panel mat. The issue of improving the thickness swell of panels like particleboard, fiberboard, and oriented strand board (OSB) has been a topic of interest for many researchers [2–4]. Briefly, there are several treatment methods that can be divided into three different means of application: pretreatment, post-treatment, and production technology. The first group includes methods that involve treatments applied to furnish before panel hot pressing, such as particle presteaming and chemical or thermal modification of particles [2–7]. The second group comprises methods applied in the consolidated panels, and thermal treatment is the most common [8,9]. Last, production technology methods involve those related to resin content improvement, mat-forming type, platen and press temperature, and water repellent application [10–17]. These days, the application of water repellents is a common practice. However, a major challenge in the manufacture of water-resistant, wood-based panels is identifying compatible combinations of water-repellent chemicals and adhesives. Waxes are commercially used to improve the water repellency of wood-based panels [18]. Oils, such as silixane systems and acrylic elastomeric coating systems, have been used to improve the weather-proofing of siding panels [19,20]. The use of silanes, silicones, and siloxanes is well established for glass fiber-reinforced plastic composites, but their use in wood-based panels is rather limited mainly due to their water insolubility and tendency to form silica deposits [21].

An attractive science, nanotechnology seems to have remarkable potential to create products of a new generation with enhanced properties [21,22]. The change in material properties is primarily due to the large interfacial area that is developed per unit of volume, since the level of added particles is reduced to nanometers. Nanomaterials enhance the properties of the original material, show great compatibility with traditional materials, and cause limited alteration of their original features [23,24]. Their use in wood has the objective of improving its physical and mechanical properties and its durability against microorganisms, since it is generally acceptable that nanosized metals and minerals interact with bacterial elements, gradually leading to cell death [23–26] or even to the disruption of enzyme function [25–27]. An excellent review of the application of nanotechnology on wood science was recently performed [21,28].

The approach of this work is to look at ways of improving the dimensional stability of OSL through nanotechnology. In particular, the aim of this study is to investigate the effect of the fortification level of nanowollastonite on urea-formaldehyde resin (UF), and its effect on mechanical and physical properties of oriented strand lumbers.

Wollastonite (a silicate mineral, $CaSiO_3$) makes bonds with cell-wall polymers of wood [29,30]. The adsorption energy of nanowollastonite (NW) on cellulose surface was reported to be as high as -6.6 eV, formed between Ca in NW and hydroxyl groups in cellulose chains [29]. In another study, the optimal adsorption distance of 2.5 Å in $OH_{cellulose}$ Ca was reported as the most important factor in the formation of the strong bond [31]. The formation of these bonds resulted in significant improvement in physical and mechanical properties in composite panels treated with NW suspension [32]. Moreover, the high thermal conductivity coefficient of wollastonite, along with its noncombustible mineral nature, was reported to ultimately cause an improvement in fire properties [30]. In addition, NW acted as a reinforcement agent in polyvinyl resin, significantly improving shear bond strength [33]. However, the application of NW in OSL panels has not yet been tested. Therefore, in the present research project nanowollastonite is used at two consumption levels to find out its effects on the physical and mechanical properties of OSL produced with two resin contents (8% and 10%).

2. Materials and Methods

2.1. Panel Production

Poplar strips were prepared from 15-year-old poplar trees (*Populus nigra*) cut from Khoy city, located in Azarbayjan Sharghi Province, Iran. The mean density of the poplar logs was 0.42 g/cm^3. The logs were first peeled and dried to a final moisture content of 6%. They were then stripped using stripper equipment produced by Iran-Randeh Co. (Tehran, Iran). Dimensions of the strips were

150 mm × 20 mm × 1 mm. The length of the strips was parallel to the longitudinal direction of the logs. Strips were kept in an oven for 48 h at 50 °C and put in sealed plastic bags to prevent moisture absorption from the air. UF, with 200–400 cP in viscosity, 47 s of gel time, and 1.277 g/cm^3 in density and containing 62% solids, was purchased from Iran Choob Co. (Ghazvin, Iran). As a hardener, 2% ammonium chloride was mixed with the resin before applying it to the strips. Ammonium chloride is a common and effective hardener used for accelerating UF resin curing. Ammonium chloride was chosen for consistency with the industry sector, as most wood-composite manufacturing factories in Iran use ammonium chloride as a hardener, although its use is regulated in some countries due to potential health hazards. The main effect of ammonium chloride on UF resin curing is that it catalyzes the reactants in UF-resin systems. The pH value and the gel time of UF resins decrease with increasing catalyst and resin solid contents and decreasing pH [34,35]. The resin was sprayed on the strips in a rotary drum, and the strips were then manually set in the forming. OSL panels were produced with two resin contents, namely, 8% and 10%. Mats were hot pressed for 10 min. The temperature of the hot-press platens was 170 °C. The pressure of the hot press was 50 kg/cm^2. Density of all OSL panels was 0.8 g/cm^3. Dimensions of panels were 45 cm × 45 cm with 16 mm in thickness. For each treatment, five replicates were produced. After manufacture, the boards were conditioned at 25 °C and 45% relative humidity. Values for mechanical properties, namely, internal bond (IB), modulus of rupture (MOR), modulus of elasticity (MOE), tension parallel to grain, shear strength parallel to grain, and impact bending were then determined according to procedures defined in the American standard for particleboards (ANSI A208.1-1998) [36]. ASTM standards were selected for the determination of mechanical properties because they are more stringent than the EN standards, in terms of mechanical properties. However, values for physical properties, namely, thickness swelling (TS) and water absorption (WA), were determined according to procedures defined in the European Union standards (EN 317-1993) [37] since they are less stringent than the ANSI standards, bearing in mind that UF resin was used as a binder. In any case, it is not the scope of this study to look at the data in light of industry standards.

2.2. Nanowollastonite Application

Nanowollastonite gel was produced in cooperation with Mehrabadi Mfg. Co. in Tehran, Iran. The size range of wollastonite nanofibers was measured as 30–110 nm. Specifications of wollastonite combination are indicated in Table 1. NW was mixed with the urea formaldehyde resin for 30 min by a magnetic stirrer for each load of the drum mixer. The mixture was then sprayed onto the wood fibers in a drum mixer 50 cm in diameter. Consumption level of wollastonite gel was 10% and 20% based on the dry weight of wood strips. The flow diagram of the experimental procedure is depicted in Figure 1.

Table 1. Composition of the nanowollastonite gel used [22,23,25].

Nanowollastonite Compounds	Content by Mass (%)
CaO	39.77
SiO$_2$	46.96
Al$_2$O$_3$	3.95
Fe$_2$O$_3$	2.79
TiO$_2$	0.22
K$_2$O	0.04
MgO	1.39
Na$_2$O	0.16
SO$_3$	0.05
Water	The rest

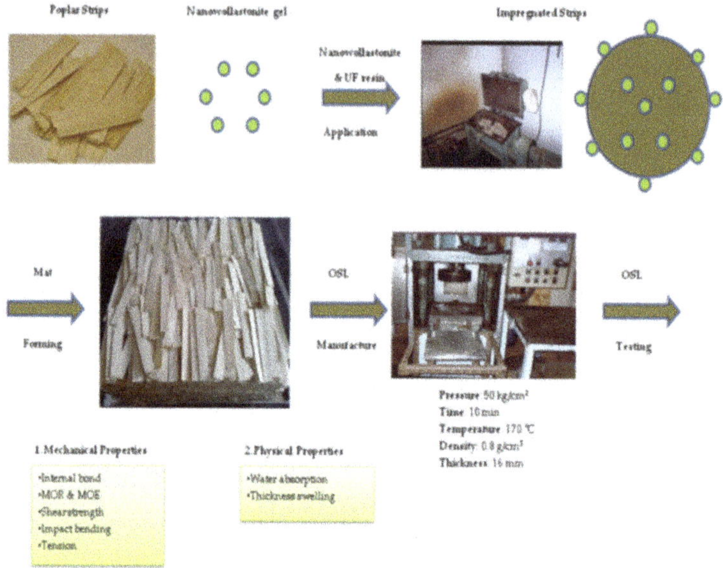

Figure 1. Flow diagram of the experimental procedure.

2.3. Thermal Conductivity Measurement

Thermal conductivity coefficient of OSL specimens was calculated based on Fourier's law for heat conduction, using an apparatus by Iranian Precise System Co. (IPS. Tehran, Iran) (Figure 2). Circular specimens were cut 30 mm in diameter and 16 mm in length (Figure 3), and all parts of the specimens were covered with silicone adhesive for better insulation. Specimens were positioned in a Teflon holder to be placed between the heating and absorbing brass rods. The heating brass bar heated the specimen from one side at 130 °C (Figure 2), while the other face of the cylindrical specimen was touched by the absorbing brass bar. The heating continued until the thermistor read a constant temperature. In order to measure the rate of heat transfer, temperature at the middle of the specimen was read and registered at 5 s intervals. Thermal conductivity was then calculated using Equations (1) and (2). Temperatures were measured with 0.1 °C precision.

$$Q = KA\frac{T_1 - T_2}{\Delta_x} \tag{1}$$

$$K = \frac{Q \times L}{A \times \Delta T} \tag{2}$$

where

k = Coefficient of thermal conductivity ($W.m^{-1}.K^{-1}$)

Q = Heat transfer (W)

L = Specimen thickness (m)

A = Cross section area of specimens (m^2)

ΔT = Temperature difference (T1–T2) (°k)

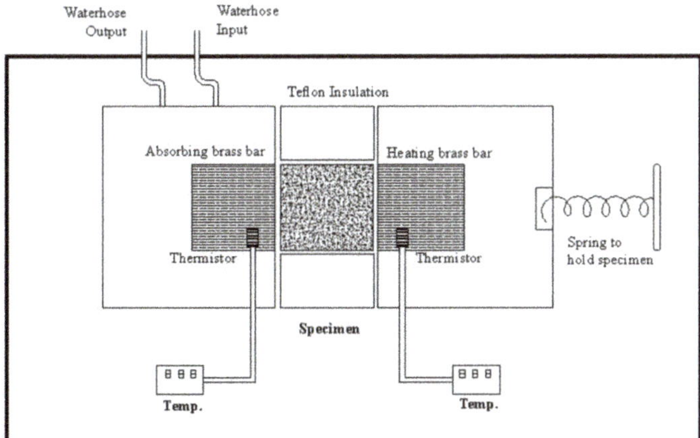

Figure 2. Schematic drawing of the apparatus for measurement of thermal conductivity [38].

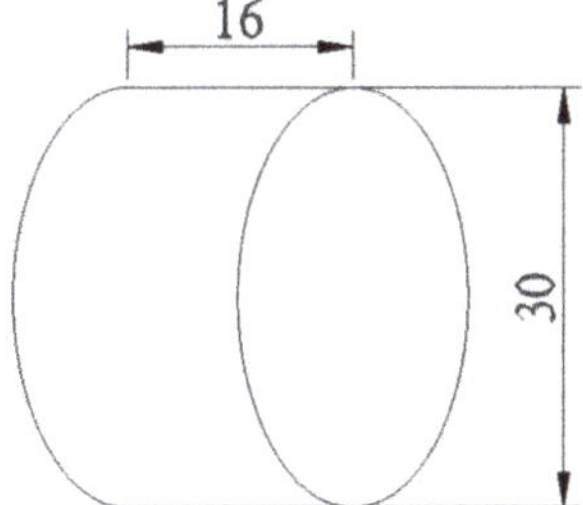

Figure 3. Schematic drawing of cylindrical specimens positioned in between the heating brass bar and absorbing brass bar (dimensions in mm) [38].

2.4. SEM Imaging

SEM imaging was carried out at the thin-film laboratory, FE-SEM lab (Field Emission), School of Electrical and Computer Engineering, University of Tehran. A field-emission cathode in the electron gun of a scanning electron microscope provided narrower probing beams at low as well as high electron energy, which improved the spatial resolution and minimized charging and damage to the specimens. As wood is a nonconductive material, a gold sputtering thickness of 6–8 nm was applied on the surface of the specimens prior to SEM imaging.

2.5. Statistical Analysis

SAS software program (Statistical Analysis System, version 9.2, 2010) was used to conduct a two-way analysis of variance (ANOVA) to discern significant differences among treatments at the 95% level of confidence. Grouping of similar treatments was made by Duncan's multiple range test. Hierarchical cluster analysis with Ward's method was performed using SPSS/20 to clarify similarities and dissimilarities among different treatments based on more than one property at the same time. In this analysis, the scaled indicator determines the degree to which the analyzed treatments are similar or different. Fitted-line and scatter plots were made using Minitab software, version 16.2.2.

3. Results and Discussion

3.1. Physical Properties

Figure 4 shows an SEM image of the surface of wood strips. Figure 5 demonstrates a wood-strip mat before being hot pressed (A), as well as specimens cut to size and ready for tests (B). Figure 6 depicts the effect of resin content on water absorption and thickness swelling of OSL. As expected, higher resin content resulted in improved properties; however, this improvement was not significant at the 0.05 probability level. Figure 6 also reveals the effect of fortification level of nanowollastonite on UF resin, in both properties. The fortification of UF resin with wollastonite gel (at 10% and 20% based on the dry weight of wood strips) did not significantly affect the water absorption, and this was true for both resin contents applied in this study, namely, 8% and 10%. This was also the case for the thickness swelling in boards manufactured with 8% resin content. However, in boards made with 10% resin content, the situation was different. The fortification of nanowollastonite on UF resin resulted in significant improvement in thickness swelling values, although the values of boards made with 20% nanowollastonite were slightly better than the ones made with 10% nanowollastonite. This requires explanation. The mechanism involved in the fortification of UF resin with nanowollastonite, which resulted in an improvement in thickness swelling values, can be attributed to the following two factors: (i) individual strips in the OSL matrix were better connected to each other through a network of bonds formed between the nanowollastonite compounds and wood-strips functional groups, mainly cellulose hydroxyl groups, putting them out of reach for bonding with the water molecules [31–33], and (ii) high thermal conductivity coefficient of wollastonite improved the transfer of heat to different layers of OSL mat [39], facilitating better and more complete resin curing. In a recent study, nanowollastonite was applied at 2, 4, 6, and 8%, based on the dry weight of wood fibers, and its effect on thermal conductivity of the medium density fiberboards was reported. It was found that nanowollastonite significantly increased the thermal conductivity of the panels, which in turn resulted in improved thickness swelling values of the MDF (Medium Density Fiberboards) boards. In particular, the use of 8% nanowollastonite contributed to better heat transfer in a way that thermal conductivity coefficient was increased by more than 29% [39].

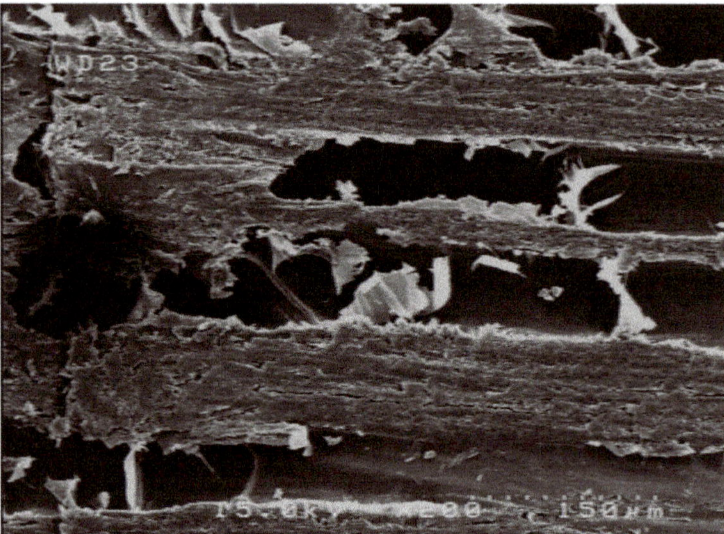

Figure 4. SEM image of the surface of wood strips to be glued and stuck together.

(A)

(B)

Figure 5. Photos of a wood-strip mat before being hot pressed (**A**), and cut-to-size and ready specimens for testing (**B**).

3.2. Mechanical Properties

Figure 7 depicts the effect of resin content on internal bond of OSL. As expected, higher resin content levels resulted in improved property; however, this improvement was not significant at the 0.05 probability level. Figure 7 also reveals the effect of fortification level of nanowollastonite on UF resin.

The fortification of nanowollastonite in boards made from 8% UF resin resulted in significant reduction in bond strength. At this resin content, it seems that nanowollastonite caused poor wetting of wood strips since the majority of failures were due to the resin and not to the wood. Another possible explanation for this behavior may be the absorption or gathering of resin molecules by wollastonite nanofibers, preventing them from being active in the process of sticking the strips together [39–41].

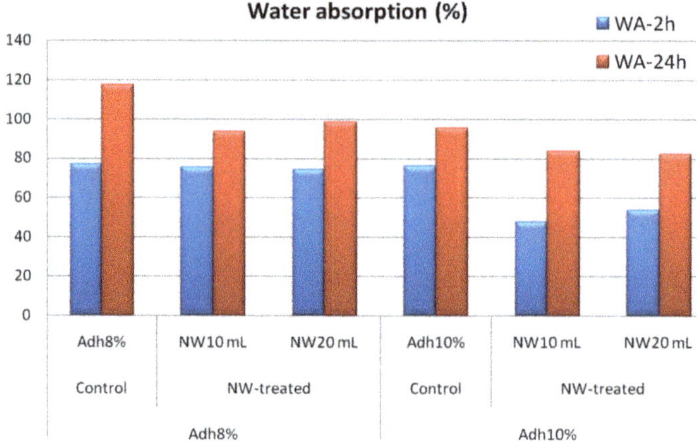

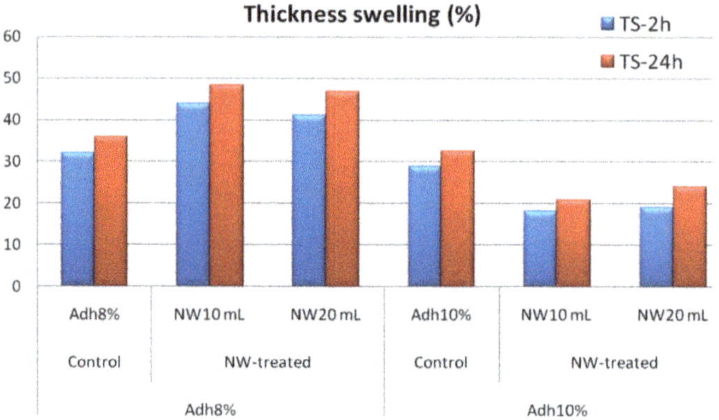

Figure 6. Water absorption and thickness swelling values of control and NW-treated OSL panels as affected by resin and nanowollastonite content (Adh = resin content; NW = nanowollastonite content).

However, at higher resin content (10%), the internal bond strength was increased. This increase was statistically significant at the 0.05 probability level, when 10% nanowollastonite was applied. Therefore, it appears that the increased resin content caused significant wetting of wood strips. Furthermore, an increase in the thermal conductivity coefficient by nanowollastonite [39] caused better curing of the resin in the core section of the mat, resulting in a higher internal bond. This observation demonstrated that 10% of nanowollastonite can be considered an optimum level. In fact, when nanowollastonite content was higher (that is, 20%), higher volume of UF resin left over from the process of sticking the strips together, and therefore was absorbed by wollastonite nanofibers. However, 10% of NW seems to be high enough to improve the properties and low enough not to interfere with the resin.

Modulus of rupture and shear strength parallel to grain at low-resin content showed similar behavior to the internal bond strength, as depicted in Figures 8 and 9, respectively. At high-resin content, the fortification with nanowollastonite did not significantly affect these two properties. The other mechanical properties, namely, modulus of elasticity, tension parallel to grain, and impact bending, were not significantly affected by either resin content or the fortification of nanowollastonite, as illustrated in Figures 10–12, respectively.

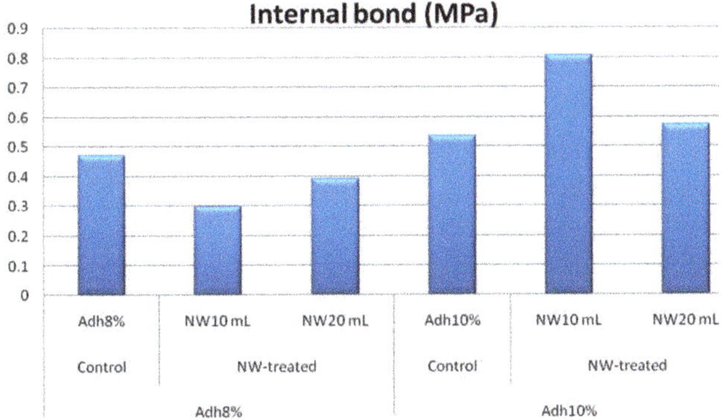

Figure 7. Internal bond strength values of control and NW-treated OSL panels as affected by resin and nanowollastonite content.

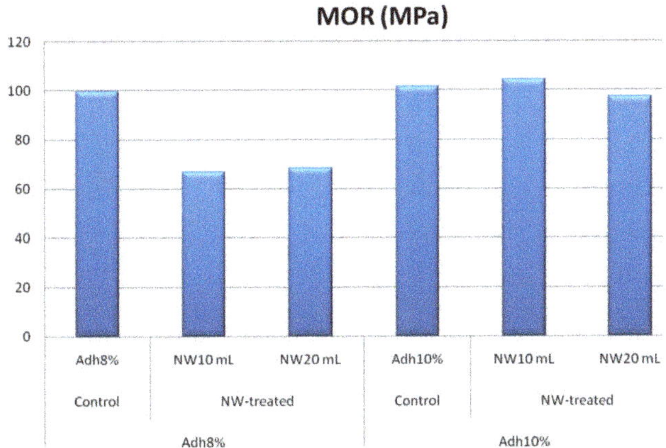

Figure 8. MOR (modulus of rupture) values of control and NW-treated OSL panels as affected by resin and nanowollastonite content.

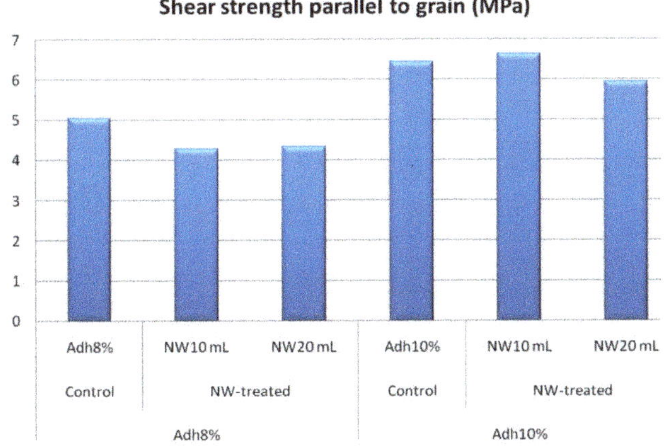

Figure 9. Shear strength values of control and NW-treated OSL panels as affected by resin and nanowollastonite content.

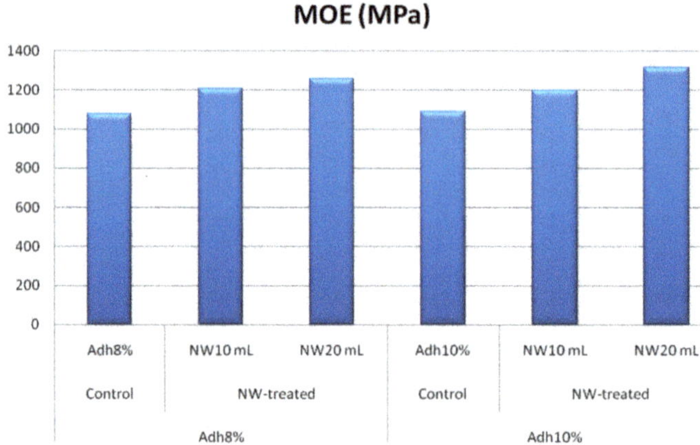

Figure 10. MOE (modulus of elasticity) values of control and NW-treated OSL panels as affected by resin and nanowollastonite content.

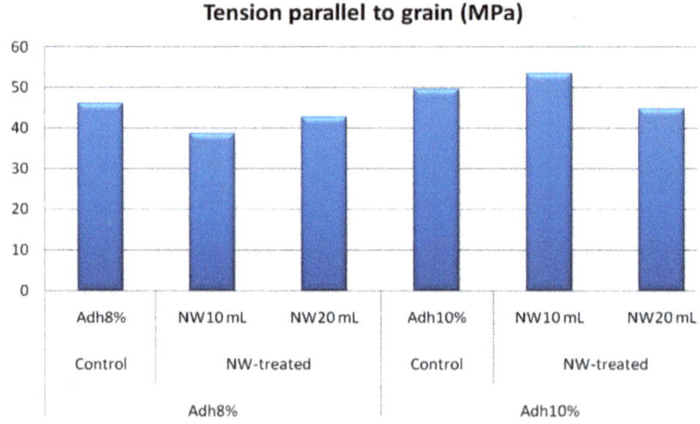

Figure 11. Tension strength values of control and NW-treated OSL panels as affected by resin and nanowollastonite content.

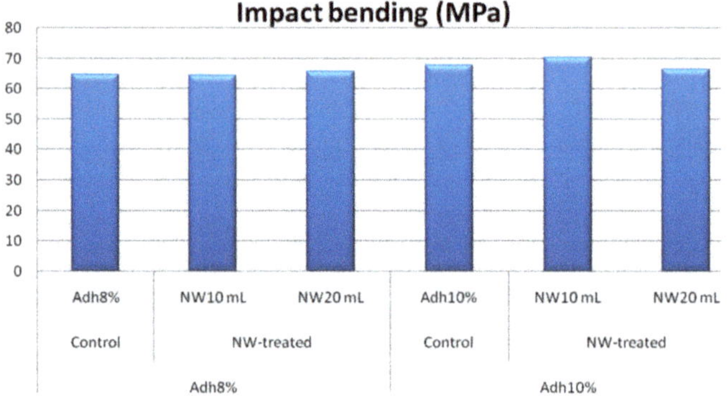

Figure 12. Impact bending values of control and NW-treated OSL panels as affected by resin and nanowollastonite content.

Cluster analysis, based on the mechanical properties, clearly showed distinctly different clustering of nanowollastonite-treated panels with 8% resin content (NW10-Adh8 and NW20-Adh8 treatments) (Figure 13A). Both control treatments with resin contents of 8% and 10% were closely clustered together. NW-treated panels with 10% resin contents were somehow clustered similarly and close to the control treatments. This clustering pattern indicated that the addition of NW at lower resin content of 8% had significantly negative effects on the overall mechanical properties, and thus is not recommended, while at the higher resin content of 10%, the addition of NW had some significantly positive effects on individual properties.

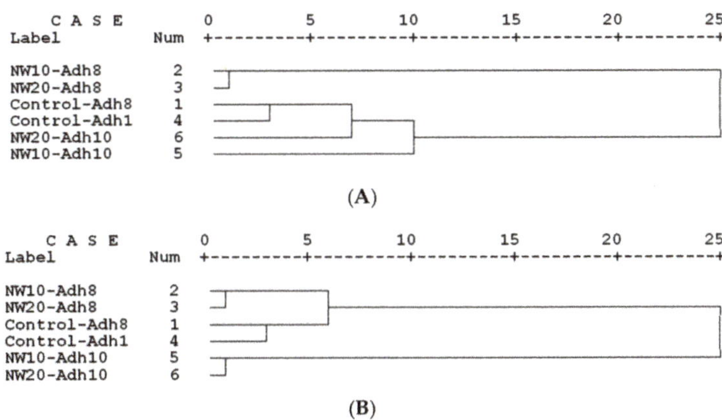

Figure 13. Cluster analysis of the six OSL panels produced based on the mechanical (**A**) and physical (**B**) properties.

On the other hand, cluster analysis based on the physical properties revealed significantly different clustering of NW-treated panels with 10% resin content (Figure 13B). Both control treatments were closely clustered to NW-treated panels with 8% resin content. This indicated that the overall physical properties of NW-treated panels with 8% resin content were similar to the control panels, although the addition of NW had negative effects on the thickness swelling. Moreover, NW can significantly improve physical properties of OSL panels when resin content is high enough (10%).

3.3. Thermal Conductivity Coefficients

Figure 14 depicts the effect of the addition of nanowollastonite on thermal conductivity coefficient of different specimens. The highest and lowest thermal coefficients were found in NW10%-Adh10% and Control-Adh8% treatments, respectively. The increase in adhesive content from 8% to 10% resulted in an increase in thermal conductivity in all treatments, although the increase was not statistically significant in any of them. The increase was attributed to the better integrity of wood strips when stuck together, as well as the higher connecting points among them.

The addition of nanowollastonite at both NW contents (10% and 20%) made thermal conductivity increase. However, the increase was statistically significant only in treatments with 10% of NW. Treatments with 20% of NW did not demonstrate significant increase in comparison to their control counterparts. Here, NW particles seemed to absorb much resin, negatively interfering with the main role of the resin, which was to stick the strips together.

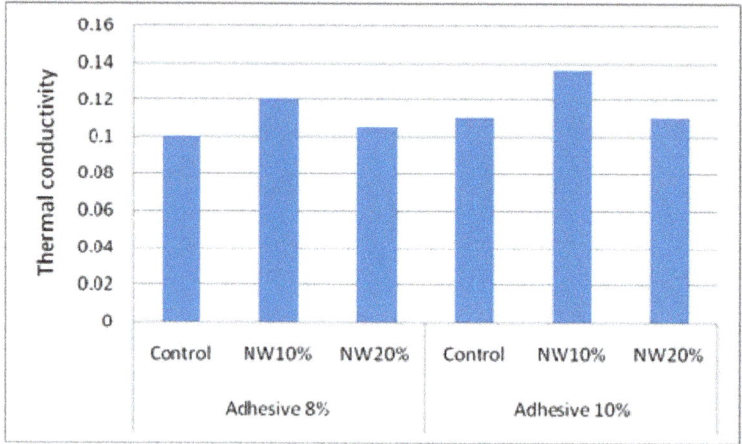

Figure 14. Thermal conductivity coefficients of control and NW-treated OSL panels as affected by resin and nanowollastonite content.

Based on the above discussion, it seems that the fortification of UF resin with 10% nanowollastonite can be considered an optimum level. When nanowollastonite content was higher (that is, 20%), a higher volume of UF resin left over from the process of sticking the strips together was absorbed by wollastonite nanofibers. However, a NW content of 10% seems to be high enough to improve the panel properties and low enough not to interfere with the resin. Therefore, in future research projects, nanowollastonite contents with intervals close to 10% are recommended to obtain a better scope of the best NW content option.

4. Conclusions

The aim of this work was to investigate the effect of fortification level of nanowollastonite on UF resin and its effect on mechanical and physical properties of oriented strand lumbers. Two resin contents were applied, namely, 8% and 10%. Nanowollastonite was mixed with the resin at two levels of 10% and 20%. It was found that the fortification of UF resin with 10% nanowollastonite can be considered as an optimum level. When nanowollastonite content was higher, a higher volume of UF resin left over from the process of sticking the strips together was absorbed by wollastonite nanofibers. Since nanowollastonite contributes to making bonds between the wood strips, which consequently improves their physical and mechanical properties, its use can be safely recommended in the OSL production process to improve the physical and mechanical properties of the panel.

Author Contributions: Methodology, H.R.T. and V.H.; Validation, V.H. and S.M.; Investigation, H.R.T., V.H. and S.M.; Writing-Original Draft Preparation, H.R.T., O.S. and A.N.P.; Writing-Review & Editing, H.R.T., O.S. and A.N.P.; Visualization, V.H. and S.M.; Supervision, H.R.T. and A.N.P.

Funding: This research received no external funding.

Acknowledgments: The authors appreciate the constant scientific support of Jack Norton (Retired, Horticulture and Forestry Science, Queensland Department of Agriculture, Forestry and Fisheries, Australia), as well as Alexander von Humboldt Stiftung, Germany.

Conflicts of Interest: The authors declare no conflict of interest.

References

1. Barbu, M.C.; Réh, R.; Irle, M. Wood–Based Composites. In *Research Developments in Wood Engineering and Technology*; Alfredo, A., Paulo, D., Eds.; IGI Global: Hershey, PA, USA, 2014.
2. Kelly, M.W. *Critical Literature Review of Relationships between Processing Parameters and Physical Properties of Particleboard*; General Technical Report; Forest Products Laboratory: Madison, WI, USA, 1977.
3. Hill, C.A.S. *Wood Modification—Chemical, Thermal and Other Processes*; John Wiley and Sons Ltd.: West Sussex, UK, 2006.
4. Rowell, R.M. Acetylation of Wood—A review. *Int. J. Lignocellul. Prod.* **2014**, *1*, 1–27.
5. Hon, D.N.S. *Chemical Modification of Lignocellulosics*; Marcel Dekker: New York, NY, USA, 2006.
6. Papadopoulos, A.N. Chemical modification of solid wood and wood raw materials for composites production with linear chain carboxylic acid anhydrides: A brief Review. *BioResources* **2010**, *5*, 499–506.
7. Mantanis, G.I. Chemical modification of wood by acetylation or furfurylation: A review of the present scaled-up technologies. *BioResources* **2017**, *12*, 4478–4489. [CrossRef]
8. Del Menezzi, C.H.S.; Tomaselli, J. Contact thermal post-treatment of oriented strandboard to improve dimensional stability: A preliminary study. *Holz Roh Werkst.* **2006**, *64*, 212–217. [CrossRef]
9. Okino, E.Y.A.; Teixeira, D.E.; Del Menezzi, C.H.S. Post-thermal treatment of oriented strandboard made from cypress. *Cienc. Tecnol.* **2007**, *9*, 199–210.
10. Bohm, M. The influence of moisture content on thickness swelling and modulus of elasticity in oriented strand board bending. *Wood Res.* **2009**, *54*, 79–90.
11. Carll, C.; Wiedenhoeft, A. Moisture-Related Properties of Wood and the Effects of Moisture on Wood and Wood Products. In *Moisture Control in Buildings: The key Factor in Mold Preventions*; Trechsel, H.R., Bomberg, M.T., Eds.; ASTM International: West Conshohocken, PA, USA, 2009.
12. Mirski, R.; Derkowski, A.; Dziurka, D. Influence of strand size, board density and adhesive type on characteristics of oriented strand lumber boards from pine strands. *Bioresources* **2019**, *14*, 6686–6696.
13. Réh, R.; Igaz, R.; Krišťák, Ľ.; Ružiak, I.; Gajtanska, M.; Božíková, M.; Kučerka, M. Functionality of Beech Bark in Adhesive Mixtures Used in Plywood and Its Effect on the Stability Associated with Material Systems. *Materials* **2019**, *12*, 1298.
14. Timsuk, P.C. An Investigation of the Moisture Sorption and Permeability Properties of Mill Fabricated Oriented Strand Board. Ph.D. Thesis, University of Toronto, Toronto, ON, Canada, 2008.
15. Leichtil, R.J.; Falk, R.H.; Laufenberg, T.L. Prefabricated wood composite I beams: A literature review. *Wood Fiber Sci.* **1990**, *22*, 62–79.
16. Saracoglou, E. Finite Element Simulations of the Influence of Cracks on the Strength of Glulam Beams. Master's Thesis, Blekinge Institute of Technology, Karlskrona, Sweden, 2011.
17. Reinprecht, L.; Svoradova, M.; Reh, R. Decay resistance of laminated veneer lumbers from European oaks. *Wood Res.* **2010**, *55*, 79–90.
18. Hsu, W.E.; Melanson, R.J.; Kozak, P.J. The effect of wax type and content on waferboard properties. In Proceedings of the 24th International Particleboard Composites Symposium, Pullman, WA, USA, 3–5 April 1990; pp. 85–96.
19. Hager, R.; Mayer, H. *Waterborne Silicones for Wood Protection*; 26th Annual Meeting; International Research Group on Wood Preservation: Elsinore, Denmark, 1995; IRG WP 30062.
20. Jusoh, E.B.; Nzokou, B.; Kamden, P. The effect of silicone on some properties of flakeboard. *Holz Roh Werkst.* **2005**, *63*, 266–271. [CrossRef]
21. Papadopoulos, A.N.; Bikiaris, D.N.; Mitropoulos, A.C.; Kyzas, G.Z. Nanomaterials and chemical modification technologies for enhanced wood properties: A review. *Nanomaterials* **2019**, *9*, 607. [CrossRef] [PubMed]
22. Roco, M. Nanotechnology's Future. *Sci. Am.* **2006**, *13*, 427–445. [CrossRef] [PubMed]
23. Taghiyari, H.R.; Schimdt, O. Nanotachnology in wood based composite panels. *Int. J. Bioinorg. Hybrid Nanomater.* **2014**, *3*, 65–73.
24. Goffredo, G.B.; Accoroni, S.; Totti, T.; Romagnoli, T.; Valentini, L.; Munafò, P. Titanium dioxide based nanotreatments to inhibit microalgal fouling on building stone surfaces. *Build. Environ.* **2017**, *112*, 209–222. [CrossRef]
25. De Filpo, G.; Palermo, A.M.; Rachiele, F.; Nicoletta, F.P. Preventing fungal growth in wood by titanium dioxide nanoparticles. *Int. Biodeterior. Biodegrad.* **2014**, *85*, 217–222. [CrossRef]

26. Civardi, C.; Schwarze, F.; Wick, P. Micronized copper wood protection: An efficiency and potential health and risk assessment for copper based nanoparticles. *Environ. Pollut.* **2015**, *200*, 20–32. [CrossRef]

27. Moya, R.; Zuniga, A.; Berrocal, A.; Vega, J. Effect of silver nanoparticles synthesized with NPsAg-ethylene glycol on brown decay and white decay fungi of nine tropical woods. *J. Nanosci. Nanotechnol.* **2017**, *17*, 1–8. [CrossRef]

28. Teng, T.; Arip, M.; Sudesh, K.; Lee, H. Conventional technology and nanotechnology in wood preservation: A review. *BioResources* **2018**, *13*, 9220–9252. [CrossRef]

29. Taghiyari, H.R.; Majidi, R.; Jahangiri, A. Adsorption of nanowollastonite on cellulose surface: Effects on physical and mechanical properties of medium-density fiberboard (MDF). *Cerne* **2016**, *22*, 215–222. [CrossRef]

30. Esmailpour, A.; Taghiyari, H.R.; Majidi, R.; Morrell, J.J.; Mohammad-Panah, B. Nano-wollastonite to improve fire retardancy in medium-density fiberboard (MDF) made from wood fibers and camel-thorn. *Wood Mater. Sci. Eng.* **2019**. [CrossRef]

31. Majidi, R.; Taghiyari, H.R.; Abdolmaleki, D. Molecular dynamics simulation evaluating the hydrophilicity of nanowollastonite on cellulose. *J. Struct. Chem.* **2019**, *6*, 1562–1569. [CrossRef]

32. Taghiyari, H.R.; Nouri, P. Effects of nano-wollastonite on physical and mechanical properties of medium-density fiberboard. *Maderas Cienc. Y Tecnol.* **2015**, *17*, 833–842. [CrossRef]

33. Esmailpour, A.; Taghiyari, H.R.; Hosseinpourpia, R.; Adamopoulos, S.; Zereshki, K. Shear strength of heat-treated solid wood bonded with polyvinyl-acetate reinforced by nanowollastonite. *Wood Res.* **2020**, in press.

34. Cheng, X.; Zhang, S.Y.; Deng, J.; Wang, S. Urea—Formaldehyde-Resin Gel Time as Affected by the pH Value, Solid Content, and Catalyst. *J. Appl. Polym. Sci.* **2007**, *103*, 1566–1569.

35. Uner, B.; Olgun, C. The effect of hardener on adhesive and fiber properties. *Wood Res.* **2017**, *62*, 27–36.

36. ANSI. *Particleboard*; American National Standard; ANSI A208.1-1998; National Particleboard Association: Corvallis, OR, USA, 1998.

37. CEN. *Particleboards and Fibreboards-Determination of Swelling in Thickness after Immersion in Water*; European Standard EN 317; CEN European Committee for Standardisation: Brussel, Belgium, 1993.

38. Taghiyari, H.R.; Norton, J.; Tajvidi, M. Effects of Nano-Materials on Different Properties of Wood-Composite Materials. In *Bio-Based Wood Adhesives: Preparation, Characterization, and Testing*; CRC Press/Taylor & Francis Group: Boca Raton, FL, USA, 2017; pp. 310–339.

39. Taghiyari, H.R.; Ghorbanali, M.; Tahir, P.M.D. Effects of improvement in thermal conductivity coefficient by nano-wollastonite on physical and mechanical properties in medium-density fiberboard (MDF). *BioResources* **2014**, *9*, 4138–4149. [CrossRef]

40. Taghiyari, H.R.; Maleki, S.; Hassani, V. Effects of nano-silane on physical and mechanical properties of oriented strand lumber (OSL). *Bois For. Trop.* **2016**, *330*, 49–55. [CrossRef]

41. Taghiyari, H.R.; Mobini, K.; Sarvari-Samadi, Y.; Doosti, Z.; Karimi, F.; Asghari, M.; Jahangiri, A.; Nouri, P. Effects of nano-wollastonite on thermal conductivity coefficient of medium-density fiberboard. *J. Nanomater. Mol. Nanotechnol.* **2013**, *2*. [CrossRef]

 *polymers*

Article

Relationships between Thermoplastic Type and Properties of Polymer-Triticale Boards

Radosław Mirski [1],*, Pavlo Bekhta [2] and Dorota Dziurka [1]

[1]   Department of Wood-Based Materials, Poznań University of Life Sciences, 60-627 Poznań, Poland; dorota.dziurka@up.poznan.pl
[2]   Department of Wood-Based Composites, Cellulose, and Paper, Ukrainian National Forestry University, 79057 Lviv, Ukraine; bekhta@nltu.edu.ua
*   Correspondence: rmirski@up.poznan.pl; Tel.: +48-061-848-7616

Received: 10 October 2019; Accepted: 23 October 2019; Published: 25 October 2019

Abstract: This study examined the effects of selected types of thermoplastics on the physical and mechanical properties of polymer-triticale boards. The investigated thermoplastics differed in their type (polypropylene (PP), polyethylene (PE), polystyrene (PS)), form (granulate, agglomerate) and origin (native, recycled). The resulting five-ply boards contained layers made from different materials (straw or pine wood) and featured different moisture contents (2%, 25%, and 7% for the face, middle, and core layers, respectively). Thermoplastics were added only to two external layers, where they substituted 30% of straw particles. This study demonstrated that, irrespective of their type, thermoplastics added to the face layers most favorably reduced the hydrophobic properties of the boards, i.e., thickness, swelling, and V100, by nearly 20%. The bending strength and modulus of elasticity were about 10% lower in the experimental boards than in the reference ones, but still within the limits set out in standard for P7 boards (20 N/mm^2 according to EN 312).

Keywords: polymer-triticale boards; thermoplastic polymers; straw; hydrophobicity; mechanical properties

---

1. Introduction

Wood plastic composites (WPCs) are usually made from forestry waste, including waste wood flour, wood shavings, and sawdust, and can be produced using different methods, e.g., extrusion, injection or compression molding [1]. Extrusion is a predominant technology for the manufacturing of WPCs, even though it only allows for predetermined dimensions of the boards. Another possibility is to produce WPCs on a flat-press [2–8], similarly as in the industrial particleboard manufacturing process. This is a new, technically more advantageous, and simple method for producing large-size WPC boards of different densities. It offers higher productivity at relatively low-pressure requirements, creates a positive environmental image of wood as a renewable material, enables the saving of wood raw material due to the lower density of composite boards, and lowers the production costs as compared with other methods [3]. Moreover, the resulting products closely compare to commercial medium-density fiberboard (MDF), particleboard, oriented strand board (OSB), and plywood.

The physical and mechanical properties of WPCs largely depend on the type of thermoplastic used, content, and properties of wood fillers, pressing parameters, and wood–plastic interactions [9–14]. The literature contains many reports on manufacturing WPCs that account for the type of thermoplastic, particle size, and influence of various technological factors [9–14].

Typical WPCs (thermoplastic content 20%–60%, wood particle content 40%–80%) are manufactured by extrusion or injection. They can also be produced using pressure methods similar to traditional board manufacturing technology involving cyclic or continuous pressing [8,15]. Additionally, chip-polymer flat-pressed composite WPCs are favorably produced by layering their mat, i.e., alternating the layers of ground thermoplastic materials and lignocellulosic particles prior to pressing [16]. Borysiuk et al. [17] demonstrated the advantages of using particles of similar shapes as this minimizes their sorting during sieving. Some authors [18,19] suggested that a reduction in the size of particles could improve the properties of particleboards made from agricultural fibers. In another work [20], binderless boards were successfully manufactured from rice straw powder with a particle size smaller than 1 mm. However, the flat-pressing method of polymer-lignocellulosic composite production requires a long pressing time and/or high temperature of the heating plates to overheat the mat and ensure thermoplastic softening. For this reason, it is better to use thermoplastics only in the layers where they can be easily melted. An important advantage of lignopolymer composites is their higher resistance to water as compared with other solid wood and wood-based materials. The polymer mechanically blocks the access of water to wood particles [21].

The demand of the wood industry—including the producers of wood-based materials—for lignocellulosic materials in the coming years will only be possible to satisfy by using the existing potential reserves. The source of additional raw material for production of wood-based materials should be plantations of fast-growing trees and annual plants. In the latter case, the board industry may use such plants and plant products as grasses, reeds, cereal straw or stems, and leaves of nettles. It also seems favorable to use commonly cultivated annual plants to minimize the risk of lignin–cellulose biomass supply. Despite the seasonality of supply, necessity of storage, low bulk density, and other negative aspects of using annual plants in board manufacturing, they still should be considered as raw materials of full value [22–25]. Cereal straw as a renewable raw material is one of the most important agricultural wastes generated in huge quantities in many countries of the world. The worldwide production of cereal straw is estimated at 1.5 billion m^3 annually [26]. The countries of the European Union produce about 140 million tons annually, of which only 2%–3 % is used by the industry as a raw material for WPCs [27]. The straw is first ground into a powder that forms a kind of reinforcement for the polymer matrix. Many studies have been carried out on the utilization of annual plant fibers in lignocellulosic composites, such as particleboard, medium–density fiberboard (MDF) and WPCs [19,21,22,26,28–34]. A review paper by Abba et al. [35] presented different methods and procedures for the production of plastic composites from agricultural waste materials. In general, utilization of lignocellulosic fibers in wood-based composites offers several advantages, such as recyclability, fewer health hazards for the operators, low density, greater deformability, less abrasiveness to equipment, biodegradability, and low cost. Their major drawback involves relatively poor compatibility with hydrophobic thermoplastics that often results in poor mechanical properties [36].

Virgin thermoplastics, such as polypropylene (PP), polyethylene (PE), polystyrene (PS), and polyvinyl chloride (PVC), are commonly used as matrices for manufacturing WPCs [37] with wood flour, but using recycled plastic as a matrix was only reported in a few studies [6,37]. A major portion of global municipal solid waste includes post-consumer plastic materials, like high-density polyethylene (HDPE), low-density polyethylene (LDPE), and PVC, which can potentially be used in WPC manufacturing. For example, plastic film is widely used for packaging daily necessities, causing a serious problem for the environment. Therefore, utilizing waste plastic is important from the perspective of environmental protection and recycling.

Considering the above, it seems that annual plant straw combined with thermoplastics could be used for manufacturing wood-based materials with high moisture resistance. For this reason, the aim of this study was to examine the effects of selected types of thermoplastics on physical and mechanical properties of polymer-triticale boards.

2. Materials and Methods

The study involved triticale straw and pine chips. The straw, obtained in agricultural bales, was ground in a laboratory mill into particles of dimensions as close as possible to pine chips. The linear dimensions of triticale particles were 17.8 mm in length, 2.55 mm in width, 0.30 mm in thickness, and for the pine chips, they were 13.73, 2.03, and 1.04 mm, respectively. A detailed dimensional analysis is presented in our previous publication [25]. The pine chips were industrial chips intended for core layers of three-ply P2 boards. Visual inspection confirmed disparate morphology of the fillers. The fillers are shown in Figure 1. The pre-selected material was sorted on a sieve with a mesh size of 0.5 mm in order to separate fine particles and silty fractions. Figure 2 presents the results of the sieve analysis.

Figure 1. The photographs of the fillers.

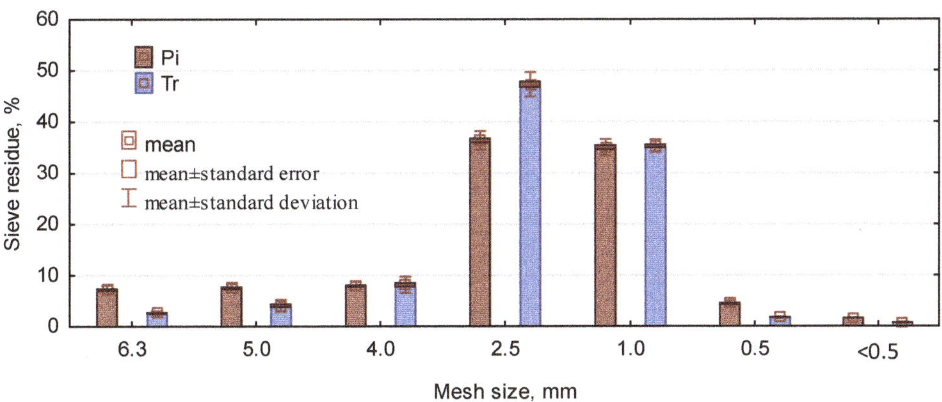

Figure 2. Sieve analysis of lignocellulose particles used in the study, (Pi—pine chips, Tr—triticale straw chips).

We produced five-ply boards with the following layer share: 0.5:0.5:1:0.5:0.5. The layers differed in the type of material and moisture content:

➢ The face layers (1, 5) were made from straw with the moisture content of 2%;
➢ The intermediate layers (2, 4) from straw with the moisture content of 25%;
➢ The core layer (3) from chips with the moisture content of 7%.

The effectiveness of a mat pressing into a board depends on numerous factors associated with the mat (raw material moisture content, degree of fineness, adhesive share) and pressing parameters (temperature, pressure, time) [38,39]. The time in which the internal layers reach the temperature of 100–110 °C largely depends on the mat moisture content [40], which is why the external layers are commonly made from materials of a higher moisture content. However, a lower moisture content finally allows for achieving a higher temperature that is crucial when using refractory additives during the mat formation. Some authors [41–43] indicate that polymers most effectively melt with lignocellulosic material when the moisture content of the material is below 3%–4%. For these reasons, our experimental boards consisted of the polymer containing layers of low moisture content, intermediate layers with high moisture content that facilitated heat transfer inside the mat, and a core layer with standard moisture content.

The study involved the following thermoplastics differing in the type, form, and softening point (Table 1):

- PP—polypropylene impact copolymer (Zakład Przetwórstwa Tworzyw Sztucznych, Czapury, Poland);
- HDPE—high-density polyethylene (Basell Orlen Polyolefins Sp. z o.o., Płock, Poland);
- LDPE—low-density polyethylene (Basell Orlen Polyolefins Sp. z o.o., Płock, Poland);
- PS—polystyrene (Zakład Przetwórstwa Tworzyw Sztucznych, Czapury, Poland);
- LDPE rec. yellow—low-density polyethylene from recycled scraps of yellow plastic (Zakład Przetwórstwa Tworzyw Sztucznych, Czapury, Poland);
- LDPE rec. pink—low-density polyethylene from recycled scraps of pink plastic (Zakład Przetwórstwa Tworzyw Sztucznych, Czapury, Poland);
- PP recycled—polypropylene from reusable packaging (Zakład Przetwórstwa Tworzyw Sztucznych, Czapury, Poland).

Table 1. Basic properties of the investigated thermoplastics.

Thermoplastic Type	Symbol	Thermoplastic Form	Density (g/cm^3)	Softening Point (Vicata (A50))	MFR (Melt Flow Index)
Polypropylene, impact copolymer	**PP**	Granulate	0.9	150 °C	4 g/10 min (230 °C/2.16 kg)
High-density polyethylene	**HDPE**	Granulate	0.952	125 °C	0.3 g/10 min (190 °C/2.16 kg)
Polystyrene	**PS**	Granulate	1.5	90 °C	6 g/min (200 °C/5 kg)
Low-density polyethylene	**LDPE**	Granulate	0.923	91 °C	1.5 g/10 min (190 °C/2.16 kg)
Polypropylene from reusable packaging	**PP_R**	Agglomerate	0.54 *	nd	nd
Low-density polyethylene from yellow plastic	**LDPE_RZ**	Agglomerate	0.37 *	nd	nd
Low-density polyethylene from pink plastic	**LDPE_RR**	Agglomerate	0.38 *	nd	nd

* bulk density; nd not determined.

The thermoplastics were added to layers 1 and 5, where they substituted 30% of straw particles. The reference boards followed the same moisture content and particle type pattern but without polymers in 1 and 5 layers. Irrespective of the layer, the lignocellulosic material was glued with pMDI in the amount of 4% of dry weight of the adhesive per dry weight of the particles. The mat was formed manually to produce 15 mm-thick boards of the target density 600 kg/m^3. The heating plate temperature was 200 °C, unit pressure 2.5 MPa, and pressing time 20 s per mm of board thickness. The mat was pressed between metal plates. Following hot pressing, the molds, together with the metal plates, were transferred onto a cold press, where they were kept under 1000 N load until their temperature dropped below 80–100 °C. The temperature was monitored with K-type thermocouples attached to the upper and lower surfaces of the plates. For each variant, we produced three molds 700 mm × 450 mm in dimension.

After conditioning (seven days, 55 ± 5% RH, 21 ± 1 °C), the boards produced this way were tested as per relevant standards and the following parameters were assessed:

- bending strength (MOR) and modulus of elasticity (MOE) according to EN 310 [44];
- internal bond (IB) according to EN 319 [45];
- internal bond after the boiling test (V-100) according to EN-1087-1 [46];
- thickness swelling (TS) after 24 h according to EN 317 [47] and water absorption (WA).

The assessments of mechanical properties and water resistance involved from 10 to 16 samples of each variant, and the remaining analyses were made in three to five replications. The results were analyzed using the STATISTICA 13.1 package (StatSoft Inc., Tulsa, OK, USA), and uncertain measurements were discarded prior to in-depth statistical analysis. The analysis was based on ANOVA, and homogeneous groups were tested with HSD Tukey's test (HSD) and the Bonferroni test. The results were analyzed for $p = 0.05$.

3. Results

Table 2 presents the board moisture content 24 h after pressing. It was relatively low, probably due to the low average moisture content of the mat (about 10.5%) and long pressing time. Nevertheless, this means that the moisture of the intermediate layer effectively transferred heat into the pressed mat.

Table 2. Board moisture content after pressing.

Board Type (Thermoplastic)	Moisture Content		
	x, % *	SD, %	υ, %
Reference Board	3.70	0.06	1.69
PP	3.13	0.09	2.99
HDPE	3.29	0.22	6.81
PS	3.07	0.04	1.36
LDPE	2.92	0.13	4.51
PP_R	3.16	0.06	1.96
LDPE_RZ	2.90	0.16	5.59
LDPE_RR	3.00	0.11	3.56

* x—mean; SD—standard deviation; υ—coefficient of variation.

The study demonstrated a drop in the bending strength and modulus of elasticity irrespective of the thermoplastic type added to the face layers (Table 3). For both parameters, the drop was the greatest for the board supplemented with high-density polyethylene, and amounted to 5.5 N/mm^2 for the bending strength and 860 N/mm^2 for the modulus of elasticity. The observed changes in bending strength depended on the type of the thermoplastic, its weight and origin, and the form in which it was added to the mat. The values for PE were considerably higher than those for the other polymers.

The analysis of the effect the origin of the thermoplastic exerted on the bending strength showed that the MOR of the boards made from secondary polymers (PP_R, LDPE_RZ, LDPE_RR) was higher than that of the boards made from virgin polymers (PP, LDPE). This was probably because the granulates of secondary polymers came in smaller sizes and the granules were more evenly spread over the board cross section. The MOR of the boards containing secondary polypropylene (PP_R) reached 24.6 MPa and was higher by 8.8% than that of the boards supplemented with virgin polypropylene (PP). The difference was statistically significant (Table 3). The MOR of the boards containing secondary polyethylene, LDPE_RZ and LDPE_RR, were 26.1 and 25.9 MPa, respectively, i.e., by 6.5% and 5.7% higher than of the boards enriched with virgin polyethylene. This time, the differences were insignificant (Table 3).

Table 3. The effects of thermoplastics on bending strength and modulus of elasticity.

Board type (Thermoplastic)	MOR			MOE		
	x, N/mm^2 *	υ, %	HSD	x, N/mm^2	υ, %	HSD
Reference board	26.5	5.2	e	3880	4.3	d
PP	22.6	5.7	a,b	3020	4.6	a,b
HDPE	21.0	4.7	a	2930	3.4	a
PS	23.4	8.9	b,c	3490	3.4	c
LDPE	24.5	6.2	c,d	3150	5.6	b
PP_R	24.6	8.6	c,d	3180	4.3	b
LDPE_RZ	26.1	3.3	d,e	3030	4.2	a,b
LDPE_RR	25.9	2.7	d,e	3020	2.5	a,b
ANOVA **	$F(7, 88) = 20.440, p = 0.00000$			$F(7, 88) = 52.715, p = 0.0000$		

* x—mean; υ—coefficient of variation, HSD—Tukey's test; ** F—distribution; p—probability value.

Further analysis demonstrated higher values of MOR for LDPE than HDPE, and for agglomerate than granulate. We also found that the lower melting point of the granulate was, the more favorable its effects were on the bending strength. However, Tukey's test identified the changes in MOR resulting from replacing some of the lignocellulosic material with recycled LDPE as insignificant. Moreover, the bending strength of the boards containing LDPE granulate was not statistically lower than of those supplemented with LDPE agglomerate. The type of polymer significantly affected the modulus of elasticity. This was probably due to the considerably lower modulus of elasticity of the polymers themselves vs. the boards made from lignocellulosic materials. Of the polymers used, only PS had the modulus of elasticity comparable to the boards composed of wood chips or annual plant particles. In this case, the MOE reached 3100–3300 N/mm^2 [48–50]. The modulus of elasticity was also clearly higher in those boards vs. the boards made from the remaining polymers, but it was still lower than the MOE of the boards made from pine chips. Nevertheless, when considering the parameters of the bending test provided in the EN 312-7 standard [51], the boards may be classified as P7 type.

The results obtained in this study are in general agreement with those reported by other researchers [52], who found that the mechanical strength of the boards made from straw and isocyanates was much lower than those made from wood under the same bonding conditions. Like with any other lignocellulosic fiber in reinforced plastic composites, the major concern when using straw is its relatively poor compatibility with hydrophobic thermoplastics, which leads to poor mechanical properties of the boards [36]. To improve the compatibility of straw and water-based synthetic resin or thermoplastic polymers, special pretreatments of straw or polymers are necessary, such as steam, cold plasma, corona, or chemical treatment, the mechanical pulverization of a wax layer, or the addition of external compatibilizers [22,33,53–56].

The thermoplastics should only slightly affect tensile strength perpendicular to the board plane, as they were added to the face layers responsible for properties determined in the bending test. However, Table 4 shows significant differences in these values. Post hoc analysis revealed two levels of tensile strength perpendicular to the board plane, with the first ranging around 0.32–0.38 N/mm^2 and the second around 0.38–0.42 N/mm^2. The values within these levels were highly variable, probably because in numerous samples, and in many cases in most of them, only the external layer was damaged. It was only in a few samples the damage reached deeper layers, mainly the core layer. Therefore, we also provided maximum values which, except for the PP board, we found for damage of the core layer. The low quality of the subsurface layers was most probably due to overheating of the external layers during the long pressing time. This should be taken into account while designing future studies. Still, the maximum values indicate a possibility of producing this type of board with a strength exceeding 0.45 N/mm^2. The improvement in the internal bond strength of WPCs vs. the control boards can be explained by the fact that, during hot pressing, some of the melted thermoplastic penetrates into the core layer. Then, the voids and spaces among the wood particles are filled with a melted thermoplastic polymer improving the adhesion between the particles.

Table 4. Effects of the thermoplastics on tensile strength perpendicular to the board plane.

Board Type (Thermoplastic)	IB				V100		
	x, N/mm^2	υ, %	Max	HSD	x, N/mm^2	υ, %	HSD
Reference board	0.33	17.8	0.43	A	0.10	10.1	A
PP	0.33	14.4	0.40	A	0.12	10.5	A,B
HDPE	0.32	10.0	0.44	A	0.12	13.1	B
PS	0.42	9.53	0.49	B	0.13	15.7	B
LDPE	0.40	9.35	0.45	B	0.13	7.6	B
PP_R	0.40	9.44	0.44	B	0.12	13.6	A,B
LDPE_RZ	0.41	12.1	0.48	B	0.13	11.5	B
LDPE_RR	0.38	12.7	0.45	A, B	0.12	13.7	A, B
ANOVA	$F(7, 88) = 9.3263, p = 0.00000$				$F(7, 88) = 5.5786, p = 0.00002$		

For tensile strength perpendicular to the board plane determined after the boiling test, the share and type of the thermoplastic should not considerably affect this value that ranged from 0.10 to 0.13 N/mm^2. Although the statistical analysis indicated some significant differences between the individual boards, they should be recognized similar.

As the thermoplastic polymers are good barriers to water due to their hydrophobic character, adding them to the outer and intermediate layers helps to significantly reduce thickness swelling and water absorption. All the boards containing the polymers in their external layers showed smaller thickness swelling than the reference ones. The differences ranged from 8.5% to nearly 22% (Table 5). Except for the HDPE boards, the improvement in thickness swelling was greater than the polymer share that was about 10%. The reduction in board swelling was greater for polymers with a lower softening point, which is why we achieved the best effects for recycled LDPE and polystyrene. The thickness swelling of boards containing these polymers was significantly lower than of the remaining ones. The degree of hydrophobic protection provided by the investigated polymers was better visible in the TS-X test, which involved immersing the boards in water for two hours while protecting their narrow surfaces with liquid foil. The results were highly variable within the experimental groups due to the degradation of the foil protecting narrow surfaces and uneven distribution of the polymers in the external layers. For the polymers containing boards, they were nearly two times lower than the control boards. Moreover, the changes were more favorable for the boards containing LDPE agglomerate, i.e., easily softening the polymer in their external layers.

Table 5. Effects of the thermoplastics on board swelling after immersion in water.

Board Type (Thermoplastic)	TS			TS-X		
	x, %	υ, %	HSD	x, %	υ, %	HSD
Reference board	24.96	4.7	A	4.51	13.5	A
PP	21.99	3.7	B,C	2.36	20.0	B
HDPE	22.83	3.7	B	2.30	36.2	B
PS	19.51	4.4	F	1.64	32.0	B
LDPE	21.29	3.3	C,D	2.34	15.9	B
PP_R	20.41	5.2	D,E	1.99	35.0	B
LDPE_RZ	19.35	4.5	F	1.77	29.8	B
LDPE_RR	19.88	3.7	F	1.51	28.0	B
ANOVA	$F(7, 88) = 55.929, p = 0.0000$			$F(7, 77) = 25.680, p = 0.0000$		

The absorbability tests showed the boards absorbed from 73% to 80% water (Figure 3). The highest absorbability was determined for the reference board and the lowest for the board containing recycled LDPE which originated from pink foil. Thus, our zero hypothesis, assuming that all produced boards absorb similar amounts of water, shall be discarded. Post hoc tests indicated that the boards enriched with the polymers in their external layers absorbed less water than the reference boards. HSD Tukey's test also indicated significant differences in absorbability for LDPE_RR and HDPE boards, but this was not confirmed by the Bonferroni test. The boards with a 30% share of polymers in their face layers absorbed only by 4%–7% less water than the control, thus indicating that this content of the polymer did not significantly limit water absorption by the lignocellulosic material.

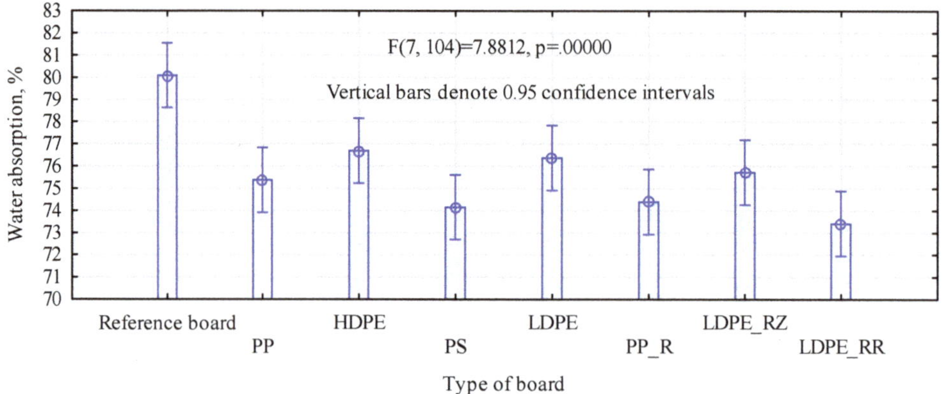

Figure 3. Water absorption in boards made from different materials.

4. Conclusions

The introduction of thermoplastics to the face layers of the mat significantly changed the mechanical properties and water resistance of the experimental boards.

The most favorable results for bending strength and modulus of elasticity were achieved for boards enriched with polystyrene and recycled polyethylene. The boards containing polystyrene showed a smaller decrease in bending strength and maintained a very high modulus of elasticity, while the boards supplemented with recycled polyethylene featured a bending strength similar to the reference boards but low modulus of elasticity.

The tensile strength perpendicular to the board plane increased considerably irrespective of the origin of the thermoplastics added to the face layers, but the best results were achieved for polyethylene recycled from yellow foil and low-density polyethylene.

The virgin thermoplastics provided more favorable tensile strength values after the boiling test than recycled thermoplastics, which is why the boards of the greatest strength were those containing polystyrene and virgin polyethylene of lower density.

Irrespective of their type, thermoplastics added to the face layers favorably reduced hydrophobic properties of the boards, i.e., thickness swelling and water absorption.

Author Contributions: Conceptualization, R.M. and P.B.; methodology, R.M. and D.D.; validation, R.M., P.B. and D.D.; investigation, R.M. and D.D.; writing—original draft preparation, R.M. and P.B.; writing—review and editing, R.M., P.B. and D.D.

Funding: This research was funded by the Ministry of Science and Higher Education program "Regional Initiative of Excellence" in the years 2019-2022, Project No. 005/RID/2018/19.

Acknowledgments: The authors express their sincere thanks to Mr. Maciej Jastrząb for his help in performing the research.

Conflicts of Interest: The authors declare no conflict of interest.

References

1. Niska, K.; Sain, M. *Wood-Polymer Composites*; Woodhead publishing limited: Cambridge, UK, 2008.
2. Wolcott, M.P. Formulation and process development of flat pressed wood-polyethylene composites. *For. Prod. J.* **2003**, *53*, 25–32.
3. Ayrilmis, N.; Jarusombuti, S. Flat-pressed wood plastic composite as an alternative to conventional wood-based panels. *J. Compos. Mater* **2011**, *45*, 103–112. [CrossRef]
4. Benthien, J.T.; Thoemen, H. Effects of raw materials and process parameters on the physical and mechanical properties of flat pressed WPC panels. *Compos. Part A Appl. Sci. Manuf.* **2012**, *43*, 570–576. [CrossRef]
5. Schmidt, H.; Benthien, J.; Thoemen, H. Processing and flexural properties of surface reinforced flat pressed WPC panels. *Eur. J. Wood Prod.* **2013**, *71*, 591–597. [CrossRef]
6. Lyutyy, P.; Bekhta, P.; Sedliačik, J.; Ortynska, G. Properties of flat-pressed wood-polymer composites made using secondary polyethylene. *Acta Fac. Xylologiae Zvolen* **2014**, *56*, 39–50.
7. Bekhta, P.; Lyutyy, P.; Ortynska, G. Effects of different kinds of coating materials on flat pressed WPC panels. *Drv. Ind.* **2016**, *67*, 113–118. [CrossRef]
8. Bekhta, P.; Lyutyy, P.; Ortynska, G. Properties of veneered flat pressed wood plastic composites by one-step process pressing. *J. Polym. Environ.* **2017**, *25*, 1288–1295. [CrossRef]
9. Stark, N.M.; Berger, M. Effect of species and particle size on properties of wood-flour-filled polypropylene composites. In Proceedings of the Conference: Functional Fillers for Thermoplastics and Thermosets, San Diego, CA, USA, 8–10 December 1997.
10. Falk, R.H.; Vos, D.; Cramer, S.M. The comparative performance of woodfiber-plastic and wood-based panels. In Proceedings of the 5th International Conference on Wood-Plastic Composites, Madison, WI, USA, 26–27 May 1999; pp. 269–274.
11. Verhey, S.A.; Laks, P. Wood particle size affects the decay resistance of wood fiber/thermoplastic composites. *For. Prod. J.* **2002**, *52*, 78–81.
12. Błędzki, A.K.; Faruk, O. Wood fiber reinforced polypropylene composites: Compression and injection molding process. *Polym. Plast. Technol.* **2004**, *43*, 871–888. [CrossRef]
13. Chen, H.C.; Chen, T.Y.; Hsu, C.H. Effects of wood particle size and mixing Ratios of HDPE on the properties of the composites. *Holz Roh Werkst.* **2006**, *64*, 172–177. [CrossRef]
14. Gozdecki, C.; Kociszewski, M.; Wilczyński, A.; Zajchowski, S. Mechanical properties of wood/polypropylene composite with coarse wood particles. Effect of specimen size. *Ann. Wars. Univ. Life Sci. For. Wood Technol.* **2009**, *68*, 283–287.
15. Mirski, R.; Dziurka, D.; Banaszak, A. Using rape particles in the production of polymer and lignocellulose boards. *BioResources* **2019**, *14*, 6736–6746. [CrossRef]
16. Synowiec, J. Sposób Wytwarzania Kompozytu z Polimerów Termoplastycznych i Cząstek Lignocelulozowych. Patent PL 183478 B1, 1996.
17. Borysiuk, P.; Mamiński, M.; Zado, A. Some comments on the manufacturing of thermoplastic-bonded particleboards. *Ann. Wars. Univ. Life Sci. For. Wood Technol.* **2009**, *68*, 50–54.
18. Hague, J.R.B.; Loxton, C.; Quinney, R.; Hobson, N. Assessment of the suitability of agri-based materials for panel products. In Proceedings of the First European Panel Products Symposium 1997, Llandudno, Wales, 9–10 October 1997; pp. 136–144.
19. Papadopoulos, A.N.; Hague, J.B.B. The potential for using flax (*Linum usitatissimum* L.) shiv as a lignocellulosic raw material for particleboard. *Ind. Crop. Prod.* **2003**, *17*, 143–147. [CrossRef]
20. Kurokochi, Y.; Sato, M. Properties of binderless board made from rice straw: The morphological effect of particles. *Ind. Crop. Prod.* **2015**, *69*, 55–59. [CrossRef]
21. Kwon, J.H.; Ayrilmis, N.; Han, T.H. Combined effect of thermoplastic and thermosetting adhesives on properties of particleboard with rice husk core. *Mater. Res.* **2014**, *17*, 1309–1315. [CrossRef]
22. Bekhta, P.; Korkut, S.; Hiziroglu, S. Effect of pretreatment of raw material on properties of particleboard panels made from wheat straw. *BioResources* **2013**, *8*, 4766–4774. [CrossRef]
23. Dziurka, D.; Mirski, R. Lightweight boards from wood and rape straw particles. *Drewno* **2013**, *56*, 19–31.
24. Mirski, R.; Dziurka, D.; Czarnecki, R. The possibility of replacing strands in the core layer of oriented strand board by particles from the stems of rape (*Brassica napus* L. *var. napus*). *BioResources* **2016**, *11*, 9273–9279. [CrossRef]

25. Mirski, R.; Dziurka, D.; Banaszak, A. Properties of particleboards produced from various lignocellulosic particles. *BioResources* **2018**, *13*, 7758–7765. [CrossRef]

26. Grigoriou, A.H. Straw-wood composites bonded with various adhesive systems. *Wood Sci. Technol.* **2000**, *34*, 355–365. [CrossRef]

27. Pawlicki, J.; Nicewicz, D. Waste thermoplastic polymers in Technologies of wood-based panels. In Proceedings of the Third International Symposium "Wood Agglomeration", Zvolen, Slovakia, 28–30 June 2000; pp. 81–84.

28. Panthapulakkal, S.; Sain, M. Injection molded wheat straw and corn stem filled polypropylene composites. *J. Polym. Environ.* **2006**, *14*, 265–272. [CrossRef]

29. Moresco, M.; Rosa, S.M.L.; Santos, E.F.; Nachtigall, S.M.B. Agrofillers in polypropylene composites: A relationship between the density and the mechanical properties. *J. Appl. Polym. Sci.* **2010**, *117*, 400–408. [CrossRef]

30. Nourbakhsh, A.; Ashori, A. Wood plastic composites from agro-waste materials: Analysis of mechanical properties. *Bioresour. Technol.* **2010**, *101*, 2525–2528. [CrossRef]

31. Hardinnawirda, K.; Aisha, I. Effect of rice husks as filler in polymer matrix composites. *J. Mech. Eng. Sci.* **2012**, *2*, 181–186. [CrossRef]

32. He, C.; Hou, R.; Xue, J.; Zhu, D. The Performance of polypropylene wood–plastic composites with different rice straw contents using two methods of formation. *For. Prod. J.* **2013**, *63*, 61–66. [CrossRef]

33. Haghdan, S.; Smith, G.D. Natural fiber reinforced polyester composites: A literature review. *J. Reinf. Plast. Comp.* **2015**, *34*, 1179–1190. [CrossRef]

34. Dukarska, D.; Czarnecki, R.; Dziurka, D.; Mirski, R. Construction particleboards made from rapeseed straw glued with hybrid pMDI/PF resin. *Eur. J. Wood Prod.* **2017**, *75*, 175–184. [CrossRef]

35. Abba, H.A.; Nur, I.Z.; Salit, S.M. Review of Agro Waste Plastic Composites Production. *J. Min. Mater. Eng.* **2013**, *1*, 271–279. [CrossRef]

36. Panthapulakkal, S.; Law, S.; Sain, M. Enhancement of Processability of Rice Husk Filled High-density Polyethylene Composite Profiles. *J. Compos.* **2005**, *18*, 445–458. [CrossRef]

37. Najafi, S.K.; Tajvidi, M.; Hamidina, E. Effect of temperature, plastic type and virginity on the water uptake of sawdust/plastic composites. *Holz Roh Werkst.* **2007**, *65*, 377–382. [CrossRef]

38. Kelly, M.W. *Critical Literature Review of Relationships Between Processing Parameters and Physical Properties of Particleboard. General Technical Report FPL-10*; USDA Forest Service, Forest Products Laboratory: Madison, WI, USA, 1977.

39. Park, B.D.; Riedl, B.; Hsu, E.W.; Shields, J. Hot-pressing process optimization by response surface methodology. *For. Prod. J.* **1999**, *49*, 62–68.

40. Dziurka, D.; Mirski, R.; Łęcka, J. The effect of pine particle moisture content on properties of particleboards resinated with PMDI. *EJPAU* **2006**, *9*, 16.

41. Borysiuk, P. *Możliwości wytwarzania płyt wiórowo-polimerowych z wykorzystaniem poużytkowych termoplastycznych tworzyw sztucznych*; Treatises and Monographs; Wydawnictwo SGGW: Warszawa, Poland, 2012; p. 393.

42. Qi, C.; Yadama, V.; Guo, K.; Wolcott, M. Thermal conductivity of sorghum and sorghum–thermoplastic composite panels. *Ind. Crop. Prod.* **2013**, *45*, 455–460. [CrossRef]

43. Research Report. *Wood Plastic Composites Study-Technologies and UK Market Opportunities. The Waste and Resources*; Action Programme (WRAP): Banbury, UK, 2003.

44. *EN 310. Wood-Based Panels—Determination of Modulus of Elasticity in Bending and of Bending Strength*; European Committee for Standardization: Brussels, Belgium, 1993.

45. *EN 319. Particleboards and fibreboards—Determination of Tensile Strength Perpendicular to the Plane of the Board*; European Committee for Standardization: Brussels, Belgium, 1993.

46. *EN-1087-1. Particleboards—Determination of Moisture Resistance—Boil Test*; European Committee for Standardization: Brussels, Belgium, 1995.

47. *EN 317. Particleboards and Fibreboards. Determination of Swelling in Thickness after Immersion in Water*; European Committee for Standardization: Brussels, Belgium, 1993.

48. Saechtling, H. *Tworzywa Sztuczne*; Wydawnictwa Naukowo-Techniczne: Warszawa, Poland, 2000.

49. Zajchowski, S.; Ryszkowska, J. Kompozyty polimerowo-drzewne—Charakterystyka ogólna oraz ich otrzymywanie z materiałów odpadowych. *Polimery* **2009**, *54*, 674–682. [CrossRef]

50. Kuciel, S.; Liber-Kneć, A.; Zajckowski, S. Kompozyty z włóknami naturalnymi na osnowie recyklatu polipropylenu. *Polimery* **2010**, *55*, 718–725. [CrossRef]

51. *EN 312. Particleboards—Specifications*; European Committee for Standardization: Brussels, Belgium, 2010.
52. Yasin, M.; Bhutto, A.W.; Bazmi, A.A.; Karim, S. Efficient Utilization of Rice-wheat Straw to Produce Value–added Composite Products. *Int. J. Chem. Environ. Eng.* **2010**, *1*, 136–143.
53. Bengtsson, M.; Oksman, K. Silane crosslinked wood plastic composites: Processing and properties. *Compos. Sci. Technol.* **2006**, *66*, 2177–2186. [CrossRef]
54. El-Kassas, A.M.; Mourad, A.-H.I. Novel fibers preparation technique for manufacturing of rice straw based fiberboards and their characterization. *Mater. Des.* **2013**, *55*, 757–765. [CrossRef]
55. Li, X.J.; Cai, Z.Y.; Winandy, J.E.; Basta, A.H. Effect of oxalic acid and steam pretreatment on the primary properties of UF-bonded rice straw particleboards. *Ind. Crop. Prod.* **2011**, *33*, 665–669. [CrossRef]
56. Panthapulakkal, S.; Sain, M.; Law, S. Effect of coupling agent on rice husk filled HDPE extruded profiles. *Polym. Int.* **2005**, *54*, 137–142. [CrossRef]

 *polymers*

Article

Combined Chemical Modification of Bamboo Material Prepared Using Vinyl Acetate and Methyl Methacrylate: Dimensional Stability, Chemical Structure, and Dynamic Mechanical Properties

Saisai Huang [1], Qiufang Jiang [1], Bin Yu [2], Yujing Nie [1], Zhongqing Ma [1,*] and Lingfei Ma [1,*]

[1] School of Engineering, Zhejiang Provincial Collaborative Innovation Center for Bamboo Resources and High-Efficiency Utilization, Zhejiang A & F University, Hangzhou 311300, China
[2] Huzhou Ruiyi Wood Industry Co., Ltd., Huzhou 313220, China
* Correspondence: mazqzafu@163.com (Z.M.); malingfei@zafu.edu.cn (L.M.);
 Tel.: +86-571-6110-0905 (Z.M. & L.M.)

Received: 24 August 2019; Accepted: 8 October 2019; Published: 11 October 2019

Abstract: Acetylation and in situ polymerization are two typical chemical modifications that are used to improve the dimensional stability of bamboo. In this work, the combination of chemical modification of vinyl acetate (VA) acetylation and methyl methacrylate (MMA) in situ polymerization of bamboo was employed. Performances of the treated bamboo were evaluated in terms of dimensional stability, wettability, thermal stability, chemical structure, and dynamic mechanical properties. Results show that the performances (dimensional stability, thermal stability, and wettability) of bamboo that was prepared via the combined pretreatment of VA and MMA (VA/MMA-B) were better than those of raw bamboo, VA single-treated bamboo (VA-B), and MMA single-treated bamboo (MMA-B). According to scanning electron microscopy (SEM) and Fourier transform infrared spectroscopy (FTIR) analyses, VA and MMA were mainly grafted onto the surface of the cell wall or in the bamboo cell lumen. The antiswelling efficiency and contact angle of VA/MMA-B increased to maximum values of 40.71% and 107.1°, respectively. From thermogravimetric analysis (TG/DTG curves), the highest onset decomposition temperature (277 °C) was observed in VA/MMA-B. From DMA analysis, the storage modulus (E') of VA/MMA-B increased sharply from 15,057 Pa (untreated bamboo) to 17,909 Pa (single-treated bamboo), and the glass transition temperature was improved from 180 °C (raw bamboo) to 205 °C (single-treated bamboo).

Keywords: bamboo; chemical modification; dimensional stability; dynamic thermodynamic; acetic anhydride; methyl methacrylate

1. Introduction

Bamboo is an important fast-growing and renewable material, and has been widely used as raw material to produce bamboo flooring, bamboo wood composite, bamboo building templates, and bamboo decorative materials because of its high mechanical properties and low strength-to-weight ratio [1,2]. However, bamboo use is highly limited by its strong hygroscopicity; specifically, the free hydroxyl groups from bamboo cell walls result in poor dimensional stability. Currently, many physical or chemical modification methods, such as heat treatment [3–6], in situ polymerization with organic monomers [7–9], acetylation treatment [10–13], and modification with 1,3-dimethylol-4,5-dihydroxyethyleneurea (DMDHEU) [14–16], have been used to improve its poor dimensional stability.

Acetylation is a conventional chemical modification method in which acetyl groups (CH_3CO-) react with hydroxyl groups (–OH) linked to the cellulose of wood/bamboo material; therefore, it improves the dimensional stability of bamboo [12,13,17–24]. The traditional reagents used in wood/bamboo acetylation modification are acetic anhydride (AA), acetyl chlorides (AC), and thioacetic acid (TA) [12,13]. However, acetylation pretreated with these reagents produces a large amount of byproducts of strong acid, and this causes undesirable odors, strength loss, and corrosion of metal fasteners [25,26].

Recently, vinyl acetate (VA) was developed as an environmentally friendly reagent used in acetylation modification [23,27–30]. The byproduct produced from VA acetylation is acetaldehyde, which has a low boiling point and is easily removed. Several researchers have reported that the dimensional stability of wood treated via VA acetylation was much better than that of AA-treated wood. For example, Jebrane et al. reported that the weight percentage gain (WPG) of maritime pine after VA acetylation was 26.2%, which was higher than that of AA-treated wood (20.5%). Additionally, the swelling of blocks treated with VA, which have a WPG above 20%, was always less puffed than AA-treated wood [23]. Furthermore, it has been reported that the bonding strength between VA and cellulose is better than that between AA and cellulose, and this results in a lower swelling coefficient [29]. More VA than AA enters into the voids of cell walls, thereby decreasing the volume of voids and limiting the overall swelling of samples [28,30].

In situ polymerization of unsaturated polymer monomers (e.g., methyl methacrylate (MMA), styrene, acrylonitrile, and acrylamide) within wood pores (e.g., vessels, tracheids, capillaries, and ray cells) to fabricate wood polymer composites (WPCs) is another effective modification method for strengthening the mechanical properties of wood or for protecting the wood matrix from being attacked by water or microorganisms [31–35]. Methyl methacrylate (MMA) is one of the most important vinyl monomers used in WPCs because it has low viscosity and is relatively inexpensive [36–40]. Matto et al. reported that pinewood samples treated via in situ polymerization of MMA resulted in a higher retention of monomers and densification, less variation of permanent swelling, and higher mechanical resistance [37]. Fu et al. grafted MMA onto a wood surface using the atom transfer radical polymerization (ATRP) method, which resulted in the wood having better hydrophobicity [39]. Shang et al. found that the macromechanical properties (bending modulus and compressive modulus) of rattan were highly improved when MMA was grafted onto the surface of rattan [38].

As previously stated, studies reported in the literature have mainly focused individually on VA acetylation or MMA in situ polymerization of wood [23,27,32,37,39,41,42]. These two methods have been documented, and it has been concluded that both lead, to different extents, to improved dimensional stability and increased decay resistance of wood. However, the combined use of these two methods with respect to the enhancement of bamboo properties has not been reported. In this study, the effects of chemical modification via single use of VA and MMA on dimensional stability, chemical properties, and thermodynamic properties of bamboo were investigated. Then, a combined chemical modification method of VA acetylation and MMA in situ polymerization was employed to achieve a synergistic improvement in the dimensional stability and mechanical properties of bamboo.

2. Materials and Methods

2.1. Materials

Bamboo, purchased from Huzhou Ruiyi Wood Industry Co., Ltd. (Huzhou, China), was cut into samples with dimensions of 20 mm × 20 mm × 5 mm (L × W × H, respectively, for dimensional stability analysis) and 35 mm × 12 mm × 2.5 mm (L × W × H, respectively, for dynamic thermodynamic analysis). Samples were oven-dried at 105 °C for 12 h until constant weight was obtained. The sizes and weights of each specimen were then measured to calculate the volume (V_0) and weight (W_0).

2.2. Bamboo Acetylation Pretreatment

Vinyl acetate (VA) acetylation pretreatment of bamboo was carried out in a stainless steel container equipped with a vacuum pump, pressure pump, and calcium chloride drying tube (Figure 1). First, the bamboo specimen and vinyl acetate/dimethylformamide solution (1:1 V/V) with 0.5% concentration of potassium carbonate as the catalyst were added to the container. The container was then put under vacuum to −0.09 MPa, and this was maintained for 12 h. The container was then put into an oven and heated to 110 °C, which was maintained for 6 h. Finally, the acetylated samples were soaked in flowing water for 48 h to remove unreacted reagents and byproducts; then, the samples were dried in a vacuum oven under 0.01 MPa at 105 °C until a constant weight was obtained. The weight percentage gain (WPG) and the volume bulking efficiency (VBE) were calculated using Equations (1) and (2), respectively.

$$WPG = \frac{W_1 - W_0}{W_0} \tag{1}$$

$$VBE = \frac{V_1 - V_0}{V_0} \tag{2}$$

where W_1 and V_1 are respectively the absolute dry mass and volume of the samples before acetylation, and W_2 and V_2 are respectively the absolute dry mass and volume of the samples after acetylation.

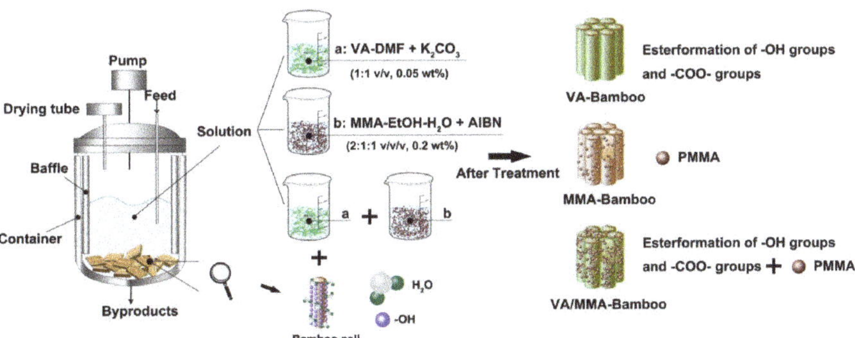

Figure 1. Schematic flow chart of bamboo acetylation, in situ polymerization of MMA on bamboo, and combined treatment of acetylation and in situ polymerization.

2.3. In Situ Polymerization of MMA on Bamboo

In situ polymerization of bamboo–MMA composite was carried out in the same container as mentioned above (Figure 1). First, MMA–EtOH solution (which was composed of MMA, ethanol, and deionized water with a volume ratio of 2:1:1) was added to the container. Azobisisobuttyronitrile (AIBN), which is an initiator with a relative molar ratio of 1:5, was then also added to the container. Next, under vacuum/pressure cycles for 12 h, VA-acetylated samples and untreated samples were impregnated with the MMA–EtOH solution in the container. All of the samples were then washed to remove the residual solvent. Finally, the samples were wrapped in aluminum foil and stored for 24 h; samples were placed separately into the oven at 80 °C for 6 h. In the end, the VA/MMA-treated bamboo and MMA-treated bamboo were vacuum-dried under 0.01 MPa at 105 °C to obtain constant weight. The conversion rate (CR) of MMA was calculated using Equation (3):

$$CR = \frac{W_3 - W_0}{W_2 - W_0} \tag{3}$$

where W_3 is the oven-dried weight after polymerization, W_2 is the wet weight of bamboo after polymerization, and W_0 is the absolute dry mass and volume of bamboo before all of the treatments.

2.4. Characterization of Raw and Pretreated Bamboo

The surface functional groups of raw and pretreated bamboo were tested using Fourier transform infrared spectroscopy (FTIR, Nicolet Is50, Thermo Fisher Scientific, Walthan, MA, USA). The surface morphologies of raw and pretreated bamboo were characterized using cold field emission scanning electron microscopy (SEM, SU8010, Hitachi, Chiyoda, Japan). The contact angles of raw and pretreated bamboo were measured using an interfacial tension tester (OCA200, DataPhysics, Filderstadt, Germany). The apparent contact angle was measured each second for the 5 s deposition on the surface of the sample. The thermal stabilities of raw and pretreated bamboo were tested using a thermogravimetric analyzer (TG-209, NETZSCH, Selb, Germany).

2.5. Dimensional Stability Analysis of Raw and Pretreated Bamboo

The dimensional stabilities of samples were evaluated using 12 specimens that were soaked in water for cyclic soaking–drying. Water soaking (at room temperature and under ambient pressure for 72 h) and oven-drying (103 °C for 24 h) were repeated three times. The weights and volumes of each sample were measured every cycle. Furthermore, the parameters of water absorption (*WA*), volume swelling efficiency (*S_w*), volume shrinking efficiency (*S_k*), and antiswelling efficiency (*ASE*) were calculated using Equations (4)–(7), respectively.

$$WA = \frac{W_{wi} - W_{w0}}{W_{w0}} \times 100\% \tag{4}$$

$$S_w = \frac{V_{wi} - V_{w0}}{V_{w0}} \times 100\% \tag{5}$$

$$S_k = \frac{V_{w0} - V_{wi}}{V_{wi}} \times 100\% \tag{6}$$

$$ASE = \frac{S_{wn} - S'_{wn}}{S'_{wn}} \times 100\% \tag{7}$$

where W_{wi} and V_{wi} are respectively the weight and size of a bamboo block after soaking it i times, and W_{w0} and V_{w0} are respectively the weight and size of a bamboo block after drying it i times. i is 1, 2, and 3. S_{wn} is the swelling efficiency of untreated bamboo, and S'_{wn} is that of reacted bamboo.

2.6. Dynamic Mechanical Analysis of Raw and Pretreated Bamboo

The storage modulus (*E'*), loss modulus (*E''*), and tan delta (tan δ) of both the raw and pretreated bamboo were recorded using a dynamic mechanical analyzer (DMA, Q800, TA Instruments, New Castle, DE, USA). The temperature was scanned from 40 to 240 °C, the heating rate was 2 °C/min, the frequency of the measurements was 1 Hz. Duplicate samples were measured to ensure the reproducibility of results.

3. Results

3.1. Weight Gain Rate, Volume Bulking Efficiency, and Conversion Rate

Table 1 shows the effects of using different modification methods on the weight gain rate (WPG) of bamboo, volume bulking efficiency (VBE) of bamboo, and conversion rate (CR) of MMA. WPG reached a maximum of 18.95% after the combined treatment of VA and MMA, and this indicates that more VA and MMA were fabricated on the bamboo. During the VA and MMA treatment process, VA has an active anhydride group that reacts with hydroxyl groups of bamboo components to create monoesters [28,29], and MMA in situ polymerizes on the surface or in a void of bamboo [23]. Ghorbani et al. reported that the WPG of poplar wood increased from 22.71% with the single treatment of maleic anhydride (MAN) and from 51.8% with the single treatment of MMA to 56.05% with the combined treatment of MAN and MMA [34].

Table 1. Weight percentage gain, volume bulking efficiency, and conversion rate of treated bamboo.

Samples	Weight Percentage Gain (%)	Volume Bulking Efficiency (%)	Conversion Rate (%)
VA-B	11.09 ± 0.56	4.25 ± 0.21	/
MMA-B	6.59 ± 0.33	3.49 ± 0.17	8.68 ± 0.43
VA/MMA-B	18.95 ± 0.95	8.66 ± 0.43	20.69 ± 1.03

Among the three chemical modification methods, the lowest value of VBE was obtained after MMA treatment because the majority of MMA entered the voids of cell walls rather than attached onto the surface of bamboo. In addition, the CR of MMA increased from 8.68% after MMA treatment to 20.69% after the combined treatment of VA and MMA, and this indicates that pretreating bamboo with VA was beneficial for increasing the CR of MMA. VA treatment enhanced interactions between the polymer and bamboo matrix, and more MMA filled the cell lumen of bamboo [41,43]. Li et al. reported that maleic anhydride successfully activated poplar through a nucleophilic substitution reaction and that this newly formed carboxyl group might act as a catalyst by providing a certain level of acidity for polymerization.

3.2. Morphology Characterization

Figure 2 shows morphology characterizations of the cross-section and longitudinal section of raw and pretreated bamboo. Compared to the raw bamboo (Figure 2A,B), the starch granules on the surface of the bamboo cells in the cross-section disappeared after VA treatment. Also, the pits on the parenchymal cells in the longitudinal section disappeared after VA treatment (Figure 2C,D), and this is because of the acetylic inflation [25]. Figure 2E,F clearly shows that MMA was in situ polymerized on the surface of the bamboo, and this is indicated by the tiny spherical granules. As seen in Figure 2G,H, more MMA was fabricated on the surface of the bamboo after the combined treatment of VA and MMA than was fabricated with just the VA treatment. In particular, abundant polymers filled most of the pores and pits in the cross-section and longitudinal section. It was expected that the use of VA would lead to penetration and swelling of the cell wall matrix and to a reduction in the number of hydroxyl groups. Pretreating the bamboo with VA could increase the adhesion amount of MMA on bamboo and create a cross-linked copolymer bonded onto the bamboo cell wall [32,34,37].

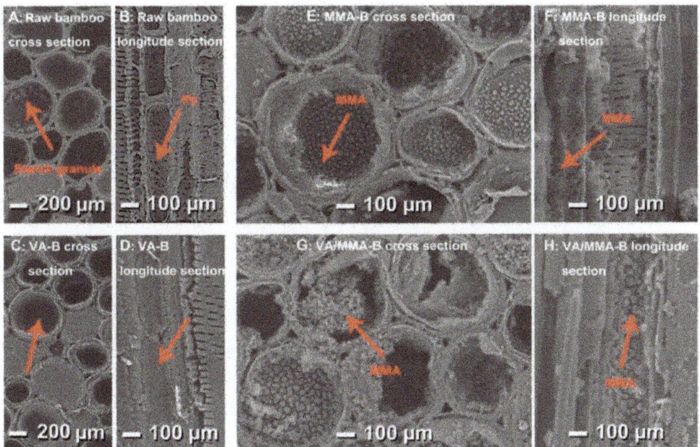

Figure 2. Morphology characterizations of the cross-section and longitudinal section of raw and pretreated bamboo: raw bamboo (**A**,**B**), VA-B (**C**,**D**), MMA-B (**E**,**F**), and VA/MMA-B (**G**,**H**).

3.3. FTIR Analysis

Figure 3 shows the effects of chemical modification on the surface functional groups of bamboo. The band at 3405 cm^{-1} is attributed to the stretching vibration of the hydroxyl group [44]. Compared to the spectrum for raw bamboo, the intensity of the hydroxyl group remarkably decreased in the spectrum for the bamboo that was pretreated using VA and MMA. Hydrophilic hydroxyl groups in bamboo are esterified by acetyl groups (CH$_3$CO–) in VA, and some of the hydroxyl groups are replaced in the polymer chains during the in situ polymerization process of MMA [37]. Among the three chemical modification methods, the lowest intensity band that corresponded to the hydroxyl group was observed in the spectrum for VA/MMA-B, and this indicates that the dimensional stability of VA/MMA-B was probably better than that of raw bamboo, VA-B, and MMA-B. The band at 2955 cm^{-1} is attributed to the stretching vibration of methyl (–CH$_3$) and methylene (–CH$_2$–) [45,46]. Obviously, the intensity of this band in the spectrum of VA/MMA-B is higher than that in the spectra of raw bamboo, MMA-B, and VA-B, and this indicates that more MMA was successfully grafted onto the bamboo cell walls after the combined treatment of VA and MMA [33,40]. The intensity of the band at 1745 cm^{-1} is ascribed to the stretching vibration of carbonyl groups (C=O) [47]. The intensity of the band for carbonyl groups in the spectrum of VA/MMA-B is much stronger than that in the spectra of raw bamboo and MMA-B, but it is slightly stronger than that in the spectrum of VA-B. This result indicates that quite a number of carbonyl groups from both VA and MMA molecules were grafted onto the bamboo cell walls of VA/MMA-B. Another slightly enhanced peak in the spectrum of VA/MMA-B is the band for the ester bond (C–O) stretching vibration at 1242 cm^{-1} [46]. This was caused by the successful reaction of bamboo hydroxyl groups with VA and MMA.

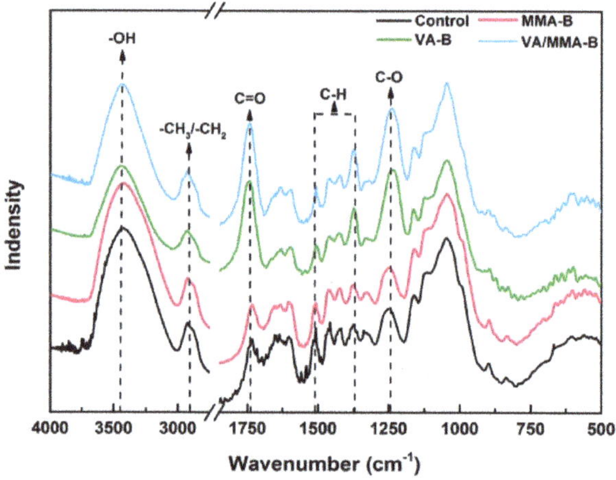

Figure 3. FTIR analysis of raw and pretreated bamboo.

All of the treated bamboo samples were also analyzed using diffuse reflectance FTIR (DRIFT) spectroscopy to study the nonadsorbing matrix. Reflectance spectra were transformed to Kubelka–Munk (K-M) units to minimize scattering contributions to the absorption measured [48,49]. Changes in the relative intensities of bands in the spectra for raw or pretreated bamboo at 1740 cm^{-1}, 1660 cm^{-1}, 1506 cm^{-1}, 1460 cm^{-1}, 1422 cm^{-1}, 1370 cm^{-1}, 1240 cm^{-1}, 1056 cm^{-1}, and 899 cm^{-1} are given in Table 2. These variations in relative intensities of various bands are because of varying quantities of cellulose, hemicellulose, and lignin present in samples that were subjected to different treatments [33,50,51]. The band intensities for VA are always higher than those of the control sample. After MMA polymerization, the bands of VA/MMA-B for the aromatic skeletal vibrations at 1740 cm^{-1}, 1660 cm^{-1}, 1506 cm^{-1}, and 1460 cm^{-1} increased in intensity. Likewise, the intensity of aromatic bending vibration at 1240 and 899 cm^{-1} increased. Regarding single MMA polymerization, the absorptions for C–H were similar with the raw sample and the band for $C–H_2$ decreased. The results from DRIFT analysis indicated that acetylation bamboo increased the weight by VA acetylation and also achieved a sufficient chemical complex by MMA polymerization.

Table 2. DRIFT spectra analysis of raw and pretreated bamboo.

Wavenumber/(cm^{-1})	Assignment	Peak Height of Associated Bands			
		Control	VA/MMA-B	VA-B	MMA-B
1740	C=O stretching vibration	0.195	0.766	0.494	0.323
1660	H–O–H deformation vibration and conjugated C=O stretching vibration	0.159	0.486	0.440	0.164
1506	Aromatic skeletal	0.171	0.446	0.377	0.171
1460	C–H deformation (asymmetric) and benzene vibration in lignin	0.190	0.421	0.457	0.176
1422	C–H deformation (asymmetric)	0.189	0.457	0.457	0.189
1370	C–H$_2$ deformation (symmetric)	0.454	0.530	0.229	0.187
1240	C–O stretching vibration in lignin, acetyl and carboxylic vibration in xylan	0.180	0.587	0.470	0.248
1056	C–O stretching	0.177	0.526	0.472	0.231
899	C1 group frequency in cellulose and hemicellulose	0.077	0.216	0.132	0.065

3.4. Dimensional Stability

Figure 4 shows the effects of chemical modification on the dimensional stability of bamboo in the three cycles of water soaking–drying experiments. Figure 4 includes data for the volume swelling ratio, volume shrinkage ratio, antiswelling efficiency, and water absorption. Compared to control samples, the volume swelling ratio and volume shrinkage ratio of bamboo treated using the three chemical modification methods all decreased, and this indicates that the dimensional stability of bamboo treated using VA and MMA was improved. Among the three chemical modification methods, bamboo treated with the combination of VA and MMA exhibited the best dimensional stability. The antiswelling efficiency of VA/MMA-B reached a maximum value of 40.71%. We suggested that VA reacts with and deactivates the hygroscopic hydroxyl groups of the cell wall polymers and thereby creates a less polar particle surface, which is for better polymerization [30,34,40].

As reported in the literature, the hydrophilic hydroxyl groups are esterified by acetyl groups (CH$_3$CO–) in VA, and this leads to a decreased availability of sites for hydrogen bonding. Furthermore, the voids of bamboo cell lumen are physically blocked via in situ polymerization of MMA, and this hinders the interaction between hydrophilic hydroxyl groups in bamboo and water in the environment [30,31,37]. With an increase in the number of times that the water soaking–drying experiment was repeated, the dimensional stability of bamboo that was treated using the three chemical modification methods decreased. This resulted in an increase in water absorption, as seen in Figure 4d. This observation is explained by the fact that part of the covering layer that formed between the bamboo surface and VA/MMA might be destroyed in the process of several rounds of the water soaking–drying experiment. A reason for this may be the partial loss of the polymer and a high physical cross-linking ratio for the treated samples during the water soaking–drying process. According to Li et al., the ASE of Poplar–PMGM–C also decreased after the third water immersion, and this is the same as our results [43]. Moreover, Zhang et al. indicated that the formed polymer also has a certain water absorption and hygroscopicity because of the high water absorption of the MMA monomer, and this may clarify our observation as well [36].

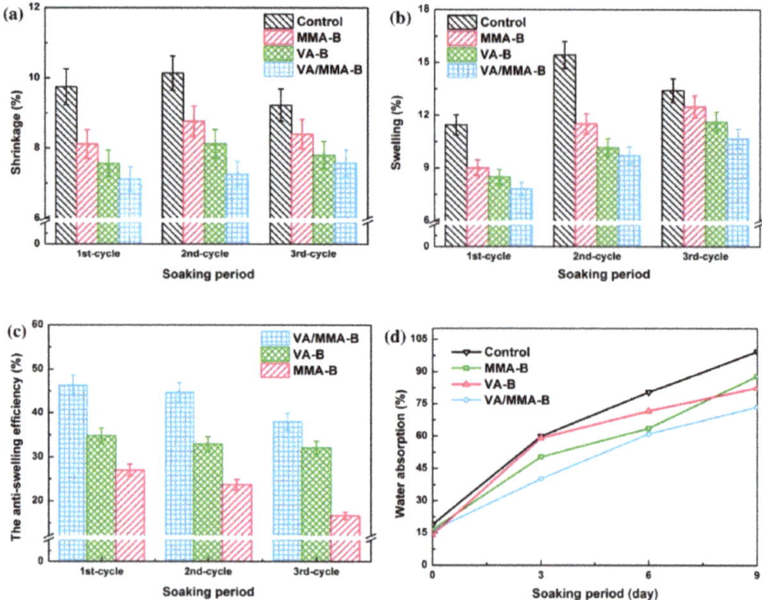

Figure 4. Dimensional stability of bamboo after different chemical modifications with three cycles of water soaking–drying experiments: volume shrinkage ratio (**a**), volume swelling ratio (**b**), antiswelling efficiency (**c**), and water absorption (**d**).

3.5. Wettability Analysis

The wettability of raw and pretreated bamboo was evaluated using contact angle analysis (CA), where a large contact angle corresponds to greater hydrophobicity and better dimensional stability [52]. Figure 5 shows the transient profiles of the CA data of raw and pretreated bamboo. For the untreated samples, the initial CA was 94.3°, and then it sharply dropped to 28.5° with an increase in contact time from 1 s to 5 s. After chemical modification, the CA values of VA-B, MMA-B, and VA/MMA-B were 93.3–43.3°, 90.7–65.2°, and 107.1–84.9°, respectively, with varying contact times. Similar to CA data for raw bamboo, the CAs of treated bamboo gradually decreased with an increase in contact time. This result might have been caused by the surface defects of bamboo, where VA and MMA were not uniformly distributed. This defect makes it hard to obtain Young's equilibrium CA, and the static CA might fluctuate within a range under real conditions [53].

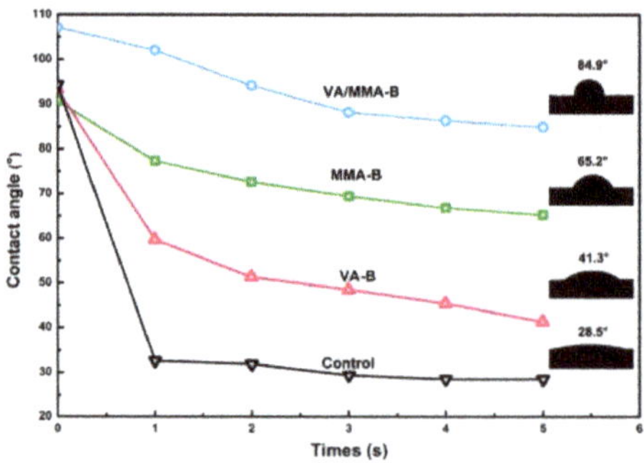

Figure 5. Contact angle profiles for the surfaces of raw and pretreated bamboo.

Among the three chemical modification methods, VA/MMA-B had the highest CA value of 84.9–107.1°, indicating that VA/MMA-B had the greatest hydrophobicity and best dimensional stability. The hydrophilic functional groups were greatly reduced by VA acetylation and in situ polymerization of MMA [23,43]. This result indicates that the combined treatment of VA and MMA is an efficient method for improving the dimensional stability and polymer coverage of the bamboo surface [39].

3.6. TG Analysis

Figure 6 shows the thermal degradation behaviors of raw and pretreated bamboo under a nitrogen atmosphere at a heating rate of 10 °C/min. According to the TG curves, the highest residual mass was observed in raw bamboo (24.5%), followed by MMA-B (21.4%), VA-B (21.3%), and VA/MMA-B (20.4%). The thermal stability of VA and MMA was lower than that of raw bamboo. Therefore, with an increase in the WPG of treated bamboo, the residual mass decreased, but the weight loss rate of the treated bamboo increased.

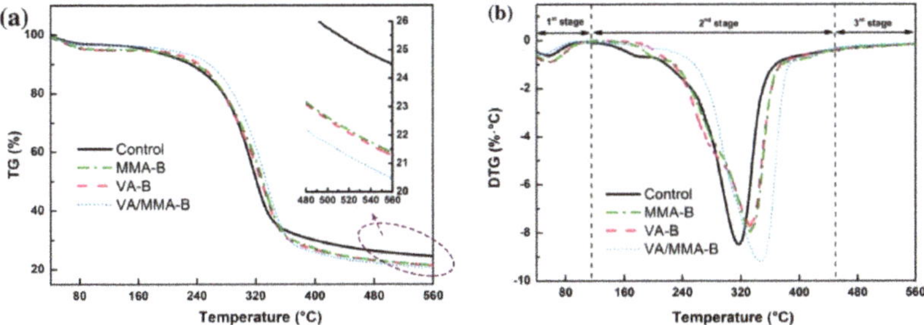

Figure 6. Thermal degradation behaviors of raw and pretreated bamboo: TG (**a**) and DTG (**b**) curves.

According to the DTG curves, the thermal degradation process of bamboo can be divided into three stages: the moisture evaporation stage (50–150 °C), fast devolatilization stage (150–450 °C), and carbonization stage (450–650 °C) [44,46,54–56]. In the fast devolatilization stage, the peak temperatures of the maximum weight loss for raw bamboo, MMA-B, VA-B, and VA/MMA-B were 317 °C, 334 °C, 333 °C, and 337 °C, respectively. This gradually moved toward the side of higher pyrolysis temperature with an increase in the extent of pretreatment via the chemical modifications. This result indicates that VA/MMA-B had enhanced thermal stability compared to all of the other samples. The esterification reaction between VA and the hydroxy groups in bamboo resulted in an increase in the degree of the crystallinity of cellulose, and this was in good agreement with the results from Wei et al. [51]. The decrease in residual mass after grafting MMA was because of the presence of MMA, which degraded more easily than bamboo [42].

3.7. Dynamic Mechanical Analysis

Variations in the dynamic storage modulus (E'), loss modulus (E''), and loss tangent (tan δ) of raw and pretreated bamboo are shown in Figure 7. The dynamic storage modulus is widely used to assess the load-bearing capability of a composite material [57]. The storage modulus of raw bamboo was about 15,057 Pa. In general, the storage modulus (E') of all of the bamboo samples decreased with an increase in temperature. It is worth noting that the value of E' for VA/MMA-B was the highest at the same temperature for each of the four samples (17,909 Pa), and this indicates that the dynamic storage modulus was enhanced after the combined pretreatment using VA and MMA. With an increase in the concentration of acetyl and methyl groups, intermolecular hydrogen bonding was broken. A certain number of hydroxyl groups were then regenerated, and the cellulose chains were consequently closer [32,33,41]. It is believed that the stiffness of the bamboo fibers was enhanced, and the initial storage modulus value of the treated samples is also reflected by this. A similar result has been reported by Jebrane et al. Some esterified material expands into the micropores or lumen of the bamboo cell wall after treatment, and this results in a higher value of E' [23].

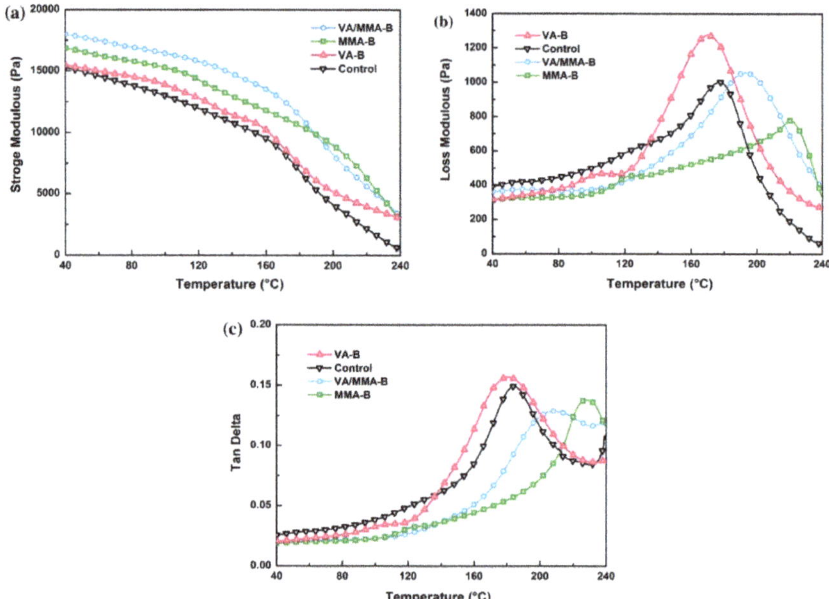

Figure 7. Storage modulus (E') (**a**), loss modulus (E'') (**b**), and loss tangent (tan δ) (**c**) curves of raw and treated bamboo.

Figure 7b illustrates the temperature spectrum of the loss modulus (E'') for raw and pretreated bamboo over the entire temperature range. The highest value of E'' was obtained for VA-B (1275.4 Pa at 170 °C), and the lowest value was for MMA-B (785.5 Pa at 220 °C). With an increase in the in situ polymerization of MMA on bamboo, intermolecular friction and energy consumption of bamboo also increased. Furthermore, a decrease in E'' is consistent with some internal plasticization occurring after polymerization, and this causes a decrease in the energy required to initiate chain mobility [58]. Researchers have reported that the thermal-softening temperature of lignin, hemicellulose, and cellulose are 30–205 °C, 150–220 °C, and 200–250 °C, respectively [59,60]. The loss peak at a temperature of 200 °C is labeled as an α relaxation process that is derived from micro-Brownian motions of bamboo cell wall polymers in the noncrystalline region [28,59]. After chemical modification, there was a remarkable difference in the α relaxation process for different samples. Compared to raw bamboo, the in situ polymerization of MMA on bamboo (MMA-B and VA/MMA-B) led to higher temperature of the α-peak because more MMA, which has amorphous chains, was grafted onto the bamboo. However, acetylation with VA did not significantly influence the temperature of the α-peak.

Figure 7c shows the glass transition temperature for raw and pretreated bamboo in terms of the mechanical loss factor (tan δ). The glass transition temperature (T_g) was approximately equal to the temperature when tan δ reached its maximum value. After in situ polymerization of MMA on bamboo, the glass transition temperature increased from 180 °C for raw bamboo to 205 °C for VA/MMA-B and 220 °C for MMA-B. The increased glass transition temperature should be ascribed to the reinforcement of polymer on bamboo as caused by the in situ polymerization of MMA in cell lumen [33]. However, compared to the glass transition temperature for raw bamboo, a slight decrease was observed in the glass transition temperature for VA-B (175 °C).

4. Conclusions

The effects of the combined treatment with VA and MMA on dimensional stability, chemical structure, and dynamic mechanical properties of bamboo were systematically investigated. Results show that the dimensional stability (i.e., antiswelling efficiency and water absorption) of bamboo after the combined treatment of VA and MMA was remarkably improved because of the decrease in hydrophilic hydroxyl groups. VA and MMA were mainly grafted onto the surface of the cell walls or in the bamboo cell lumen. From TG analysis, an increase in the extent of pretreatment via chemical modifications resulted in the peak temperatures of the maximum weight loss gradually moving toward the side of higher pyrolysis temperature. From DMA analysis, compared to untreated and single-treated bamboo, the storage modulus (E') of VA/MMA-B sharply increased by about 3 MPa, and the glass transition temperature increased from 180 °C to 205 °C. At the same time, the glass transition temperature of MMA-B was the highest (220 °C).

Author Contributions: Conceptualization and experiment design was done by Z.M. and L.M.; material preparation and methodology were done by S.H., Q.J., B.Y., and Y.N.; software was done by S.H. and Q.J.; review and editing were done by Z.M. and L.M.; data analysis and discussion were done by all of the authors.

Funding: This research was funded by the Natural Science Foundation of China (31270592), the Fund of Zhejiang Provincial Collaborative Innovation Center for Bamboo Resources and High-Efficiency Utilization (2017ZZY2-02), and the Young Elite Scientists Sponsorship Program by CAST (2018QNRC001).

Conflicts of Interest: The authors declare no conflict of interest.

References

1. Yeasmin, L.; Ali, M.N.; Gantait, S.; Chakraborty, S. Bamboo: An overview on its genetic diversity and characterization. *3 Biotech* **2015**, *5*, 1–11. [CrossRef] [PubMed]
2. Archila, H.; Kaminski, S.; Trujillo, D.; Zea Escamilla, E.; Harries, K.A. Bamboo reinforced concrete: A critical review. *Mater. Struct.* **2018**, *51*, 102. [CrossRef]
3. Bekhta, P.; Niemz, P. Effect of high temperature on the change in color, dimensional stability and mechanical properties of spruce wood. *Holzforschung* **2003**, *57*, 539–546. [CrossRef]
4. Hakkou, M.; Pétrissans, M.; Zoulalian, A.; Gérardin, P. Investigation of wood wettability changes during heat treatment on the basis of chemical analysis. *Polym. Degrad. Stab.* **2005**, *89*, 1–5. [CrossRef]
5. Rowell, R.M.; Ibach, R.E.; McSweeny, J.; Nilsson, T. Understanding decay resistance, dimensional stability and strength changes in heat-treated and acetylated wood. *Wood Mater. Sci. Eng.* **2009**, *4*, 14–22. [CrossRef]
6. He, Z.; Qu, L.; Wang, Z.; Qian, J.; Yi, S. Effects of zinc chloride–silicone oil treatment on wood dimensional stability, chemical components, thermal decomposition and its mechanism. *Sci. Rep.* **2019**, *9*, 1601. [CrossRef] [PubMed]
7. Bhaskar, J.; Haq, S.; Yadaw, S.B. Evaluation and testing of mechanical properties of wood plastic composite. *J. Thermoplast. Compos. Mater.* **2011**, *25*, 391–401. [CrossRef]
8. Krehula, L.K.; Katančić, Z.; Siročić, A.P.; Hrnjak-Murgić, Z. Weathering of high-density polyethylene-wood plastic composites. *J. Wood Chem. Technol.* **2013**, *34*, 39–54. [CrossRef]
9. Seki, M.; Tanaka, S.; Miki, T.; Shigematsu, I.; Kanayama, K. Forward extrusion of bulk wood containing polymethylmethacrylate: Effect of polymer content and die angle on the flow characteristics. *J. Mater. Process. Technol.* **2017**, *239*, 140–146. [CrossRef]
10. Hung, K.-C.; Wu, T.-L.; Chen, Y.-L.; Wu, J.-H. Assessing the effect of wood acetylation on mechanical properties and extended creep behavior of wood/recycled-polypropylene composites. *Constr. Build. Mater.* **2016**, *108*, 139–145. [CrossRef]
11. Kumar, S. Chemical modification of wood. *Wood Fiber Sci.* **2007**, *26*, 270–280.
12. Kurimoto, Y.; Sasaki, S. Preparation of acetylated wood meal and polypropylene composites ii: Mechanical properties and dimensional stability of the composites. *J. Wood Sci.* **2013**, *59*, 216–220. [CrossRef]
13. Rowell, R.M. 14 chemical modification of wood. In *Handbook of Wood Chemistry and Wood Composites*; CRC Press: Boca Raton, FL, USA, 2005; pp. 381–413.

14. Xie, Y.; Krause, A.; Militz, H.; Turkulin, H.; Richter, K.; Mai, C. Effect of treatments with 1,3-dimethylol-4,5-dihydroxy-ethyleneurea (dmdheu) on the tensile properties of wood. *Holzforschung* **2007**, *61*, 43–50. [CrossRef]

15. Xie, Y.; Xiao, Z.; Grüneberg, T.; Militz, H.; Hill, C.A.S.; Steuernagel, L.; Mai, C. Effects of chemical modification of wood particles with glutaraldehyde and 1,3-dimethylol-4,5-dihydroxyethyleneurea on properties of the resulting polypropylene composites. *Compos. Sci. Technol.* **2010**, *70*, 2003–2011. [CrossRef]

16. Jiang, T.; Gao, H.; Sun, J.; Xie, Y.; Li, X. Impact of dmdheu resin treatment on the mechanical properties of poplar. *Polym. Polym. Compos.* **2014**, *22*, 669–674. [CrossRef]

17. Rowell, R.M.; Simonson, R.; Hess, S.; Plackett, D.V.; Cronshaw, D.; Dunningham, E. Acetyl distribution in acetylated whole wood and reactivity of isolated wood cell-wall components to acetic anhydride. *Wood Fiber Sci.* **2007**, *26*, 11–18.

18. Obataya, E.; Furuta, Y.; Gril, J. Dynamic viscoelastic properties of wood acetylated with acetic anhydride solution of glucose pentaacetate. *J. Wood Sci.* **2003**, *49*, 152–157. [CrossRef]

19. Kumari, R.; Ito, H.; Takatani, M.; Uchiyama, M.; Okamoto, T. Fundamental studies on wood/cellulose-plastic composites: Effects of composition and cellulose dimension on the properties of cellulose/pp composite. *J. Wood Sci.* **2007**, *53*, 470–480. [CrossRef]

20. Li, J.; Zhang, L.-P.; Peng, F.; Bian, J.; Yuan, T.-Q.; Xu, F.; Sun, R.-C. Microwave-assisted solvent-free acetylation of cellulose with acetic anhydride in the presence of iodine as a catalyst. *Molecules* **2009**, *14*, 3551–3566. [CrossRef]

21. Huang, X.; Kocaefe, D.; Kocaefe, Y.; Pichette, A. Combined effect of acetylation and heat treatment on the physical, mechanical and biological behavior of jack pine (pinus banksiana) wood. *Eur. J. Wood Wood Prod.* **2017**, *76*, 525–540. [CrossRef]

22. Hill, C.A.S.; Jones, D.; Strickland, G.; Cetin, N.S. Kinetic and mechanistic aspects of the acetylation of wood with acetic anhydride. *Holzforschung* **1998**, *52*, 623–629. [CrossRef]

23. Jebrane, M.; Sèbe, G. A novel simple route to wood acetylation by transesterification with vinyl acetate. *Holzforschung* **2007**, *61*, 143–147. [CrossRef]

24. Huang, S.; Ma, Z.; Nie, Y.; Lu, F.; Ma, L. Comparative study of the performance of acetylated bamboo with different catalysts. *BioResources* **2018**, *14*, 44–58.

25. Li, J.-Z.; Furuno, T.; Katoh, S.; Uehara, T. Chemical modification of wood by anhydrides without solvents or catalysts. *J. Wood Sci.* **2000**, *46*, 215–221. [CrossRef]

26. Hill, C. Modifying the properties of wood. In *Wood Modification: Chemical, Thermal and Other Processes*; Wiley: Hoboken, NJ, USA, 2006; pp. 19–44.

27. Çetin, N.S.; Özmen, N.; Birinci, E. Acetylation of wood with various catalysts. *J. Wood Chem. Technol.* **2011**, *31*, 142–153. [CrossRef]

28. Jebrane, M.; Harper, D.; Labbé, N.; Sèbe, G. Comparative determination of the grafting distribution and viscoelastic properties of wood blocks acetylated by vinyl acetate or acetic anhydride. *Carbohydr. Polym.* **2011**, *84*, 1314–1320. [CrossRef]

29. Jebrane, M.; Pichavant, F.; Sèbe, G. A comparative study on the acetylation of wood by reaction with vinyl acetate and acetic anhydride. *Carbohydr. Polym.* **2011**, *83*, 339–345. [CrossRef]

30. Özmen, N.; Çetin, N.S.; Mengeloğlu, F.; Birinci, E.; Karakuş, K. Effect of wood acetylation with vinyl acetate and acetic anhydride on the properties of wood-plastic composites. *BioResources* **2013**, *8*, 753–767. [CrossRef]

31. Huang, J.; Schols, H.A.; Jin, Z.; Sulmann, E.; Voragen, A.G. Characterization of differently sized granule fractions of yellow pea, cowpea and chickpea starches after modification with acetic anhydride and vinyl acetate. *Carbohydr. Polym.* **2007**, *67*, 11–20. [CrossRef]

32. Cetin, N.S.; Tingaut, P.; Özmen, N.; Henry, N.; Harper, D.; Dadmun, M.; Sebe, G. Acetylation of cellulose nanowhiskers with vinyl acetate under moderate conditions. *Macromol. Biosci.* **2009**, *9*, 997–1003. [CrossRef] [PubMed]

33. YongFeng, L.; XiaoYing, D.; ZeGuang, L.; WanDa, J.; YiXing, L. Effect of polymer in situ synthesized from methyl methacrylate and styrene on the morphology, thermal behavior, and durability of wood. *J. Appl. Polym. Sci.* **2013**, *128*, 13–20. [CrossRef]

34. Ghorbani, M.; Nikkhah Shahmirzadi, A.; Amininasab, S.M. Physical and morphological properties of combined treated wood polymer composites by maleic anhydride and methyl methacrylate. *J. Wood Chem. Technol.* **2017**, *37*, 443–450. [CrossRef]

35. Yildiz, Ü.C.; Yildiz, S.; Gezer, E.D. Mechanical properties and decay resistance of wood–polymer composites prepared from fast growing species in turkey. *Bioresour. Technol.* **2005**, *96*, 1003–1011. [CrossRef] [PubMed]

36. Zhang, Y.; Zhang, S.Y.; Yang, D.Q.; Wan, H. Dimensional stability of wood–polymer composites. *J. Appl. Polym. Sci.* **2006**, *102*, 5085–5094. [CrossRef]

37. Mattos, B.D.; de Cademartori, P.H.; Missio, A.L.; Gatto, D.A.; Magalhães, W.L. Wood-polymer composites prepared by free radical in situ polymerization of methacrylate monomers into fast-growing pinewood. *Wood Sci. Technol.* **2015**, *49*, 1281–1294. [CrossRef]

38. Shang, L.; Jiang, Z.; Tian, G.; Ma, J.; Yang, S. Effect of modification with methyl methacrylate on the mechanical properties of plectocomia kerrana rattan. *Bioresources* **2016**, *11*, 2071–2082. [CrossRef]

39. Fu, Y.; Li, G.; Yu, H.; Liu, Y. Hydrophobic modification of wood via surface-initiated arget atrp of mma. *Appl. Surf. Sci.* **2012**, *258*, 2529–2533. [CrossRef]

40. Yu, F.; Yang, W.; Song, J.; Wu, Q.; Chen, L. Investigation on hydrophobic modification of bamboo flour surface by means of atom transfer radical polymerization method. *Wood Sci. Technol.* **2013**, *48*, 289–299. [CrossRef]

41. Li, Y.F.; Liu, Y.X.; Wang, X.M.; Wu, Q.L.; Yu, H.P.; Li, J. Wood–polymer composites prepared by the in situ polymerization of monomers within wood. *J. Appl. Polym. Sci.* **2011**, *119*, 3207–3216. [CrossRef]

42. Stolf, D.O.; Bertolini, M.d.S.; Christoforo, A.L.; Panzera, T.H.; Ribeiro Filho, S.L.M.; Lahr, F.A.R. Pinus caribaeavar. Hondurensis wood impregnated with methyl methacrylate. *J. Mater. Civ. Eng.* **2017**, *29*, 05016004. [CrossRef]

43. Li, Y.; Wu, Q.; Li, J.; Liu, Y.; Wang, X.-M.; Liu, Z. Improvement of dimensional stability of wood via combination treatment: Swelling with maleic anhydride and grafting with glycidyl methacrylate and methyl methacrylate. *Holzforschung* **2012**, *66*, 59–66. [CrossRef]

44. Ma, Z.; Wang, J.; Li, C.; Yang, Y.; Liu, X.; Zhao, C.; Chen, D. New sight on the lignin torrefaction pretreatment: Relevance between the evolution of chemical structure and the properties of torrefied gaseous, liquid, and solid products. *Bioresour. Technol.* **2019**, *288*, 121528. [CrossRef] [PubMed]

45. Ma, Z.; Chen, D.; Gu, J.; Bao, B.; Zhang, Q. Determination of pyrolysis characteristics and kinetics of palm kernel shell using tga–ftir and model-free integral methods. *Energy Convers. Manag.* **2015**, *89*, 251–259. [CrossRef]

46. Chen, D.; Wang, Y.; Liu, Y.; Cen, K.; Cao, X.; Ma, Z.; Li, Y. Comparative study on the pyrolysis behaviors of rice straw under different washing pretreatments of water, acid solution, and aqueous phase bio-oil by using tg-ftir and py-gc/ms. *Fuel* **2019**, *252*, 1–9. [CrossRef]

47. Ma, Z.; Zhang, Y.; Li, C.; Yang, Y.; Zhang, W.; Zhao, C.; Wang, S. N-doping of biomass by ammonia (nh3) torrefaction pretreatment for the production of renewable n-containing chemicals by fast pyrolysis. *Bioresour. Technol.* **2019**, *292*, 122034. [CrossRef] [PubMed]

48. Zeng, Y.; Yang, X.; Yu, H.; Zhang, X.; Ma, F. The delignification effects of white-rot fungal pretreatment on thermal characteristics of moso bamboo. *Bioresour. Technol.* **2012**, *114*, 437–442. [CrossRef] [PubMed]

49. Nuopponen, M.H.; Birch, G.M.; Sykes, R.J.; Lee, S.J.; Stewart, D. Estimation of wood density and chemical composition by means of diffuse reflectance mid-infrared fourier transform (drift-mir) spectroscopy. *J. Agric. Food Chem.* **2006**, *54*, 34–40. [CrossRef] [PubMed]

50. Weiland, J.-J.; Guyonnet, R. Study of chemical modifications and fungi degradation of thermally modified wood using drift spectroscopy. *Holz Als Roh-Und Werkst.* **2003**, *61*, 216–220. [CrossRef]

51. Ye, X.; Wang, H.; Wu, Z.; Zhou, H.; Tian, X. The functional features and interface design of wood/polypropylene composites based on microencapsulated wood particles via adopting in situ emulsion polymerization. *Polym. Compos.* **2018**, *39*, 427–436. [CrossRef]

52. Xu, J.; Zhang, Y.; Shen, Y.; Li, C.; Wang, Y.; Ma, Z.; Sun, W. New perspective on wood thermal modification: Relevance between the evolution of chemical structure and physical-mechanical properties, and online analysis of release of vocs. *Polymers* **2019**, *11*, 1145. [CrossRef]

53. Nystrom, D.; Lindqvist, J.; Ostmark, E.; Antoni, P.; Carlmark, A.; Hult, A.; Malmstrom, E. Superhydrophobic and self-cleaning bio-fiber surfaces via atrp and subsequent postfunctionalization. *ACS Appl. Mater. Interfaces* **2009**, *1*, 816–823. [CrossRef]

54. Chen, Y.; Liu, B.; Yang, H.; Wang, X.; Zhang, X.; Chen, H. Generalized two-dimensional correlation infrared spectroscopy to reveal the mechanisms of lignocellulosic biomass pyrolysis. *Proc. Combust. Inst.* **2019**, *37*, 3013–3021. [CrossRef]

55. Ma, Z.; Zhang, Y.; Shen, Y.; Wang, J.; Yang, Y.; Zhang, W.; Wang, S. Oxygen migration characteristics during bamboo torrefaction process based on the properties of torrefied solid, gaseous, and liquid products. *Biomass Bioenergy* **2019**, *128*, 105300. [CrossRef]
56. Ma, Z.; Wang, J.; Zhou, H.; Zhang, Y.; Yang, Y.; Liu, X.; Ye, J.; Chen, D.; Wang, S. Relationship of thermal degradation behavior and chemical structure of lignin isolated from palm kernel shell under different process severities. *Fuel Process. Technol.* **2018**, *181*, 142–156. [CrossRef]
57. Das, M.; Chakraborty, D. Influence of mercerization on the dynamic mechanical properties of bamboo, a natural lignocellulosic composite. *Ind. Eng. Chem. Res.* **2006**, *45*, 6489–6492. [CrossRef]
58. Su, C.; Zong, D.; Xu, L.; Zhang, C. Dynamic mechanical properties of semi-interpenetrating polymer network-based on nitrile rubber and poly (methyl methacrylate-co-butyl acrylate). *J. Appl. Polym. Sci.* **2014**, *131*, 40217. [CrossRef]
59. Jiang, J.; Lu, J. Dynamic viscoelastic behavior of wood under drying conditions. *Front. For. China* **2009**, *4*, 374–379. [CrossRef]
60. Hristov, V.; Vasileva, S. Dynamic mechanical and thermal properties of modified poly (propylene) wood fiber composites. *Macromol. Mater. Eng.* **2003**, *288*, 798–806. [CrossRef]

Article

Ammonium Lignosulfonate Adhesives for Particleboards with pMDI and Furfuryl Alcohol as Crosslinkers

Venla Hemmilä, Stergios Adamopoulos *, Reza Hosseinpourpia and Sheikh Ali Ahmed

Department of Forestry and Wood Technology, Linnaeus University, Lückligs plats 1, 351 95 Växjö, Sweden; venla.hemmila@lnu.se (V.H.); reza.hosseinpourpia@lnu.se (R.H.); sheikh.ahmed@lnu.se (S.A.A.)
* Correspondence: stergios.adamopoulos@lnu.se

Received: 12 September 2019; Accepted: 4 October 2019; Published: 10 October 2019

Abstract: Tightening formaldehyde emission limits and the need for more sustainable materials have boosted research towards alternatives to urea-formaldehyde adhesives for wood-based panels. Lignin residues from biorefineries consist of a growing raw material source but lack reactivity. Two crosslinkers were tested for ammonium lignosulfonate (ALS)—bio-based furfuryl alcohol (FOH) and synthetic polymeric 4,4′-diphenylmethane diisocyanate (pMDI). The addition of mimosa tannin to ALS before crosslinking was also evaluated. The derived ALS adhesives were used for gluing 2-layered veneer samples and particleboards. Differential Scanning Calorimetry showed a reduction of curing temperature and heat for the samples with crosslinkers. Light microscopy showed that the FOH crosslinked samples had thicker bondlines and higher penetration, which occurred mainly through vessels. Tensile shear strength values of 2-layered veneer samples glued with crosslinked ALS adhesives were at the same level as the melamine reinforced urea-formaldehyde (UmF) reference. For particleboards, the FOH crosslinked samples showed a significant decrease in mechanical properties (internal bond (IB), modulus of elasticity (MOE), modulus of rupture (MOR)) and thickness swelling. For pMDI crosslinked samples, these properties increased compared to the UmF. Although the FOH crosslinked ALS samples can be classified as non-added-formaldehyde adhesives, their emissions were higher than what can be expected to be sourced from the particles.

Keywords: biorefinery lignin; wood panels; sustainable adhesives; adhesive penetration; particleboard properties; formaldehyde emissions

---

1. Introduction

Wood-based panel industry uses almost solely petrol-based adhesives, such as urea-formaldehyde and melamine-urea-formaldehyde. There are two main drivers to replace these synthetic systems with formaldehyde-free bio-based alternatives; lowered formaldehyde limits and the need for sustainability. Formaldehyde has been classified as a human cancer-causing carcinogen (Group 1) since 2004 [1]. The formaldehyde limits have steadily been lowered since, the latest example being the Federal Ministry for the Environment, Nature Conservation and Nuclear Safety of Germany changing their legislation, and effectively lowering the European E1 emission level of 0.10 ppm to 0.05 ppm (EN 717–1) in Germany [2]. These low limits can be reached by modifying the existing systems, as an example by adding formaldehyde scavengers, such as urea, or by using natural compounds, such as waste wood bark floors [3]. Active powdered bark has been found to be able to reduce the free formaldehyde release to about 75% on plywood [4]. Bio-based adhesives can be fully formaldehyde-free and have the additional benefit of increased sustainability. However, to reach 100% bio-based formulations is challenging, and the focus has been on increasing the content of bio-based material with a stepwise approach. The properties and reaction time are enhanced by using synthetic, reactive crosslinkers, such as polyethyleneimines (PEI) for proteins [5], aldehydes for lignins and tannins [6], and the universal crosslinker, polymeric diphenylmethane diisocyanate (pMDI) for most bio-based materials [7,8].

The most researched biomaterials for wood panel adhesives are proteins (e.g., soy and wheat), starches, tannins, and lignins [9–13]. Out of these, lignin is the one with high volumes and only a few value-added applications. Out of the 50 million tons of lignin produced worldwide annually, the majority is used as a combustion fuel in pulp mills, and only 2 wt % is used commercially [14,15]. Lignin consists of randomly crosslinked phenylpropanoid units linked to each other by C-O-C and C-C bonds. The word "lignin" refers both to natural lignin found in plants, as well as to the by-product from pulping, papermaking, and biorefinery processes that generate energy, fuels, and materials [16]. The increasing amounts of lignin available from biorefinery processes make it important to find suitable value-adding applications [17]. Lignins from sulfite processes, lignosulfonates, are water-soluble almost over the entire pH range [18]. In the sulfite pulping process, aqueous sulfur dioxide is combined with counter ions, such as ammonium (NH^{4+}) or sodium (Na^+). Sulfonic acid is introduced to the α-carbon atoms of lignin, leading to hydrolysis, which increases the hydrophilicity, and thus the solubility of lignin in the pulping process. This is due to the cleavage of the linkages between phenylpropane units that frees phenolic hydroxyl groups, which are later available for further reactions [19].

One of the main uses of lignin residues is lignin-phenol-formaldehyde (LPF) resins, where lignin is used to partially replace phenol. The phenol replacement amounts are typically below 50%, as the addition of lignin lowers the reactivity of the resin, leading to increased reaction times [20]. Lignin modifications, such as methylolation and phenolation, can be used to increase lignin reactivity with industrially acceptable cost [11,21]. However, this is not enough, especially if unpurified sulfur-containing technical lignins (Kraft lignins and lignosulfonates) are used. In order to use these lignins as main components in adhesive formulations, the low reactivity needs to be compensated by using a suitable crosslinker, especially for particleboard and fibreboard production, where the production speed is a critical parameter [11].

Furfuryl alcohol (furan-2-ylmethanol) is a bio-sourced alcoholic heterocyclic compound from the furan family and can be produced by the reduction of furfural. In the presence of a catalyst at elevated temperatures, furfuryl alcohol polymerizes into a dark-colored, insoluble polymer through a complex polycondensation process that is influenced by many factors, such as the strength and type of the acid catalyst, temperature, and presence of water [22–24]. Furfuryl alcohol requires very acid conditions for curing, which is problematic for wood-based materials as the low pH and high residual acidity hydrolyzes the holocellulose of wood at the interface. However, there are some studies suggesting that it might also be possible to have reactive furfuryl alcohol at alkaline conditions [25]. When blended in plasticized form into furfuryl alcohol during its polymerization, lignin is capable of creating linkages to furfuryl alcohol to form thermoset resins. In a study by Guigo et al. [26], two addition rates of straw and grass lignin from the soda pulping process were tested, 20 and 30 wt %. Zhang et al. found that furfuryl alcohol could be reacted with glyoxal before reacting with lignin. The addition of epoxy resin (3, 6, and 9%) further improved the performance of the lignin-glyoxal-furfuryl alcohol particleboard adhesives [27]. Lignins are also proposed to react with furfuryl alcohol precursor, furfural, at high acidic conditions, where lignin and furfural are proposed to replace phenol and formaldehyde in PF resin formulations, respectively [28].

Polymeric 4,4′-diphenylmethane diisocyanate (pMDI) is commonly used as a sole adhesive in the wood panel industry, especially in the production of oriented strand board (OSB) and medium-density fibreboard (MDF) [29]. For particleboards, it is the only industrial formaldehyde-free adhesive used in large industrial scale [30], standing approximately 1% of the annual European panel production [31]. However, pMDI is very expensive and needs to be used in stabilized form due to its high reactivity towards moisture [9]. It can be used as a crosslinker for bio-based materials, such as proteins [32] and mixtures of tannins and lignins [33,34]. As pMDI is a strong adhesive on its own, the final test results of these combinations can be more dependent on the pMDI amount, and the contribution of the bio-based material can remain unclear [30]. The previous study on the performance of one-layer particleboard bonded with glyoxalated lignin combined with pMDI showed superior internal bond (IB) strength in dry and boiled conditions at a mixture level of 60/40 wt % glyoxalated lignin/pMDI [7]. Lei et al. (2008) reported that the internal bond strength of particleboards produced with glyoxalated lignin, mimosa tannin, and pMDI, up to 80 wt % natural polymers, met the requirements for interior grade panels.

The utilization of proper crosslinker is an important factor for developing a lignin-based adhesive that can meet the reaction speed and adhesion strength requirements of the wood panel industry. High molecular weight ammonium lignosulfonates (ALS) have high absorptivity, making them suitable to be used as dispersants. The high amount of hydroxyl group makes ALS suitable for adhesive applications, such as reaction with isocyanates to form urethanes [35]. Thus, ALS has the potential to be used as an adhesive with additional dispersive properties to enhance the spreading of the other components. Therefore, this work was conducted with the aim to study ALS adhesives with synthetic (pMDI) and bio-based (furfuryl alcohol) crosslinkers for particleboard making. The effect of tannin addition into the adhesive mixture was also studied for both crosslinkers. The adhesives were initially evaluated by means of differential scanning calorimetry (DSC) and veneer gluing to get details on their curing behavior, interaction with wood tissues (penetration), and shear bond strength. Three-layered particleboards were then tested in terms of their physical, mechanical, and formaldehyde emission properties, and the results were compared with those bonded with a commercial adhesive.

2. Materials and Methods

2.1. Chemicals and Adhesive Formulations

Commercial ammonium lignosulfonate (ALS) and mimosa tannin (m) were kindly provided by Borregaard (Neuss, Germany) and Silvachimica S.r.l. (San Michele Mondovì, Italy), respectively. ALS was derived from fermented Norway spruce (*Picea abies*) wood sulfite liquor. Sulfuric acid (H_2SO_4, 98%) and furfuryl alcohol (FOH, 98%) were purchased from Sigma Aldrich (Saint Louis, MO, USA). Melamine reinforced urea-formaldehyde (UmF) with a solid content of 68% was purchased from AkzoNobel (Stockholm, Sweden). Ammonium sulfite, with a concentration of 30%, was purchased from Yara (Köping, Sweden). Polymeric 4,4′-diphenylmethane diisocyanate (pMDI) adhesive (I-Bond PBEM 4352) was purchased from Huntsman (Everberg, Belgium).

Table 1 shows details on the various adhesive formulations for testing of 2-layered veneer samples and particleboards based on ALS and mimosa tannin (m) with FOH and pMDI as crosslinkers. For 2-layered veneer testing, the same application amount (100 g solids/m^2) was used. For particleboard tests, a base application amount of 12% (wt % on dry particles) was used for the UmF and ALS references. For the formulations with crosslinkers, half of the base (6% of ALS or ALS + m) was replaced with the crosslinker. As a typical application amount of these crosslinkers is around 4%, half of this, i.e., 2 wt % on dry fibers (25% of the total adhesive amount), was used as the crosslinker amount, leading to a final application amount of 8%.

For all formulations, the dry ALS powder, with or without tannin (m) powder, was first dissolved in water to form the base solution. Lignin (or lignin and tannin) to water ratio was 1:2, and the final solid content for all formulations with crosslinkers was 48% ± 2%. For the formulations with the bio-based crosslinker FOH (ALS-FOH and ALS-FOH + m), the base solution consisting of ALS or ALS + m was first heated to 70 ± 3 °C under constant magnetic stirring, after which the FOH was added. The pH was adjusted to 2 with 30% H_2SO_4 (2% *w/w* of solids), and the reaction proceeded for 45 min for adhesive evaluations on veneer samples. An identical procedure was applied for the preparation of the same adhesives for particleboard manufacturing. A shorter reaction time (30 min) was required to keep the viscosity of the adhesive at a spray-able range (<300 mPa·s) for the lab-scale semi-manual blender used in this study. Viscosity at the end of the reaction was controlled, using a TQC rotational portable viscometer (DV1400, Capelle aan den IJssel, The Netherlands). Spindle 2 was used for values below 400 and spindle 7 for values around 2000. For the adhesives with the synthetic crosslinker ALS-pMDI and ALS-pMDI + m, pMDI was added to the ALS-water solution (concentration

50%) just before the application of the adhesive for veneer gluing. It was possible to apply these adhesives containing pMDI within 25 min. For particleboard production, the base solutions (ALS and ALS + m) were sprayed first on the wood particles, and then pMDI was sprayed separately to avoid a preliminary reaction. As references for the various tests (2-layered veneer samples, particleboards), industrial UmF resin, with a melamine content of 5%, and 50% ALS-water solution alone was used at 12% (based on the dry weight of wood). For UmF, 3% (*w/w* of dry resin) of ammonium sulfite with a solid content of 45% was added as a hardener.

Table 1. Adhesive compositions for testing of 2-layered veneer samples and particleboards based on ammonium lignosulfonate (ALS), furfuryl alcohol (FOH), and polymeric 4,4′-diphenylmethane diisocyanate (pMDI) as crosslinkers, and with or without mimosa tannin (m).

Adhesive ID	Base [1]: Lignin to Tannin Ratio	Crosslinker Amount in Relation to Base [1] (%)	Application Amount for 2-Layered Veneer (g solids/m^2)	Application Amount for Particleboard (wt % to dry particles)	Final Viscosity for Veneer/Particleboard (mPa·s)
UmF [2]			100	12	370/370
ALS	10:0	-	100	12	80/90
ALS-FOH	10:0	25	100	8	2100/210
ALS-FOH + m	9:1	25	100	8	2350/190
ALS-pMDI	10:0	25	100	8	120/110
ALS-pMDI + m	9:1	25	100	8	110/100

[1] Base refers to lignin or the lignin-tannin mixture, [2] Melamine-urea formaldehyde with 3% (*w/w* of dry resin) of ammonium sulfite as a hardener.

2.2. Differential Scanning Calorimetry (DSC)

Curing behaviors of the adhesive formulations (except for UmF) were determined using a NETZSCH STA 409PC (Bayern, Germany) instrument. A total of 5–10 mg of freshly blended mixtures was heated from 25 to 225 °C at a heating rate of 20 °C min^{-1} under a nitrogen flow of 10 mL/min, as explained previously [36].

2.3. Shear Bond Strength

The shear bond strength of all the adhesive formulations shown in Table 1 was evaluated by tensile shear strength using a lap joint test according to EN 205 [37] with Instron universal testing machine (Buckinghamshire, England). Beech veneers with a moisture content of 12 ± 1% were used to produce 2-layered samples of dimensions 100 × 20 × 3.2 (L × W × H) mm^3. Adhesives' application amount was 100 g of solids per m^2 to an area of 20 × 20 mm^2. The samples were pressed in a single daylight press at a temperature of 150 °C for 90 s and conditioned before testing. The loading speed (constant) was set to 2 mm/min, and a total of ten repetitions of each adhesive formulation were tested.

2.4. Fourier Transform Infrared (FTIR) Spectroscopy

The chemical structure of the adhesive formulations crosslinked with pMDI was analyzed with a P-Elmer FTIR Spectrometer (Seer Green, United Kingdom). The evaluation was performed between 4000 and 800 cm^{-1} at room temperature, accumulating 10 scans with a resolution of 4 cm^{-1}, as described previously [38]. The spectra were collected from bondlines of tensile shear strength test samples in order to confirm the incorporation of ALS in the cured adhesives by chemical reaction with pMDI, i.e., the formation of urethane linkages.

2.5. Adhesive Penetration

To observe the bondline thickness and adhesive penetration into the porous wood structure, 30–40 μm thick sections exposing a bondline with a cross-sectional surface at random longitudinal positions of the 2-layered veneer samples were cut using a sliding microtome (WSL, Birmensdorf, Switzerland). The sections were placed unstained on glass slides and viewed under a motorized

BX63F light microscope equipped with the DP73 color CCD cooled camera (max. 17.28 megapixel) and the software program cellSens DIMENSION 1.18 (all Olympus, Lund, Sweden). For each adhesive formulation, a minimum of 10 microtome slices were cut in order to obtain microscopy pictures. Random regions of interest with dimensions of 1100×600 (W × H) μm^2 were selected to measure the bondline thickness and the maximum penetration (MP), as explained in previous studies [39,40]. A total number of 30 MP measurements were performed. The used software was able to distinguish between the darker adhesive parts from the lighter colored wood tissues, and hence to calculate the desired statistical parameters. The average of 5 deepest detected adhesive objects in the interphase region was selected to measure the MP using the equation:

$$MP = \frac{\sum_i^5 (y_i + r_i - y_0)}{5}$$

where MP is the maximum depth of penetration (μm), y_i is the centroid of adhesive object i (μm), r_i is the mean radius of adhesive object i (μm), and y_0 is the reference y-coordinate of the bondline interface (μm).

In addition, for each of the 30 selected regions, a bondline thickness was measured as the thickness of the region around the center of the bondline, where the adhesive was continuously present.

2.6. Particleboard Manufacturing and Testing

Standard industry core and surface layer wood chips from softwood species (Scots pine and Norway spruce) were oven-dried to $1.0 \pm 0.3\%$ moisture content. Dried chips were then blended with the adhesive formulations, shown in Table 1, in a drum fitted with a gun triggered by an air compressor. No wax or any other additives were applied for the manufacturing of the exploratory laboratory boards. The resinated chips were placed in a mold with dimensions of 500×450 mm^2 to produce 3-layer particleboards (65% core share) with a thickness of 12 mm and a target density of 620 kg/m^3. All boards were pressed at a temperature of 190 °C for 270 s using 12 mm metal stops in a single daylight hot press (Ake, Mariannelund, Sweden). After cooling to room temperature, the boards were cut to 400×400 mm^2 dimensions and conditioned at $65 \pm 3\%$ relative humidity and 20 ± 2 °C before testing. Two particleboard panels were used to evaluate the physical and mechanical properties for each adhesive system. For internal bond (IB) measurements, eight squares with a side length of 50 mm were cut and tested according to EN 319 [41]. For moduli of elasticity (MOE) and rupture (MOR), 6 pieces with a thickness of 50 mm and length of 210 mm were cut and tested according to EN 310 [42]. Mechanical tests were performed with an Instron universal testing machine (Buckinghamshire, England). Eight squares with a side length of 50 mm were cut and immersed in water with a temperature of 20 °C for 2 h and 24 h, and the respective thickness swelling values were calculated according to EN 317 [43].

2.7. Formaldehyde Emission Testing

Formaldehyde emission measurements with 0.044 m^3 Dynamic Micro Chamber (DMC) were performed according to ASTM D 6007–14 [44] standard method (10.39 m^2/m^3 of particleboard at 25 °C, 50% relative humidity, and an air exchange rate of 12.19 h^{-1}), as described previously [45]. The ASTM D 6007–14 method was chosen for being fast and for having a good correlation to EN 717–1 chamber method [46].

2.8. Statistical Analysis

Mean values and standard deviations were calculated with Microsoft Excel. Standard deviations (sd) were calculated using the formula:

$$sd = \sqrt{\sum \frac{(x_i - \bar{x})^2}{(n-1)}}$$

where x_i is the observed value, $\bar{x}$ is the mean of the values, and n is the sample size. The statistical differences between the means were assessed by ANOVA and Tukey's honestly significant difference (HSD) using a *p*-value of under 0.05 as the threshold of statistical significance. In one case (maximum penetration), the statistical differences between two sample data sets were assessed with Welch Two Sample t-test.

3. Results

3.1. Curing Behavior of the Adhesives

Differential Scanning Calorimetry (DSC) was used to analyze the curing behavior of the ALS adhesives crosslinked with pMDI and FOH, with and without tannin (Figure 1, Table 2). For particleboard adhesives, the curing temperature determines if the adhesive fully cures during pressing. The ALS sample had an exothermic curing peak T_{max1} of 129 °C. The addition of both crosslinkers—FOH and pMDI—lowered the curing temperature, indicating higher reactivity [47]. ALS-FOH and ALS-FOH + m had T_{max1} of 119 °C and 120 °C, respectively. The presence of tannin, however, caused an additional peak to appear at 115 °C for the ALS-FOH + m sample, indicating two separate reactions happening during the heating. The additional reaction can be connected to the reaction of FOH with tannin, which has been reported previously [48]. An additional reaction peak was also detectable for both pMDI crosslinked ALS samples. However, there was a big difference between the samples; the curing of ALS-pMDI happened at the lowest temperature of all the samples (107 °C), while the T_{max1} of ALS-Pmdi + m was the highest of the crosslinked samples at 125 °C. Lignin-based adhesives have varying curing temperatures depending on the crosslinker used. As an example, experimental lignin-formaldehyde resin was found to have curing temperature around 203 °C [49], lignin-phenol-formaldehyde adhesives reported to have curing temperature of 168 °C (75% lignin) and 144 °C (40% lignin) [50], and between 145.7–150.9 °C in another study [47]. Lignin-based epoxy resins had curing temperatures ranging from 90–145 °C, depending on the catalyst and heating rate [51].

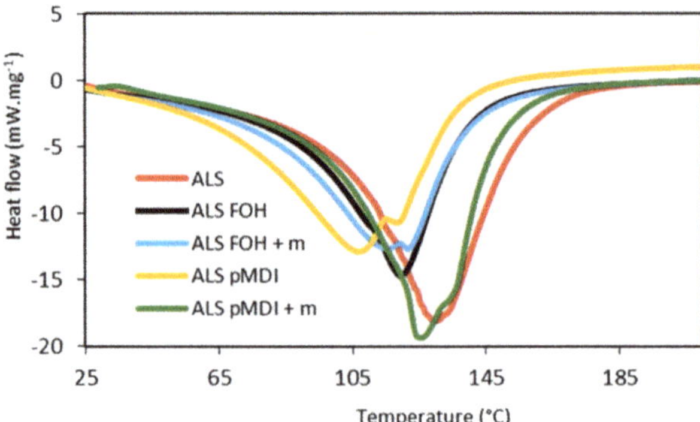

Figure 1. DSC heat of reaction thermograms of adhesive systems based on ammonium lignosulfonate (ALS), furfuryl alcohol (FOH), and polymeric 4,4′-diphenylmethane diisocyanate (pMDI) as crosslinkers, and with or without mimosa tannin (m).

Table 2. Details on the DSC thermograms of adhesive systems based on ammonium lignosulfonate (ALS), furfuryl alcohol (FOH), and pMDI as crosslinkers, and with or without mimosa tannin (m).

Adhesive identification	Onset (°C)	T_{max1} (°C)	T_{max2} (°C)	Curing Heat (J/g)
ALS	104	129	-	172
ALS-FOH	98	119	-	117
ALS-FOH + m	84	120	115	126
ALS-pMDI	75	107	118	123
ALS-pMDI + m	113	125	135	167

The curing heat of ALS was 172 J/g, which is lower than found for lignin extracted from different fiber sources using laboratory organosolv methods [52]. The ALS-FOH had a curing heat of 117 J/g, and the tannin reinforced sample ALS-FOH + m had a curing heat of 126 J/g. The reactivity of furfuryl alcohol-based systems is dependent on the type and amount of catalyst used, as the catalyst needs to reach all the reactive sites in the resin network. In this case, a fixed amount of catalyst (2% of the total solids) was used, and increasing this might have a positive impact on the curing heat based on previous investigation [53]. The ALS-pMDI + m had the highest curing heat (167 J/g), while the ALS-pMDI had the lowest at 123 J/g. This indicated that it took more total energy to cure the ALS-pMDI + m sample than any other sample. This was surprising since pMDI-cured samples typically required lower energy, as in the case of the ALS-pMDI sample. It should be noted that measuring pMDI-containing samples on DSC was challenging, as the reaction with water of the adhesive starts before the temperature was raised. In the case of ALS-pMDI + m, the reaction might have been so fast that the curing had partially happened before the measurement. The curing heats of all of the pMDI and FOH crosslinked ALS samples were lower than that of un-crosslinked, being in the range of 167–117 J/g. They were higher than those reported by Kalami et al. [49] for their experimental lignin-formaldehyde resin (90 J/g) but lower than the commercial phenol-resorcinol-formaldehyde resin used in the same study (171 J/g).

3.2. Tensile Shear Strength and Adhesive Penetration

As shown in Figure 2 for the 2-layered veneers, gluing with sole ALS resulted in poor tensile shear strength (0.56 N/mm^2). The addition of crosslinkers significantly increased the tensile shear strength of the ALS sample to the level of the UmF cured samples (UmF 1.30 N/mm^2, crosslinked 1.47–1.72 N/mm^2). There was no significant difference between the crosslinked samples or the UmF bonded sample ($p > 0.05$).

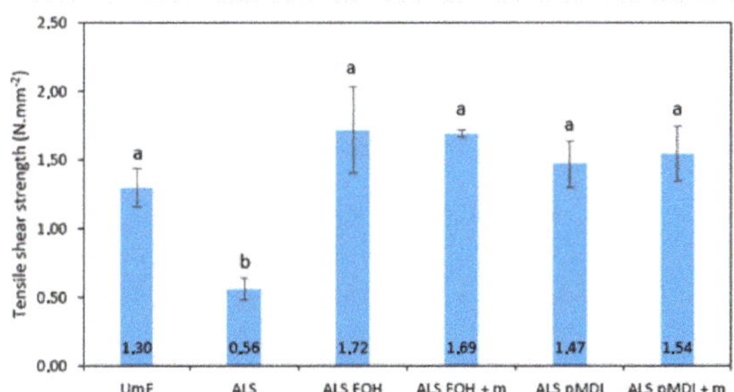

Figure 2. Tensile shear strength of 2-layered beech samples bonded with urea-melamine-formaldehyde (UmF), ammonium lignosulfonate (ALS), ALS crosslinked with furfuryl alcohol (ALS FOH), ALS FOH with mimosa tannin (ALS FOH + m), ALS crosslinked with pMDI (ALS pMDI) and ALS pMDI with mimosa tannin (ALS pMDI + m). Each column is the average of 10 determinations, and error bars represent standard deviations. Values labeled with different letters (a and b) are statistically different at an error probability of $\alpha = 0.05$ (ANOVA and Tukey's HSD tests). UmF, melamine reinforced urea-formaldehyde; HSD, honestly significant difference.

Although the tensile shear strength values were not statistically different among the crosslinked samples, there were differences in the bondline thickness, as seen in Figure 3. The adhesive formulations containing FOH as crosslinker exhibited significantly thicker ($p < 0.0001$) bondlines (ALS-FOH with 52 µm and ALS-FOH + m with 67 µm) than the ones crosslinked with pMDI (ALS-pMDI with 12 µm and ALS-pMDI + m with 26 µm). However, the different bondline morphologies did not seem to play any role in the adhesion, and systems with both crosslinkers were able to bridge the gap between the wood substrates [54]. The two pMDI crosslinked ALS samples did not fill cell lumens beyond the bondline, and thus only the bondline thickness was determined. Adhesive penetration could be observed in adhesive formulations crosslinked with FOH. The MP for ALS-FOH was significantly higher ($p > 0.05$) than for the ALS-FOH + m sample (178 µm and 87 µm, respectively). This could partially be due to the higher viscosity of the ALS-FOH + m sample. Higher viscosities have previously shown to decrease the average penetration depth of adhesives, such as urea-formaldehyde [55]. However, the difference in viscosities was not large enough to fully explain the different penetration behavior. Another explanation can be that tannin increases the crosslinking and contributes to forming larger molecules that prohibit penetration. It should be noted that ALS bonded samples could not be evaluated since wetting of the surfaces with a brush for microtome sectioning resulted in complete delamination. Also, determinations were not performed for the UmF samples as it was aimed to evaluate the behavior of the crosslinked ALS adhesives and not to make any such comparisons.

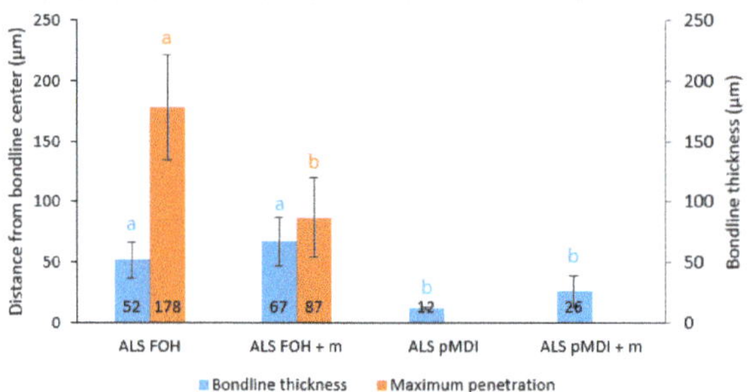

Figure 3. Bondline thickness (blue bars) and maximum penetration (orange bars) in 2-layered beech samples bonded with ammonium lignosulfonate (ALS) crosslinked with furfuryl alcohol (ALS FOH), ALS FOH with mimosa tannin (ALS FOH + m), ALS crosslinked with pMDI (ALS pMDI) and ALS pMDI with mimosa tannin (ALS pMDI + m). Each bondline thickness and maximum penetration column is the average of 15 and 30 determinations, respectively. Error bars represent standard deviations. Values labeled with different letters (a and b) are statistically different at an error probability of $\alpha = 0.05$ (bondline thickness ANOVA and Tukey's HSD tests, maximum penetration Welch Two Sample t-test).

Microscopic observation of wood-adhesive interphases provided additional information on the penetration of the adhesive formulations (Figure 4a–d). Distinct and continuous bondlines with some voids were observed for the ALS adhesive crosslinked with FOH (ALS-FOH, Figure 4a). Large interphases with continuous and distinct bondlines were observed for the ALS adhesive crosslinked with FOH in the presence of tannin (ALS-FOH + m, Figure 4b). Imperfect and starved bondlines were observed in the case of ALS adhesives crosslinked with pMDI (Figure 4c,d). In addition, the bondlines were lighter in color compared to the FOH crosslinked ALS samples, partially due to the lighter color of pMDI compared to FOH. The lighter color could also be due to the poor crosslinking of pMDI with ALS, leading the lignin to dissolve into the wooden structure when water was added for the sampling. The reaction of isocyanate groups with moisture (in wood or ALS adhesive) results in the production of CO_2 gas, which causes an inner vapor pressure that drives more adhesive from the bondline towards the wood structure. This phenomenon was reported previously by Bastani and colleagues [40,56]. It has been shown previously that penetration is inversely related to molecular weight, viscosity, and solids content [57]. In this study, the deeper penetration of the adhesive crosslinked with the bio-based crosslinker could be attributed to the lower molecule weight of FOH as compared with pMDI, although the viscosity of the pMDI crosslinked samples was lower. For hardwoods, such as beech, adhesive penetration mainly occurs in longitudinal vessels [56]. In the ALS samples crosslinked with FOH, the penetration mainly occurred through vessels with some filling of fiber lumens. No penetration through rays could be detected. Finding a meaningful relationship between the adhesion penetration results and shear bond performance can be challenging [58]. By looking the microscopy (Figure 3) and tensile strength results (Figure 2), a statistically significant difference in adhesion penetration between the FOH and pMDI crosslinked samples occurred ($p < 0.0001$), but no significant difference ($p > 0.05$) in tensile strength could be detected.

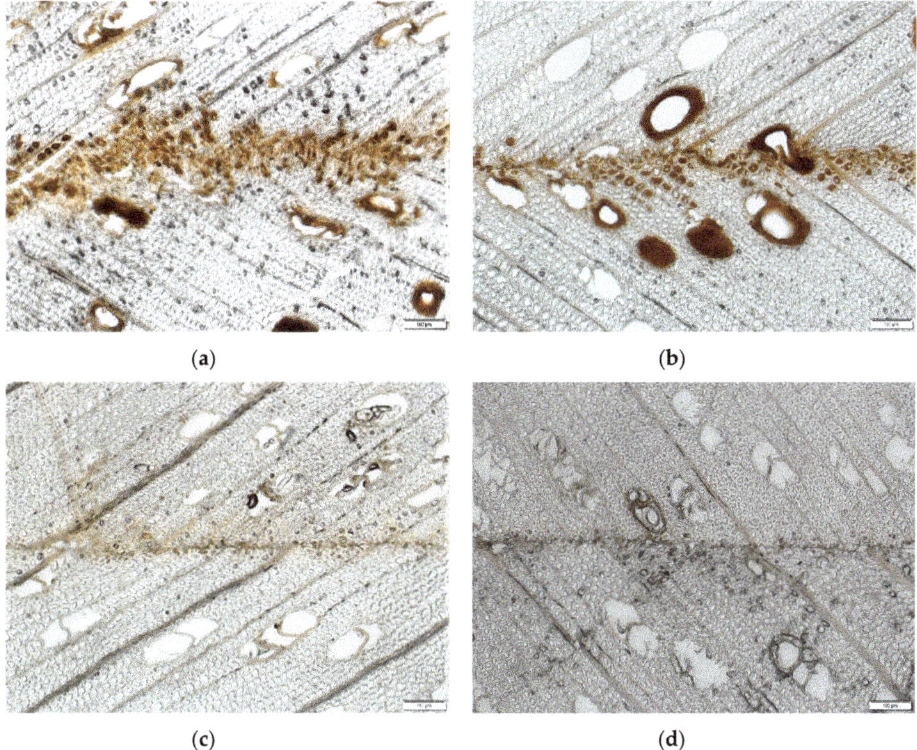

Figure 4. Transverse visible-light view of horizontal bondlines in 2-layered veneer samples: (**a**) ALS-FOH, (**b**) ALS-FOH +m, (**c**) ALS-pMDI, and (**d**) ALS-pMDI + m.

3.3. Chemical Characterization of Adhesive Bondline

FTIR spectroscopy analysis was used to monitor changes in the chemical structure of adhesive formulations crosslinked with pMDI upon curing in wood veneers (Figure 5). The stretching vibrations at 2960 and 3410 cm^{-1} were contributed to the aromatic and aliphatic OH groups. ALS alone illustrated two distinct peaks at 1705 and 1720 cm^{-1}, which are related to the unconjugated carbonyl stretching vibration [35]. The absorption bands at 1141 and 1180 cm^{-1} were attributed to the aromatic CH bonds in guaiacyl units and stretching of SO groups, respectively. Obvious changes across all regions of the spectra were observed for ALS formulations with the synthetic crosslinker and in combination with tannin, i.e., ALS-pMDI and ALS-pMDI+m. The appearance of peaks at 1712 cm^{-1} and shoulder at 1660 cm^{-1} in the formulations containing pMDI, i.e., ALS-pMDI and ALS-pMDI + m, confirmed the formation of urethane linkages in the bondlines [36,59]. In addition, strong stretching bonds at 1510–1530 cm^{-1} in the ALS-pMDI and ALS-pMDI + m formulations were attributed to urea linkages, which could be due to the reaction of pMDI and water in the adhesive formulations [60]. Formation of urethane and urea bonds proved that all isocyanate groups in pMDI (2270–2250 cm^{-1}) were consumed.

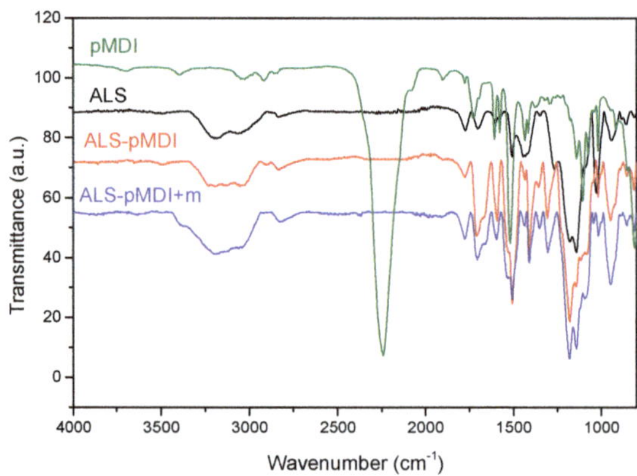

Figure 5. Fourier transform infrared spectroscopy (FTIR) spectra of pure pMDI and cured adhesive formulations with ALS, ALS-pMDI, and ALS-pMDI + m.

3.4. Mechanical and Physical Properties of Particleboards

Figure 6 illustrates the internal bond (IB) strength for the particleboards bonded with ALS-based adhesives for synthetic and bio-based crosslinkers, and with or without tannin. It should be mentioned that the panels with sole ALS as binder delaminated after the opening of the press and could not be tested. It was observed that the IB strength in the panels containing pMDI crosslinker was significantly higher than in the panels crosslinked with FOH as well as in controls bonded with UmF ($p < 0.0001$). Crosslinking of ALS with FOH with or without tannin resulted in poor IB results as compared with the control panels, and the differences were statistically significant ($p < 0.0001$). The addition of tannin reduced the IB strength of the panels bonded with ALS-pMDI + m adhesive ($p < 0.05$). However, the IB strength remained unchanged by adding tannin in the panels bonded with ALS-FOH (ALS-FOH + m) ($p > 0.05$). According to EN 312 (2010) [61], the minimum requirement of IB strength of particleboards for indoor applications (P2) in thickness range 6–13 mm is 0.40 N·mm^{-2}. The panels bonded with ALS-based adhesives crosslinked with pMDI (ALS-pMDI and ALS-pMDI + m) were above this limit (0.62 N/mm^2 and 0.49 N/mm^2, respectively). These values are in line with other pMDI crosslinked lignin adhesives, where 2.5–4% (% on dry wood mass) of pMDI has been used [30]. However, the contribution of lignin into the final strength can be questioned and needs to be further evaluated.

Similar to IB strength, modulus of rupture (MOR) in static bending for the panels was considerably higher for the pMDI crosslinked ALS samples compared to the control UmF (Figure 7, $p < 0.05$). The modulus of elasticity (MOE) of the panels with pMDI was at the same level as the control ($p > 0.05$). However, the panels bonded with ALS-based adhesive crosslinked with FOH showed significantly lower MOR and MOE as compared with the formulation containing pMDI ($p < 0.0001$) and with the UmF control ($p < 0.0001$). The addition of tannin did not change the MOR and MOE in the particleboards bonded with adhesives having FOH or pMDI as crosslinkers. The highest bending properties were obtained from panels bonded with ALS-pMDI (MOR: 11.1 N/mm^2, MOE: 2056 N/mm^2) and ALS-pMDI + m (MOR: 10.9 N/mm^2, MOE: 1856 N/mm^2). According to EN 312 [61], the respective minimum requirements for MOR and MOE of particleboards in thickness range 6 to 13 mm for indoor applications are 11 N/mm^2 and 1800 N/mm^2. No difference in mechanical properties could be detected with the addition of 10% of tannin to the lignin-FOH copolymer, and that was also true for the tensile shear strength of the 2-layered beech samples. This is in contradiction to the pre-mentioned study by Luckender et al. (2016) [62], where for FOH copolymers, the best results could be obtained by using spent liquor lignin and tannin together with FOH, in oppose to using only lignin or only tannin with FOH. Best results were obtained with spent liquor lignin: tannin ratios of 50:10 and 40:20 (IB 0.55 N/mm^2 and 0.53 N/mm^2, respectively, with adhesive addition amount of 15%).

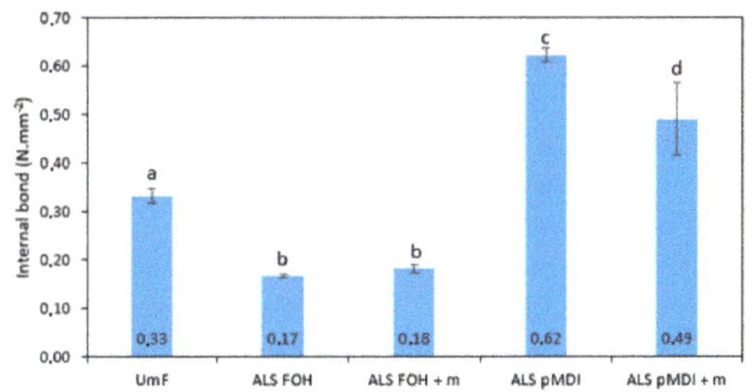

Figure 6. Internal bond strength of particleboard panels produced with urea-melamine-formaldehyde (UmF), ammonium lignosulfonate (ALS) crosslinked with furfuryl alcohol (ALS FOH), ALS FOH with mimosa tannin (ALS FOH + m), ALS crosslinked with pMDI (ALS pMDI) and ALS pMDI with mimosa tannin (ALS pMDI + m), and with or without mimosa tannin (m). Error bars represent standard deviations. Values labeled with different letters (a, b, c and d) are statistically different at an error probability of $\alpha = 0.05$ (Tukey's HSD test).

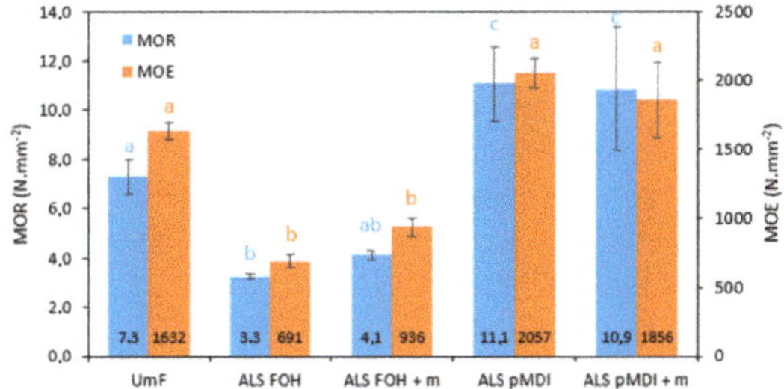

Figure 7. MOR (modulus of rupture, blue bars) and MOE (modulus of elasticity, orange bars) of particleboards produced with urea-melamine-formaldehyde (UmF), ammonium lignosulfonate (ALS) crosslinked with furfuryl alcohol (ALS FOH), ALS FOH with mimosa tannin (ALS FOH + m), ALS crosslinked with pMDI (ALS pMDI) and ALS pMDI with mimosa tannin (ALS pMDI + m). Error bars represent standard deviations. Values labeled with different letters (a, b, and c) are statistically different at an error probability of $\alpha = 0.05$ (ANOVA and Tukey's HSD tests). Value marked ab is not statistically different from a or b. Different colored letters refer to the matching mechanical property.

All mechanical results of particleboards were poor for the FOH crosslinked ALS samples. However, in the case of 2-layered beech samples, the tensile shear strength was at the same level for reference and all the crosslinked ALS samples. One explanation could be the difference in the FOH pre-polymerization between the veneer and particleboard samples. The increase of viscosity of furfuryl copolymers happens very rapidly and is highly affected by small differences in the amount of hardener, temperature, and time [62]. In order to keep the viscosity at a constant level suitable for the spray nozzle to reach even distribution in the chip blender, FOH was only reacted for 30 min in the adhesive formulation before resination. This was shorter than the reaction time of 45 min used for gluing the 2-layered beech samples. Thus, FOH had less time for pre-polymerization before board pressing compared to the veneer pressing. In a study by Dao et al. [63], the effect of polymerization time on the final furfuryl alcohol strength was shown. Different polymerization times of polyfurfuryl alcohol were tested for wood powder composites, and increasing the polymerization time from 30 min to 110 min resulted in an increase of the dry tensile shear strength from 2450 psi (16.9 N/mm^2) to 3730 psi (25.7 N/mm^2), and the wet tensile shear strength from 1270 psi (8.7 N/mm^2) to 1960 psi (13.5 N/mm^2). Another explanation for the low strength of the FOH crosslinked ALS samples could be the low addition amount for gluing the particleboards. A previous study reported that the bonding quality of the panels is directly related to the amount of FOH and catalyst in the adhesive formulation. In this study, a total amount of 40% of FOH was used in making the particleboards, and the total adhesive amounts were 10% and 15% [62]. In the current study for particleboard manufacturing, it was aimed to keep the amounts of pMDI and FOH crosslinker the same (25% of total adhesive) at the industrially relevant level of total added adhesive (8 wt % to dry particles). Therefore, both the total adhesive amount and the FOH share in it were lower than in the previously mentioned study.

The thickness swelling (TS) of the particleboards bonded with UmF and different ALS-based adhesive systems are presented in Figure 8. The reference boards showed TS of 40% and 52% measured after 2 h and 24 h, respectively. The panels bonded with ALS-FOH and ALS-FOH + m were disintegrated after 2 h and thus could not be measured. There was no significant difference between the ALS-FOH and ALS-FOH + m samples ($p > 0.05$). The sensitivity of FOH towards the water was reported previously by Luckender et al. [62]. The samples with the synthetic crosslinker pMDI had significantly lower TS compared to the reference UmF ($p < 0.0001$ for 2 h swelling and $p < 0.05$ for 24 h swelling). The isocyanate groups in pMDI are highly unsaturated and can react with a number of active hydroxyl groups in wood chips' surface and ALS structure, as well as with the moisture contained in the chips and the glue system. That will result in the formation of polyurethane and polyurea linkages, respectively [64]. The presence of excess primary amino-groups at elevated temperature will also lead to allophanate and/or biuret bonds [65]. The mentioned reactions provide durable linkages for adhesive systems containing pMDI crosslinker under moist conditions. One possible way to increase the water-resistance of the lignin-FOH adhesives is the addition of glyoxal to the lignin-FOH copolymer. Zhang et al. [27] found that furfuryl alcohol does react with glyoxal, and the addition of glyoxal has a positive effect on the dry strength and water resistance of lignin-FOH copolymers. The addition of epoxy further increased the properties to a level where lignin-FOH glyoxal adhesive with 9% of epoxy had IB of 0.41 MPa and IB after 2 h in boiling water of 0.21 MPa.

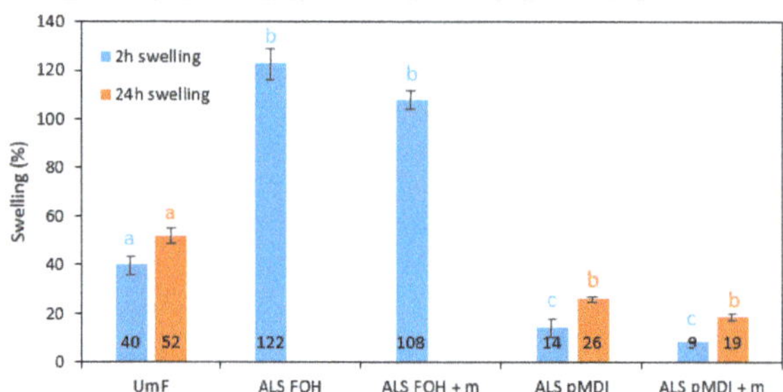

Figure 8. Thickness swelling of particleboards produced with urea-melamine-formaldehyde (UmF), ammonium lignosulfonate (ALS) crosslinked with furfuryl alcohol (ALS FOH), ALS FOH with mimosa tannin (ALS FOH + m), ALS crosslinked with pMDI (ALS pMDI) and ALS pMDI with mimosa tannin (ALS pMDI + m) after two hours (blue bars) and after 24 hours (orange bars). Error bars represent standard deviations. Values labeled with different letters (a, b and c) are statistically different at an error probability of $\alpha = 0.05$ (ANOVA and Tukey's HSD tests). Different colored letters refer to swelling property.

3.5. Formaldehyde Emissions

In this work, the previously evaluated method based on a dynamic microchamber (DMC), according to ASTM D 6007 [44], after 1-day conditioning [45] was used to determine the extreme formaldehyde emission values of the particleboard samples (Figure 9).

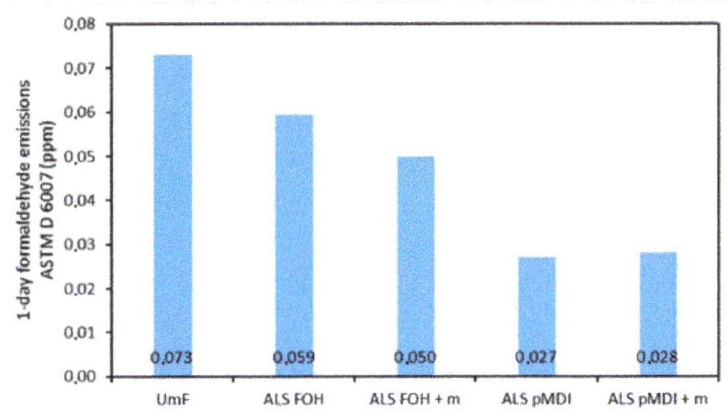

Figure 9. Formaldehyde emissions of particleboards produced with urea-melamine-formaldehyde (UmF), ammonium lignosulfonate (ALS) crosslinked with furfuryl alcohol (ALS FOH), ALS FOH with mimosa tannin (ALS FOH + m), ALS crosslinked with pMDI (ALS pMDI) and ALS pMDI with mimosa tannin (ALS pMDI + m).

The panels bonded with commercial adhesive (UmF) had a formaldehyde emission value of 0.073 ppm. The emissions were lower for all ALS-based adhesive systems. The lowest emissions were obtained for the boards bonded with ALS-pMDI (0.027 ppm). The panels bonded with ALS-FOH exhibited a considerably higher formaldehyde emission value of 0.059. This can be explained by the polycondensation mechanism of FOH, where the reaction between two methylol groups in the FOH monomer leads to the formation of ether bridge that can transform into methylene bridge by releasing formaldehyde molecule [23,48] (Figure 10). The formaldehyde emissions of the particleboards crosslinked with pMDI remained unchanged by tannin addition. However, emissions decreased to a minor extent in the boards crosslinked with FOH. The formaldehyde scavenger effect of tannin was reported previously [66].

Figure 10. One of furfuryl polycondensation pathways: the reaction between the methylol groups of two furfuryl alcohol molecules, leading to dimethyl ether bridge that can be transformed into a methylene bridge by the release of formaldehyde [48].

4. Conclusions

In this work, two crosslinkers for ALS, with and without mimosa tannin, were evaluated—furfuryl alcohol (FOH) and polymeric 4,4′-diphenylmethane diisocyanate (pMDI). Curing of the adhesive systems was performed using DSC, and properties of the crosslinked ALS adhesives were tested on 2-layered beech veneers and 3-layer particleboards. The conclusions can be summarized as follows:

- All samples with crosslinkers had lower curing temperature and curing heat than the un-crosslinked ALS sample. The ALS-pMDI sample cured at the lowest temperature (107 °C) and had the lowest curing heat (123 J/g).
- For those formulations crosslinked with pMDI, FTIR results pointed out for chemical reaction of ALS with pMDI in the cured adhesives, i.e., the formation of urethane linkages.
- The FOH crosslinked ALS samples had thick dark bondlines, while the pMDI cured ALS samples had thin light bondlines, possibly due to the higher molecular weight of pMDI and vapor pressure created by the isocyanate crosslinking.
- The penetration of the FOH crosslinked ALS occurred mainly through vessels with some filling of the fiber lumens. The FOH crosslinked ALS without tannin had deeper penetration than the one with tannin, suggesting that tannin prohibits penetration, allowing more glue to remain near the bondline.
- The mechanical properties of the particleboards produced using FOH crosslinked ALS, with and without tannin (IB: 0.17 N/mm^2 and 0.18 N/mm^2, respectively), were inferior to those produced using pMDI crosslinked ALS, with and without tannin (IB: 0.62 N/mm^2 and 0.49 N/mm^2, respectively). A similar trend could be seen in thickness swelling where the FOH crosslinked ALS samples disintegrated after 2 h. The contribution of ALS to the final strength of the ALS pMDI samples needs to be further evaluated.
- Although particleboard properties were worse for the FOH crosslinked samples, the tensile shear strengths of both FOH and pMDI crosslinked ALS were at the same level as the UmF reference for the 2-layered veneer samples. The poor performance of FOH crosslinked ALS samples in particleboards could be due to the shorter pre-polymerization time and low FOH and total glue amount.
- The addition of mimosa tannin (10% to ALS amount) had no effect on any of the mechanical properties for both particleboard and 2-layered veneer samples. However, it lowered the emissions of FOH crosslinked ALS samples from 0.059 ppm of ALS FOH to 0.050 ppm of ALS FOH + m.
- The formaldehyde emissions of pMDI crosslinked particleboards were at the level of natural wood (0.027 ppm and 0.028 ppm). Interestingly, the FOH crosslinked boards emitted beyond the level of natural wood (0.059 ppm and 0.050 ppm). This can be due to one of the polycondensation mechanisms of FOH, where two methylol groups of two FOH molecules react, leading to ether bridge that can transform into methylene bridge by the release of formaldehyde.

No particleboards could be produced using the ALS without crosslinker, highlighting the need for a crosslinker if the lignosulfonate is used without modification. More work in finding suitable crosslinkers for bio-based materials is needed, but until then, pMDI is a promising crosslinker that works for most bio-based adhesives. This study demonstrated the potential to combine ALS and pMDI in particleboard manufacturing; however, the chemical interaction between the polymers needs to be further elucidated for optimum usage. Thus, further work in lowering the pMDI amount, optimizing pressing parameters, and modifying the lignosulfonate formula should be done as the next step to confirm that the ALS is truly contributing to the final adhesion strength.

Author Contributions: Conceptualization, V.H. and S.A.; methodology, V.H. and R.H.; formal analysis, V.H., S.A., and R.H.; investigation, V.H., R.H., and S.A.A.; writing—original draft preparation, V.H.; writing—review and editing, R.H. and S.A.; supervision, S.A.

Funding: Venla Hemmilä and Stergios Adamopoulos would like to thank the Knowledge Foundation for the financial support (project titled "New environment-friendly board materials", 2015–2019).

Conflicts of Interest: The authors declare no conflict of interest.

References

1. International Agency for Research on Cancer (IARC). *IARC Classifies Formaldehyde as Carcinogenic to Humans*; International Agency for Research on Cancer (IARC): Lyon, Franc, 2004; Press Release June 15, No. 153.

2. Bundesministerium für Umwelt Naturschutz und nukleare Sicherheit, Bekanntmachung analytischer Verfahren für Probenahmen und Untersuching für die in Anlage 1 der Chemikalien-Verbotsverordnung genannten Stoffe und Stoffgruppen. 2018, Bundesanzeiger, BAnz AT 26.11.2018 B2. Available online: https://www.umwelt-online.de/recht/gefstoff/chemverb.vo/chvvanalyt18.htm (accessed on 9 October 2019).

3. Aydin, I.; Demirkir, C.; Colak, S.; Colakoglu, G. Utilization of bark flours as additive in plywood manufacturing. *Eur. J. Wood Wood Prod.* **2017**, *75*, 63–69. [CrossRef]

4. Réh, R.; Igaz, R.; Krišťák, Ľ.; Ružiak, I.; Gajtanska, M.; Božíková, M.; Kučerka, M. Functionality of beech bark in adhesive mixtures used in plywood and Its effect on the stability associated with material systems. *Materials* **2019**, *12*, 1298.

5. Gu, K.; Li, K. Preparation and evealuation of particleboard with a soy flour-polyethylenimine-maleic anhydride adhesive. *J. Am. Oil Chem. Soc.* **2011**, *88*, 673–679. [CrossRef]

6. Da Silva, C.G.; Grelier, S.; Pichavant, F.; Frollini, E.; Castellan, A. Adding value to lignins isolated from sugarcane bagasse and Miscanthus. *Ind. Crops Prod.* **2013**, *42*, 87–95. [CrossRef]

7. El Mansouri, N.E.; Pizzi, A.; Salvadó, J. Lignin-based wood panel adhesives without formaldehyde. *Holz Roh-Werkst. (1937–2008)* **2006**, *65*, 65. [CrossRef]

8. Tan, H.; Zhang, Y.; Weng, X. Preparation of the Plywood Using Starch-based Adhesives Modified with blocked isocyanates. *Procedia Eng.* **2011**, *15*, 1171–1175. [CrossRef]

9. Hemmilä, V.; Adamopoulos, S.; Karlsson, O.; Kumar, A. Development of sustainable bio-adhesives for engineered wood panels: A Review. *RSC Adv.* **2017**, *7*, 38604–38630. [CrossRef]

10. Frihart, C.; Birkeland, M. *Soy Properties and Soy Wood Adhesives*; American Chemical Society: Midland, MI, USA, 2014; pp. 167–192.

11. Pizzi, A. Recent developments in eco-efficient bio-based adhesives for wood bonding: Opportunities and issues. *J. Adhes. Sci. Technol.* **2006**, *20*, 829–846. [CrossRef]

12. Norström, E.; Demircan, D.; Fogelström, L.; Khabbaz, F.; Malmström, E. Chapter 4: Green Binders for Wood Adhesives. In *Applied Adhesive Bonding Science and Technology*; Ozer, H., Ed.; InTech Open: Rijeka, Croatia, 2018.

13. Ferdosian, F.; Pan, Z.; Gao, G.; Zhao, B. Bio-based adhesives and evaluation for wood composites application. *Polymers* **2017**, *9*, 70. [CrossRef]

14. Costello, H. *Lignin Products Global Market Size, Sales Data 2017–2022 & Applications in Animal Feed Industry 2017: (Market Analysis No. 188450)*; Orbis Research: Dallas, TX, USA, 2017.

15. Farag, S.; Kouisni, L.; Chaouki, J. Lumped approach in kinetic modeling of microwave pyrolysis of Kraft lignin. *Energy Fuels* **2014**, *28*, 1406–1417. [CrossRef]

16. Fatehi, P.; Chen, J. Extraction of Technical Lignins from Pulping Spent Liquors, Challenges and Opportunities. In *Production of Biofuels and Chemicals from Lignin*; Fang, Z., Smith, J.R.L., Eds.; Springer: Singapore, 2016; pp. 35–54.

17. Ragauskas, A.J.; Beckham, G.T.; Biddy, M.J.; Chandra, R.; Chen, F.; Davis, M.F.; Davison, B.H.; Dixon, R.A.; Gilna, P.; Keller, M.; et al. Lignin valorization: Improving lignin processing in the biorefinery. *Science* **2014**, *344*, 1246843. [CrossRef] [PubMed]

18. Azadi, P.; Inderwildi, O.R.; Farnood, R.; King, D.A. Liquid fuels, hydrogen and chemicals from lignin: A critical review. *Renewable Sustainable Energy Rev.* **2013**, *21*, 506–523. [CrossRef]

19. Chakar, F.S.; Ragauskas, A.J. Review of current and future softwood kraft lignin process chemistry. *Ind. Crops Prod.* **2004**, *20*, 131–141. [CrossRef]

20. Danielson, B.; Simonson, R. Kraft lignin in phenol formaldehyde resin. Part 1. Partial replacement of phenol by kraft lignin in phenol formaldehyde adhesives for plywood. *J. Adhes. Sci. Technol.* **1998**, *12*, 923. [CrossRef]

21. Vázquez, G.; González, J.; Freire, S.; Antorrena, G. Effect of chemical modification of lignin on the gluebond performance of lignin-phenolic resins. *Bioresour. Technol.* **1997**, *60*, 191–198. [CrossRef]

22. Choura, M.; Belgacem, N.; Gandini, A. Acid-Catalyzed Polycondensation of Furfuryl Alcohol: Mechanisms of Chromophore Formation and Cross-Linking. *Macromolecules* **1996**, *29*, 3839–3850. [CrossRef]

23. Bertarione, S.; Bonino, F.; Cesano, F.; Jain, S.; Zanetti, M.; Scarano, D.; Zecchina, A. Micro-FTIR and Micro-Raman Studies of a Carbon Film Prepared from Furfuryl Alcohol Polymerization. *J. Phys. Chem.* **2009**, *113*, 10571–10574. [CrossRef]

24. Bertarione, S.; Bonino, F.; Cesano, F.; Damin, A.; Scarano, D.; Zecchina, A. Furfuryl Alcohol Polymerization in H-Y Confined Spaces: Reaction Mechanism and Structure of Carbocationic Intermediates. *J. Phys. Chem.* **2008**, *112*, 2580–2589. [CrossRef]

25. Abdullah, U.H.B.; Pizzi, A. Tannin-furfuryl alcohol wood panel adhesives without formaldehyde. *Eur. J. Wood Wood Prod.* **2013**, *71*, 131–132. [CrossRef]

26. Guigo, N.; Mija, A.; Vincent, L.; Sbirrazzuoli, N. Eco-friendly composite resins based on renewable biomass resources: Polyfurfuryl alcohol/lignin thermosets. *Eur. Polym. J.* **2010**, *46*, 1016–1023. [CrossRef]

27. Zhang, J.; Wang, W.; Zhou, X.; Liang, J.; Du, G.; Wu, Z. Lignin-based adhesive crosslinked by furfuryl alcohol–glyoxal and epoxy resins. *Nord. Pulp Pap. Res. J.* **2019**, *34*, 228–238. [CrossRef]

28. Dongre, P.; Driscoll, M.; Amidon, T.; Bujanovic, B. Lignin-Furfural Based Adhesives. *Energies* **2015**, *8*, 7897–7914. [CrossRef]

29. Papadopoulos, A.; Hill, C.; Traboulay, E.R.B.; Hague, J. Isocyanate resins for particleboard: PMDI vs EMDI. *Holz Roh- Werkst. (1937-2008)* **2002**, *60*, 81–83. [CrossRef]

30. Solt, P.; Konnerth, J.; Gindl-Altmutter, W.; Kantner, W.; Moser, J.; Mitter, R.; van Herwijnen, H.W.G. Technological performance of formaldehyde-free adhesive alternatives for particleboard industry. *Int. J. Adhes. Adhes.* **2019**, *94*, 99–131. [CrossRef]

31. Kutnar, A.; Burnard, M.D. The past, present, and future of EU wood adhesive research and market. In Proceedings of the International Conference on Wood Adhesives, Toronto, ON, Canada, 9–11 October 2014; pp. 22–35.

32. Amaral-Labat, G.A.; Pizzi, A.; Gonçalves, A.R.; Celzard, A.; Rigolet, S.; Rocha, G.J.M. Environment-friendly soy flour-based resins without formaldehyde. *J. Appl. Polym. Sci.* **2008**, *108*, 624–632. [CrossRef]

33. Lei, H.; Pizzi, A.; Du, G. Environmentally friendly mixed tannin/lignin wood resins. *J. Appl. Polym. Sci.* **2008**, *107*, 203–209. [CrossRef]

34. Ballerini, A.; Despres, A.; Pizzi, A. Non-toxic, zero emission tannin-glyoxal adhesives for wood panels. *Holz Roh- Werkst. (1937-2008)* **2005**, *63*, 477–478. [CrossRef]

35. Hemmilä, V.; Hosseinpourpia, R.; Adamopoulos, S.; Eceiza, A. Characterization of Wood-based Industrial Biorefinery Lignosulfonates and Supercritical Water Hydrolysis Lignin for Value-Added Applications. Under review for Waste and Biomass Valorization. 2019.

36. Hosseinpourpia, R.; Echart, A.S.; Adamopoulos, S.; Gabilondo, N.; Eceiza, A. Modification of Pea Starch and Dextrin Polymers with Isocyanate Functional Groups. *Polymers* **2018**, *10*, 939. [CrossRef]

37. EN 205, Adhesives. Wood adhesives for non-structural applications. In *Determination of Tensile Shear Strength of Lap Joints*; European Committee for Standardization: Brussels, Belgium, 2016; Available online: http://www.rivacommerce.com/images/EN205-204.pdf (accessed on 9 October 2019).

38. Hosseinpourpia, R.; Adamopoulos, S.; Parsland, C. Utilization of different tall oils for improving the water resistance of cellulosic fibers. *J. Appl. Polym. Sci.* **2018**, *136*, 47303. [CrossRef]

39. Sernek, M.; Resnik, J.; Kamke, F.A. Penetration of liquid urea-formaldehyde adhesive into beech wood. *Wood Fiber Sci.* **1999**, 41–48.

40. Bastani, A.; Adamopoulos, S.; Militz, H. Effect of open assembly time and equilibrium moisture content on the penetration of polyurethane adhesive into thermally modified wood. *J. Adhes.* **2017**, *93*, 575–583. [CrossRef]

41. EN 319, Particleboards and fibreboards—Determination of tensile strength perpendicular to the plane of the board. 1993, European Standard. Available online: https://apawood-europe.org/official-guidelines/european-standards/individual-standards/en-319/ (accessed on 9 October 2019).

42. EN 310, Wood-based panels—Determination of modulus of elasticity in bending and of bending strength. 1993, European Standard. Available online: https://www.iso.org/standard/32836.html (accessed on 9 October 2019).

43. EN 317, Particleboards and fibreboards—Determination of swelling in thickness after immersion in water. 1993, European Standard. Available online: https://www.scienceopen.com/document?vid=38d0d5f2-5ac7-45ff-89ff-e33e6831ea7f (accessed on 9 October 2019).

44. ASTM D 6007, Standard test method for determining formaldehyde concentrations in air from wood products using a small-scale chamber. 2014, ASTM International. Available online: https://www.astm.org/Standards/D6007.htm (accessed on 9 October 2019).

45. Hemmilä, V.; Zabka, M.; Adamopoulos, S. Evaluation of Dynamic Microchamber as a Quick Factory Formaldehyde Emission Control Method for Industrial Particleboards. *Adv. Mater. Sci. Eng.* **2018**, 1–9. [CrossRef]

46. Hemmilä, V.; Meyer, B.; Larsen, A.; Schwab, H.; Adamopoulos, S. Influencing factors, repeatability and correlation of chamber methods in measuring formaldehyde emissions from fiber- and particleboards. *Int. J. Adhes. Adhes.* **2019**, *95*, 102420. [CrossRef]

47. Solt, P.; Jääskeläinen, A.-S.; Lingenfelter, P.; Konnerth, J.; van Herwijnen, H. Impact of Molecular Weight of Kraft Lignin on Adhesive Performance of Lignin-Based Phenol-Formaldehyde Resins. *For. Prod. J.* **2018**, *68*, 365–371.

48. Szczurek, A.; Fierro, V.; Thébault, M.; Pizzi, A.; Celzard, A. Structure and properties of poly(furfuryl alcohol)-tannin polyHIPEs. *Eur. Polym. J.* **2016**, *78*, 195–212. [CrossRef]

49. Kalami, S.; Arefmanesh, M.; Master, E.; Nejad, M. Replacing 100% of phenol in phenolic adhesive formulations with lignin. *J. Appl. Polym. Sci.* **2017**, *134*, 45124. [CrossRef]

50. Wang, M.; Leitch, M.; Xu, C. Synthesis of phenol–formaldehyde resol resins using organosolv pine lignins. *Eur. Polym. J.* **2009**, *45*, 3380–3388. [CrossRef]

51. Ferdosian, F.; Yuan, Z.; Anderson, M.; Xu, C. Sustainable lignin-based epoxy resins cured with aromatic and aliphatic amine curing agents: Curing kinetics and thermal properties. *Thermochim. Acta* **2015**, *618*, 48–55. [CrossRef]

52. Watkins, D.; Nuruddin, M.; Hosur, M.; Tcherbi-Narteh, A.; Jeelani, S. Extraction and characterization of lignin from different biomass resources. *J. Mater. Res. Technol.* **2015**, *4*, 26–32. [CrossRef]

53. Domínguez, J.C.; Grivel, J.C.; Madsen, B. Study on the non-isothermal curing kinetics of a polyfurfuryl alcohol bioresin by DSC using different amounts of catalyst. *Thermochim. Acta* **2012**, *529*, 29–35. [CrossRef]

54. Chen, H.; Yan, N. Application of Western red cedar (Thuja plicata) tree bark as a functional filler in pMDI wood adhesives. *Ind. Crops Prod.* **2018**, *113*, 1–9. [CrossRef]

55. Gavrilovic-Grmusa, I.; Dunky, M.; Miljkovic, J.; Djiporovic-Momcilovic, M. Influence of the viscosity of UF resins on the radial and tangential penetration into poplar wood and on the shear strength of adhesive joints. *Holzforschung* **2012**, *66*, 849–856. [CrossRef]

56. Bastani, A.; Adamopoulos, S.; Koddenberg, T.; Militz, H. Study of adhesive bondlines in modified wood with fluorescence microscopy and X-ray micro-computed tomography. *Int. J. Adhes. Adhes.* **2016**, *68*, 351–358. [CrossRef]

57. Kamke, F.A.; Lee, J.N. Adhesive penetration in wood—a review. *Wood Fiber Sci.* **2007**, *39*, 205–220.

58. Kariz, M.; Sernek, M. Bonding of Heat-Treated Spruce with Phenol-Formaldehyde Adhesive. *J. Adhes. Sci. Technol.* **2010**, *24*, 1703–1716. [CrossRef]

59. Gómez-Fernández, S.; Ugarte, L.; Calvo-Correas, T.; Peña-Rodríguez, C.; Corcuera, M.A.; Eceiza, A. Properties of flexible polyurethane foams containing isocyanate functionalized kraft lignin. *Ind. Crops Prod.* **2017**, *100*, 51–64. [CrossRef]

60. Duong, L.D.; Nam, G.-Y.; Oh, J.-S.; Park, I.-K.; Luong, N.D.; Yoon, H.-K.; Lee, S.-H.; Lee, Y.; Yun, J.-H.; Lee, C.-G.; et al. High Molecular-Weight Thermoplastic Polymerization of Kraft Lignin Macromers with Diisocyanate. *BioResources* **2014**, *9*, 13. [CrossRef]

61. EN 312:2010 Particleboards: Specifications. 2010, Brussels, European standard. Available online: https://infostore.saiglobal.com/en-au/standards/une-en-312-2010-16459_SAIG_AENOR_AENOR_36532/ (accessed on 9 October 2019).

62. Luckeneder, P.; Gavino, J.; Kuchernig, R.; Petutschnigg, A.; Tondi, G. Sustainable phenolic fractions as basis for furfuryl alcohol-based co-polymers and their use as wood adhesives. *Polymers* **2016**, *8*, 396. [CrossRef]

63. Dao, L.T.; Zavarin, E. Chemically activated furfuryl alcohol-based wood adhesives. 1. The role of furfuryl alcohol. *Holzforschung* **1996**, 470–476. [CrossRef]

64. Hosseinpourpia, R.; Adamopoulos, S.; Mai, C.; Taghiyari, H.R. Properties of medium-density fibreboards bonded with dextrin-based wood adhesive. *Wood Res.* **2019**, *64*, 185–194.

65. Sonnenschein, M.F. Introduction to polyurethane chemistry. In *Polyurethanes: Science, Technology, Markets, and Trends*; John Wiley & Sons, Inc.: Hoboken, NJ, USA, 2015.

66. Boran, S.; Usta, M.; Ondaral, S.; Gümüşkaya, E. The efficiency of tannin as a formaldehyde scavenger chemical in medium density fiberboard. *Composites B* **2012**, *43*, 2487–2491. [CrossRef]

Article

Physical and Mechanical Properties of Thermally-Modified Beech Wood Impregnated with Silver Nano-Suspension and Their Relationship with the Crystallinity of Cellulose

Siavash Bayani [1], Hamid R. Taghiyari [2,*] and Antonios N. Papadopoulos [3,*]

[1]   Department of Wood and Paper Science and Technology, College of Agriculture and Natural Resources, Science and Research Branch, Islamic Azad University, Tehran, Iran

[2]   Wood Science and Technology Department, Faculty of Materials Engineering & New Technologies, Shahid Rajaee Teacher Training University, Tehran, Iran

[3]   Laboratory of Wood Chemistry and Technology, Department of Forestry and Natural Environment, International Hellenic University, GR-661 00 Drama, Greece

*   Correspondence: htaghiyari@sru.ac.ir (H.R.T.); antpap@teiemt.gr (A.N.P.)

Received: 25 July 2019; Accepted: 17 September 2019; Published: 20 September 2019

Abstract: The aim of this study was to investigate the physical and mechanical properties of thermally modified beech wood impregnated with silver nano-suspension and to examine their relationship with the crystallinity of cellulose. Specimens were impregnated with a 400 ppm nanosilver suspension (NS); at least, 90% of silver nano-particles ranged between 20 and 100 nano-meters. Heat treatment took place in a laboratory oven at three temperatures, namely 145, 165, and 185 °C. Physical properties and mechanical properties of treated wood demonstrated statistically insignificant fluctuations at low temperatures compared to control specimens. On the other hand, an increase of temperature to 185 °C had a significant effect on all properties. Physical properties (volumetric swelling and water absorption) and mechanical properties (MOR and MOE) of treated wood demonstrated statistically insignificant fluctuations at low temperatures compared to control specimens. This degradation ultimately resulted in significant decrease in MOR, impact strength, and physical properties. However, thermal modification at 185 °C did not seem to cause significant fluctuations in MOE and compression strength parallel to grain. As a consequence of the thermal modification, part of amorphous cellulose was changed to crystalline cellulose. At low temperatures an increased crystallinity caused some of the properties to be improved. Crystallinity also demonstrated a decrease in NS-HT185 in comparison to HT185 treatment. TCr indices in specimens thermally treated at 145 °C revealed a significant increase as a result of impregnation with nanosilver suspension. This improvement in TCr index resulted in a noticeable increase in MOR and MOE values. Other properties did not show significant fluctuations, suggesting that the effect of the increased crystallinity and cross-linking in lignin was more than the negative effect of the low cell-wall polymer degradation caused by thermal modification. Change of amorphous cellulose to crystalline cellulose, as well as cross-linking in lignin, partially ameliorated the negative effects of thermal degradation at higher temperatures and therefore, compression parallel to grain and modulus of elasticity did not decrease significantly. Overall, it can be concluded that increased crystallinity and cross-linking in lignin can compensate for some decreased properties caused by thermal modification, but it would be significantly dependent on the temperature under which modification is carried out. Impregnating specimens with silver nano-suspension prior to thermal modification enhanced the effects of thermal modification as a result of improved thermal conductivity.

Keywords: thermal modification; nanocompounds; mechanical and physical properties; cellulose; crystallinity

---

1. Introduction

The main drawbacks of wood, namely dimensional instability and biological durability are mainly due to the nature of the cell wall main polymers and in particular due to their high abundance of hydroxyl groups (OH) [1–3]. The protection of wood without toxic chemicals has been the subject of numerous studies. The application of chemical or thermal modification technique can be considered as a solution to this, since species of inferior properties, mainly softwoods, can be modified and transformed to entirely new green products with superior properties [4–6]. An interesting article, reviewed recently the state of the art in the area of chemical and thermal modification. The article was focused on the two most promised techniques, namely acetylation and furfurylation [7].

Though thermal modification has been successful in improving dimensional stability and resistance against fungal attack [8,9], it reduces mechanical properties in wood. The reduction in mechanical properties limits the industrial applications in which the strength of wood is of prime importance. It is to be kept in mind that there is always a trade-off between the improved dimensional stability and fungal resistance versus the reduced mechanical properties. Degradation of main wood components (cell-wall polymers including cellulose, hemicellulose, and lignin) caused by thermal modification is a well-known phenomenon [3]. The outcome of this degradation is usually materialized by increased weight loss as the temperature for thermal modification increases. However, under relatively mild thermal conditions (<200 °C), some semi-crystalline cellulose regions change to crystalline regions [10] and cross-linking and polycondensation reactions in lignin structure occur [11,12]. These can have significant improving impact on different properties of wood, though weight loss may show very little degradation of cell-wall polymers.

Nanotechnology, an attractive science, seems to have a remarkable potential to create products of a new generation with enhanced properties [13]. The change in material properties is primarily due to the large interfacial area that is developed per unit of volume, since the level of added particles is reduced to nanometers. Nanomaterials enhance the properties of the original material, shows great compatibility with the traditional materials and cause limited alteration of their original features [14,15]. Their use in wood has the objective to improve its physical and mechanical properties and its durability against microorganisms, since it is generally acceptable that nanosized metals and minerals interact with the bacterial elements, leading gradually to the cell death [14–17] or even to the disruption of the enzyme function [15,18]. Moreover, wood has a low thermal conductivity [14] and therefore, it is expected that the outer layer and inner part of individual specimens would not be similarly modified by thermal modification. Impregnating specimens with a nano-metal suspension, such as silver, would improve the thermal conductivity of specimens, eventually making both outer and inner parts to be thermally modified to nearly same degree [19,20]. The cited authors studied physical properties (water absorption and thickness swelling), and mechanical properties (MOR, MOE, compression parallel to grain, brittleness, and pull-off strength), however, little or no studies were carried out on the effects of thermal modification at different temperatures on crystallinity and cellulose content of nanosilver-impregnated specimens.

The sustainability, excellent mechanical properties and interesting physical properties of cellulose nanomaterials have recently received a great deal of attention for modifying polymers and acting as both active and passive components in a wide range of potential products. The applications for which cellulose nanomaterials have been investigated include medical and pharmaceutical applications, paper and paperboard additives, paints and coatings, security papers, food packaging, electronic devices and displays, and energy harvesting and storage devices [21–23].

As cellulose is broken down into nanometer-scaled particles and fibers, the materials begin to exhibit some interesting and unique properties depending on their size, morphology, charge, and other characteristics [23,24]. Typically, cellulose nanomaterials, especially from plant sources, can be categorized as either cellulose nanocrystals (CNCs) or cellulose nanofibrils (CNFs). CNCs are discrete, rod-shaped cellulose particles typically with high crystallinity and are primarily produced through acid hydrolysis of native cellulose. CNCs have nanometer-scaled diameters with lengths of hun dreds to thousands of nanometers, and they often contain surface charge groups and can thus be colloidally stable [24–26]. Conversely, CNFs are typically network structured nanoscaled fibers that can be produced from countless methods usually involving some kind of chemical treatment followed by mechanical refining. The morphology of CNFs widely varies from process to process, and they can often have a wide distribution of diameters and lengths [26–28].

Cellulose nanomaterials are often touted as potential polymer reinforcements because of their high strength and stiffness, and these good mechanical properties also lend these materials for use in functional devices such as electronics. The cellulose crystal has been measured to have extremely high strength and stiffness with values of about 10 GPa and up to about 200 GPa, respectively, but bulk materials, such as films, made from nanocellulose have considerably lower mechanical properties [28–30]. Nonetheless, cellulose nanomaterial films and composites have mechanical properties that rival or exceed many polymers. The optical properties of cellulose nanomaterials are also distinctive from bulk cellulose and can be favorable for electronic applications. Cellulosic materials have low thermal expansion, which is desirable in electronic components that generate heat, and the coefficient of thermal expansion (CTE) for cellulose nanomaterials and their compos ites has been measured and reported for various CNF films [31,32]. Because of all these properties, including sustainability, high strength, flexibility, transparency, and low thermal expansion, cellulose nanomaterials have been explored for their use in electronics and functional devices. In some cases, cellulose nanomaterials can provide advantages over traditional materials [26–28,30,32]. The aim of this study, therefore, was to investigate the physical and mechanical properties of thermally modified beech wood impregnated with silver nano-suspension and to examine their relationship with the crystallinity of cellulose.

2. Materials and Methods

2.1. Sample Preparation

Beech trees (*Fagus orientalis*) were cut from Shafaroud district at the Rezvanshahr city in Gilan province (Iran). One-meter log from each tree was cut at the breast height to be air-dried for more than five months. From each log, twenty-one specimens for each of the properties studied in the present project were prepared; these specimens were randomly divided into seven groups of control, heat-treated (HT) at 145, 165, and 185 °C, as well as nanosilver-impregnated heat-treated (NS-HT) at the same temperatures. The chemical composition of beech wood, as far as the main cell wall polymers is concerned, is as follows: 69% holocellulose, 47.6% cellulose, and 25.5% lignin. Once impregnated, all specimens were again air-dried for three more months to the final moisture content of 8% before any test was carried out on them, because wood has a thermo-hygromechanical behavior and its properties depend on the combined action of temperature, relative humidity, and mechanical load variations [33].

2.2. Nanosilver Impregnation

In this study, silver nano-suspension was applied due to its easy of application and its effectiveness in increasing thermal conductivity in wood and wood-composites [14,19,20]. Specimens were impregnated with a 400 ppm nanosilver suspension (NS). At least 90% of silver nano-particles ranged between 20 and 100 nano-meters. A pressure of 2.5 bars for 10 minutes was applied in a sealed tank. Once impregnated, the specimens were randomly placed under the same conditions (30 °C, 40%–45% relative humidity) along with the un-treated specimens for two months. For checking the depth of impregnation, separate specimens were prepared to be simultaneously impregnated with NS specimens. After the impregnation, these specimens were cut in half to observe that NS-suspension has penetrated in the core section of specimens.

2.3. Thermal Modification

Specimens were heat-treated in a laboratory oven (model Memmert UFE 700), filled with atmospheric air. Un-treated and nanosilver-impregnated specimens were randomly arranged on the middle trays. Wooden strips of 3-mm thickness were put under specimens to avoid direct contact with metal trays of the oven. For HT-145 and NS-HT-145 specimens, thermal treatment was carried out at 145 °C for 24 h. For HT-165 and NS-HT-165 specimens, thermal treatment was carried out at two stages; the first stage was heating at 145 °C for 24 h continued with heating at 165 °C for 4 h. The flow diagram of the experimental procedure is depicted in Figure 1.

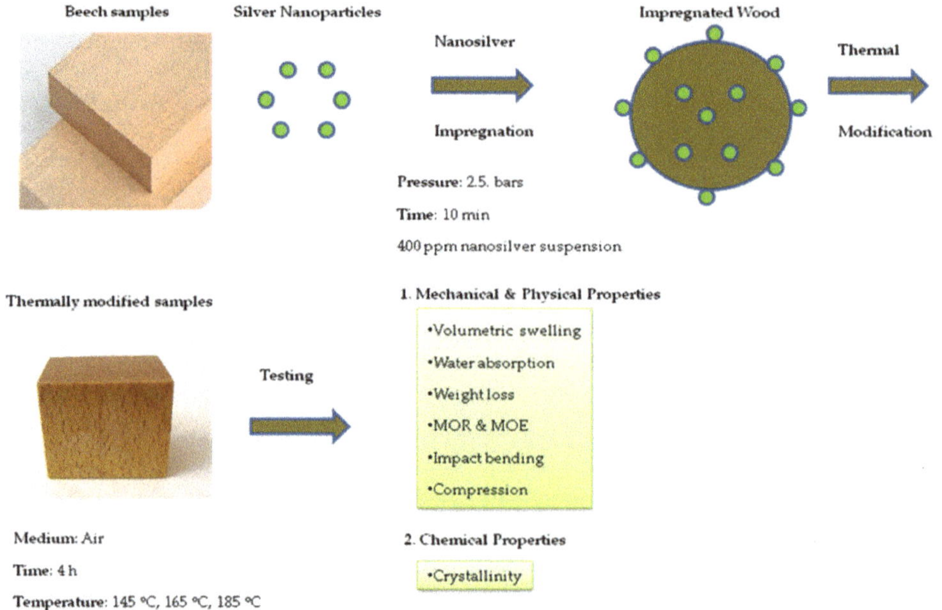

Figure 1. Flow diagram of the experimental procedure.

2.4. Total Crystillanity Index (TCrI)

Fourier Transform Infrared Spectroscopy (FTIR) was used to measure crystallinity of cellulose in the present project. In this method, specific infrared bands in FTIR spectra are related to the crystallinity and the type of cellulose (amorphous or crystalline) in each specimen [34]. The total crystallinity index (TCrI or TCI) was used to measure and report the amount of crystallinity in different treatments. In this index, the band (at 1372 cm^{-1}) is attributed to C–H deformation, and the band at 2900 cm^{-1} is attributed to C–H and CH$_2$ stretching. The areas under the curve of the above-mentioned bands were measured using OMNIC 6.1a software (Thermo Nicolet Corporation, USA). The areas were measured based on the on-set and off-set points for each band. Once the area under each band was measured, TCr index for each treatment was defined and calculated as a$_{1372\text{cm}}{}^{-1}$ divided by a$_{2900\text{cm}}{}^{-1}$ [34]. It is to be noted that a linear relationship was found between the results from TCrI with those obtained from X-ray diffraction (XRD) [35].

2.5. Physical and Mechanical Properties

Physical and mechanical properties, namely volumetric swelling, water absorption, modulus of rupture (MOR), modulus of elasticity (MOE), impact bending, compression strength parallel to grain, were determined in accordance with ASTM D0143-94 [36].

2.6. Statistical Analysis

Statistical analysis was conducted using SAS software, version 9.2 (Cary, NC, USA). Two-way analysis of variance (ANOVA) was performed on the mean data to determine significant differences at the 95% level of confidence. Duncan's multiple range groupings were carried out among treatments to elucidate significant difference among groups. Hierarchical cluster analysis, including dendrograms and Ward methods with squared Euclidean distance intervals, was conducted with SPSS/18, version 18 (IBM; NY, USA). Cluster analysis was performed to find similarities and dissimilarities between treatments based on more than one property simultaneously. The scaled indicator in each cluster analysis shows similarities and differences between treatments; lower scale numbers show more similarities while higher ones show dissimilarities. Fitted-line, contour, and surface plots were done in Minitab software, version 16.2.2 (Minitab Inc.; NY, USA).

3. Results and Discussion

Results showed an increasing trend in weight losses as the temperature of thermal modification increased from 145 °C to 185 °C (Figure 2). The highest and lowest weight losses were observed in NS-HT185 and HT145 specimens, respectively. Weight losses caused by thermal modification at 145 °C, were very low, both for untreated and NS-impregnated specimens. This indicated that degradation of cell-wall polymers was very low at this temperature. It is expected that mostly hemi-cellulose and lignin were degraded at this temperature as to the fact that cellulose tends to have higher resistance to thermal degradation in comparison to hemi-cellulose and lignin [11,12,37].

Physical properties (volumetric swelling and water absorption) (Figure 3A–D) and mechanical properties (MOR and MOE) (Figure 4A,B) of treated wood demonstrated statistically insignificant fluctuations at low temperatures compared to control specimens. These are considered corroborating evidence of low degradation of cell-wall polymers at this low temperature of 145 °C for thermal modification. On the other hand, an increase of temperature to 185 °C had a significant effect on all properties. There was a significant weight loss, indicating degradation in cell-wall polymers. This degradation ultimately resulted in significant decrease in MOR, impact strength, and physical properties. However, thermal modification at 185 °C did not seem to cause significant fluctuations in MOE and compression strength parallel to grain (Figure 4B,D). In fact, compression strength parallel to grain showed an increase, though the increase was not statistically significant. In this connection, it was reported that some semi-crystalline regions in wood changes to crystalline regions as a result of heat treatment [38]. The crystalline regions in wood structure act as a framework that supports the whole wood structure and therefore, increase in this part had an improving effect on MOE and compression strength parallel to grain values, though cell-wall polymers were degraded to some extent. However, crystallinity measurement showed a decrease when specimens were thermally modified at 185 °C (Figure 5) and therefore, other factors appeared to be involved in the process too. Thermal modification in the present study was carried out at three temperatures which are all more than glass transition of lignin in most wood species [3,38,39]. Therefore as a binding agent, cross-linking in lignin could occur, making new bonds and connecting points between lignin with cellulose and hemicellulose fibrils that were thermally degraded [10]. This reasoning can also partly explain the increases in MOR and MOE values in N-HT145 specimens, though the increase was not statistically significant (Figure 4A,B). In fact, the increased crystallinity as well as cross-linking in lignin put the small thermal degradation in perspective, eventually MOR and MOE increased. That is, these could significantly ameliorate the negative effects of thermal degradation. Attenuated Total Reflectance (ATR) graph of crystallinity was in agreement with this hypothesis, showing the highest crystallinity in N-HT145 treatment (Figure 5). Water absorption and volumetric swelling values also illustrated significant decreases when specimens were heat-treated at 185 °C, indicating thermal degradation of cellulose and hemicellulose, change of amorphous cellulose to crystalline cellulose, and cross-linking in lignin binding cellulose molecules and inhibiting their hydroxyl groups to make bonds with water molecules. Moreover, the reduction in the physical properties can also be partially attributed to irreversible hydrogen bonding in the course of water movements within the pore system of the cell walls, eventually decreasing hygroscopicity in wood [40]. As to the fact that one of the consequences of irreversible hydrogen bonding is reported to be hornification, the unexpected high values of compression strength parallel to grain and MOE in HT185 and NS-HT185 treatments can also partially be attributed to hornification.

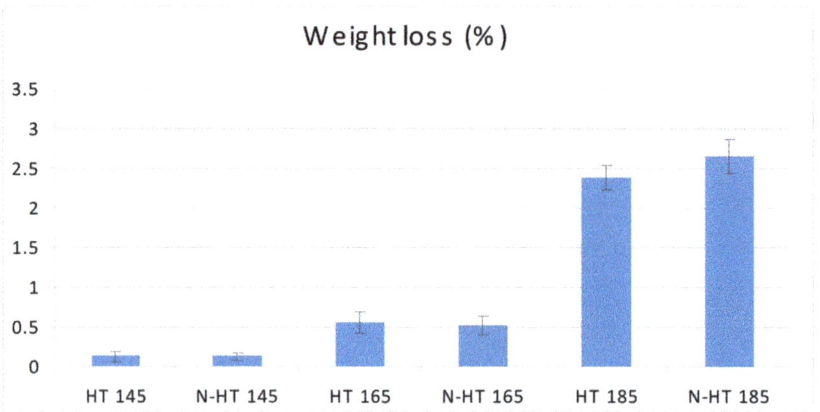

Figure 2. Weight losses in beech wood heat-treated at 145, 165, and 185 °C and impregnated with silver nano-suspension (HT = heat treatment at a determined temperature; N = nanosilver impregnated).

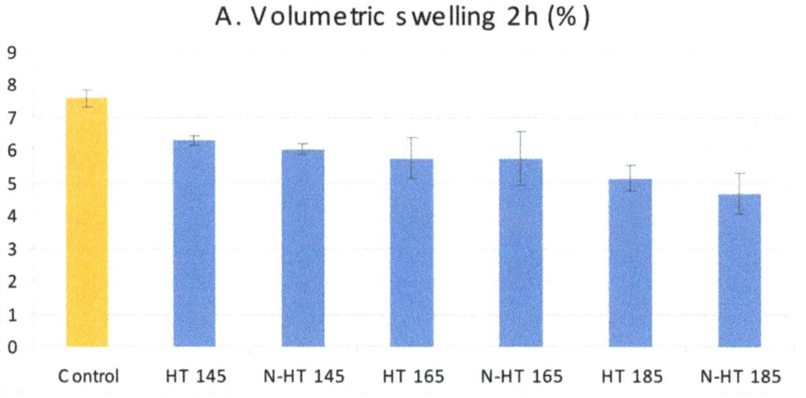

Figure 3. *Cont.*

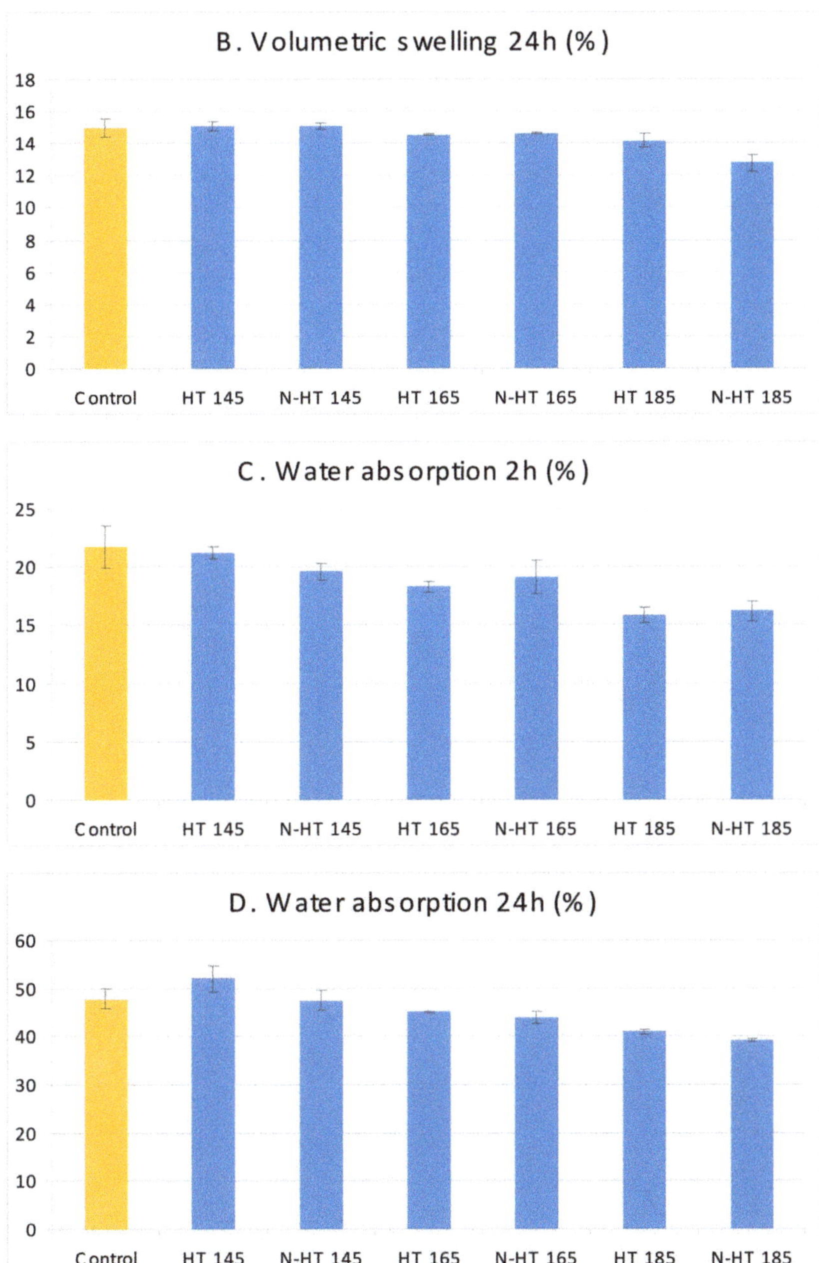

Figure 3. Physical properties in beech wood heat-treated at 145, 165, and 185 °C and impregnated with silver nano-suspension (HT = heat treatment at a determined temperature; N = nanosilver impregnated).

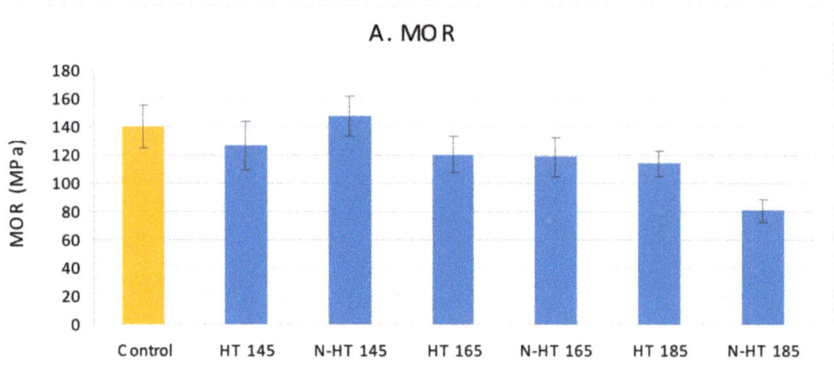

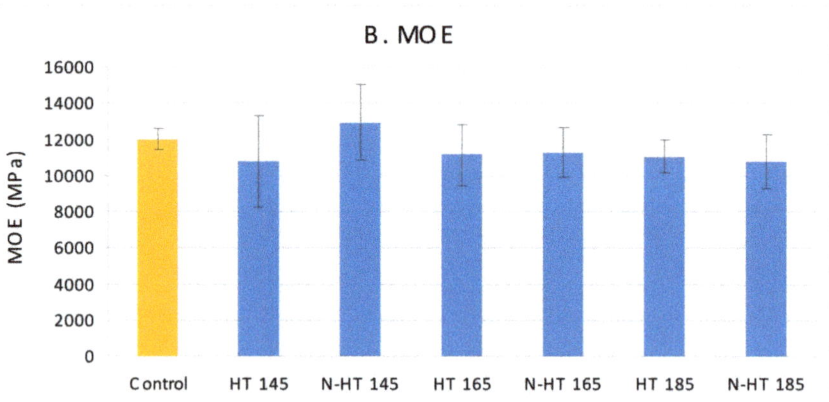

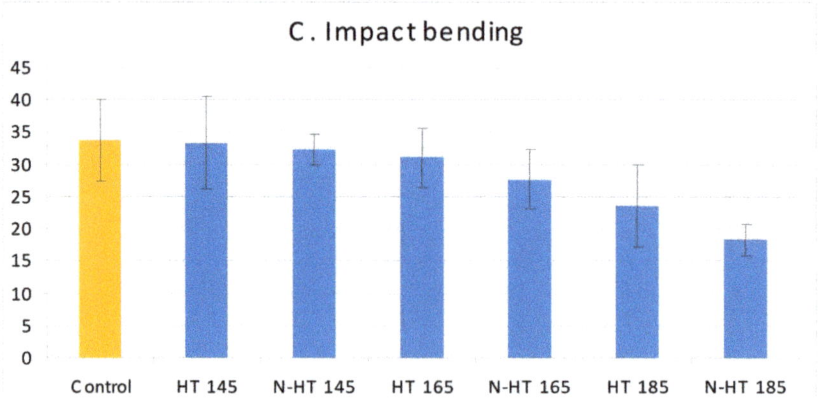

Figure 4. *Cont.*

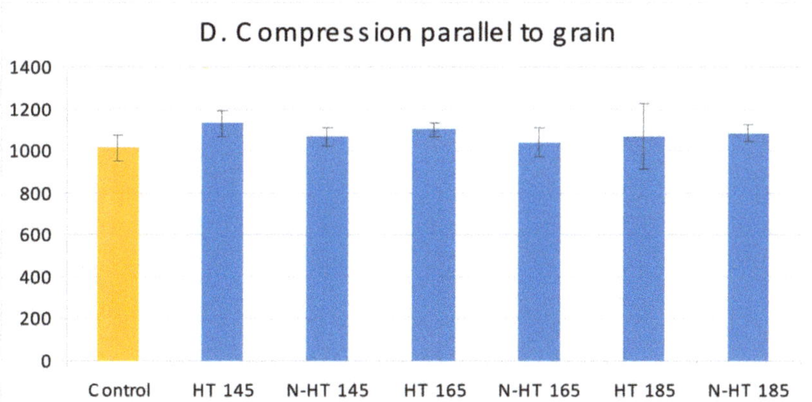

Figure 4. Mechanical properties in beech wood heat-treated at 145, 165, and 185 °C and impregnated with silver nano-suspension (HT=Heat Treatment at a determined temperature; N = nanosilver impregnated).

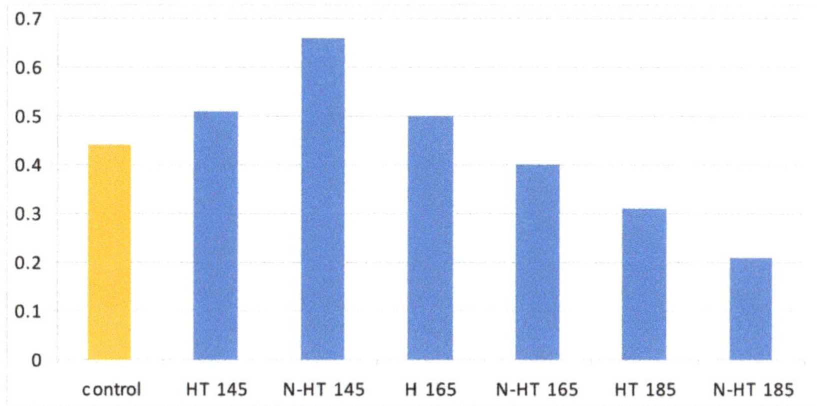

Figure 5. Crystallinity in the seven treatments studied in the present project (HT = heat treatment at a determined temperature; N = nanosilver impregnated).

Infrared spectra were in agreement with the above-mentioned changes in physical and mechanical properties as a result of thermal modification at different temperatures (Figure 6 A,B). Clear fluctuations and differences were observed in wave numbers 3300–3000 cm^{-1}, relating to hydroxyl groups of cell-wall polymers which are greatly influential on both physical and mechanical properties. Significant differences were also observed in the intensities of fingerprint regions of wave number 1700–1500 cm^{-1} (C=O stretching) in both un-impregnated and NS-impregnated treatments. The peak of 1733 cm^{-1} was reduced in intensity in almost all thermally modified specimens; only NS-HT165 showed an increased intensity. This wave number is assigned to the stretching vibrations of carbonyl groups which belong mainly to hemicellulose. The wave number 3700–3600 cm^{-1} (relating to nonbonded hydroxy groups, OH stretch) showed similarity between the control and specimens heat-treated at 145 °C, both nanosilver-impregnated and untreated. Specimens heat-treated at 165 and 185 illustrated significant differences with that of control specimens, showing the effects of thermal modification on hydroxy groups.

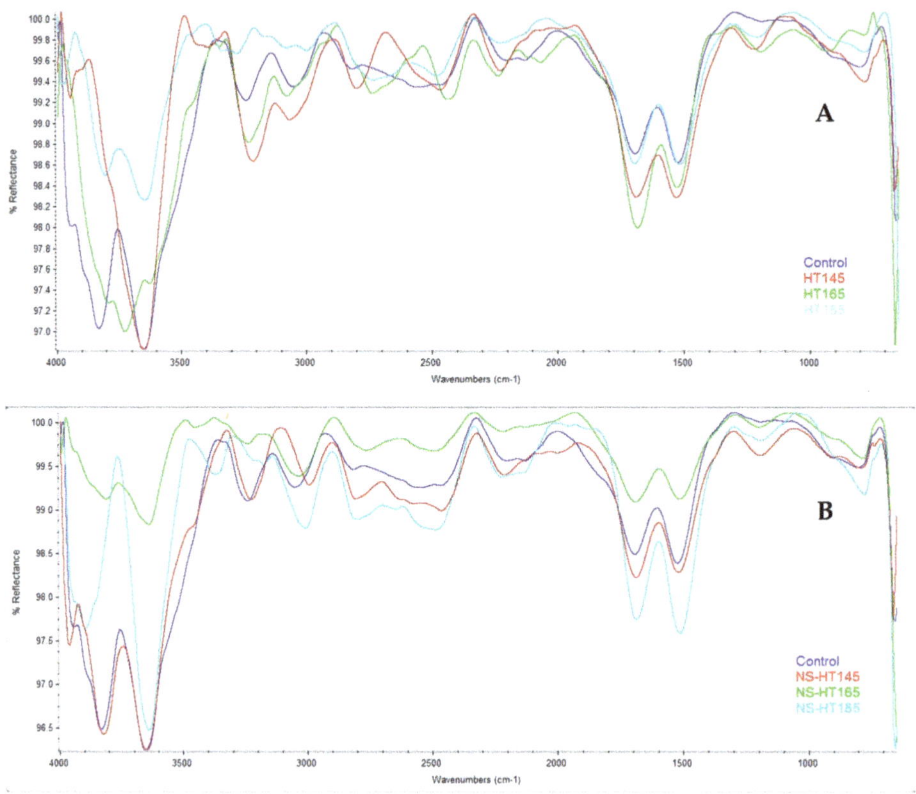

Figure 6. Infrared spectra of the control and thermally-modified (**A**) and the control and nanosilver suspension (NS)-impregnated thermally modified (**B**) beech specimens (HT = heat treatment at a determined temperature; NS = nanosilver impregnated).

Previous studies, reported increased thermal conductivity in solid wood and wood-based composites with addition of nano-metals and nano-minerals [41,42]. Physical and mechanical properties of composite panels were improved as a result of addition of nano-copper in particleboards [41,42]; and an increased thermal conductivity of 29% by addition of nano-wollastonite was associated with improved physical and mechanical properties [42]. In the present study, thermal conductivity coefficient was not measured however, the effects of impregnating specimens with nanosilver suspension were observed in some of the properties. Weight loss in NS-HT185 specimens was higher, though the amount of increase was not statistically significant. This had a decreasing effect on volumetric swelling, MOR, and compression strength parallel to grain. Crystallinity also demonstrated a decrease in NS-HT185 in comparison to HT185 treatment. TCr indices in specimens thermally treated at 145 °C revealed a significant increase as a result of impregnation with nanosilver suspension. This improvement in TCr index resulted in a noticeable increase in MOR and MOE values. Other properties did not show significant fluctuations, suggesting that the effect of the increased crystallinity and cross-linking in lignin was more than the negative effect of the low cell-wall polymer degradation caused by thermal modification. The same hypothesis concurs with NS-HT165 and HT165 treatments. That is, no significant difference was observed between these two treatments on the basis of any of the properties, with the exception of impact bending. Impact bending (impact strength) was previously reported to be one the most sensitive property in solid wood species to be affected by thermal modification [10].

Fitted-line plot between different properties versus crystallinity showed rather low and statistically insignificant R-square. For instance, R-square between MOR versus ATR-crystallinity was only 68% (Figure 7). This implied the interaction and involvement of a variety of factors besides crystallinity. For example, and with regard to the physical properties of water absorption and volumetric swelling, degradation of cell-wall polymers and change in lignin may have caused alteration of chemical structure of the polymers and therefore, the number of hydroxyl groups to absorb water molecules differed. As to the mechanical properties, degradation of cell-wall polymers had negative effects on most of the mechanical properties, while increased crystallinity acted as a supporting framework, eventually ameliorating part of the negative effects. Overall, it can be concluded that increased crystallinity and cross-linking in lignin can compensate for some decreased properties caused by thermal modification, but it would be significantly dependent on the temperature under which modification is carried out. Contour plot and surface analysis illustrated the same trends. That is, a general direct and smooth relationship was observed among some of the properties, like MOR and MOE versus impact bending (Figure 8), indicating that thermal modification had the same increasing or decreasing effect on all these properties. However, distortions happened in the plots for some other properties, indicating significant different effect of thermal modification on the properties.

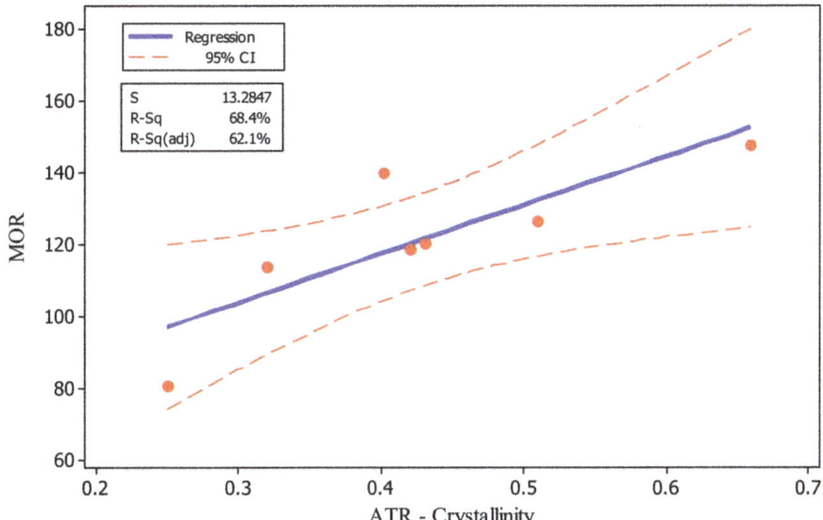

Figure 7. Fitted-line plot between Attenuated Total Reflectance (ATR)-crystallinity versus mechanical properties (MOR) in the seven treatments studied in the present project.

Cluster analysis based on all properties measured in the present project showed rather close clustering of control, HT145, and NS-HT145 treatments (Figure 9). This indicated that the overall alterations of different properties caused by thermal modification at 145 °C only slightly changed the overall properties of beech wood, though some of properties revealed improvements, such as MOR and MOE. The other four treatments at 165 °C and 185 °C temperatures showed distinct remote clustering from the control treatment, demonstrating a great impact on the overall properties of beech wood. NS-HT185 clustered rather remotely from HT165, NS-HT165, and HT185 which were closely clustered. This can indicate the influential effect of an increased thermal conductivity caused by impregnation with NS-suspension. However, close clustering of HT146 with NS-HT145 can indicate that at lower temperature of about 145 °C, an increased thermal conductivity may not have as great impact in small woody specimens as it has at higher temperature of 185 °C.

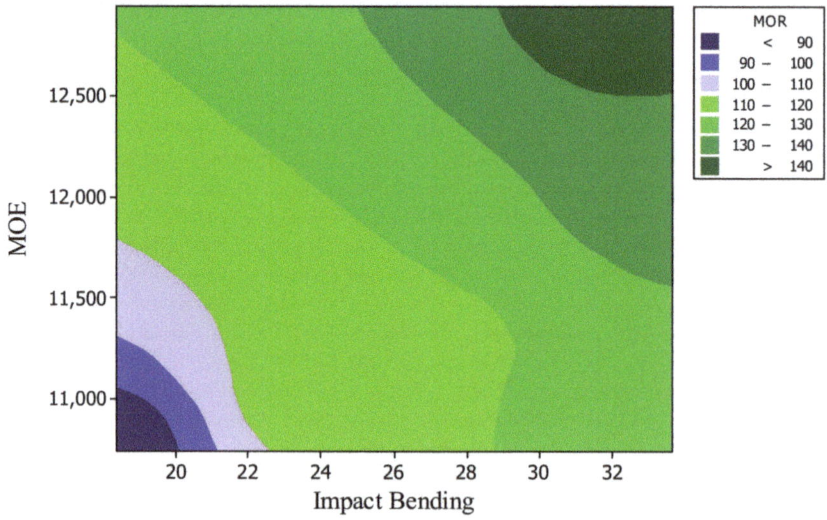

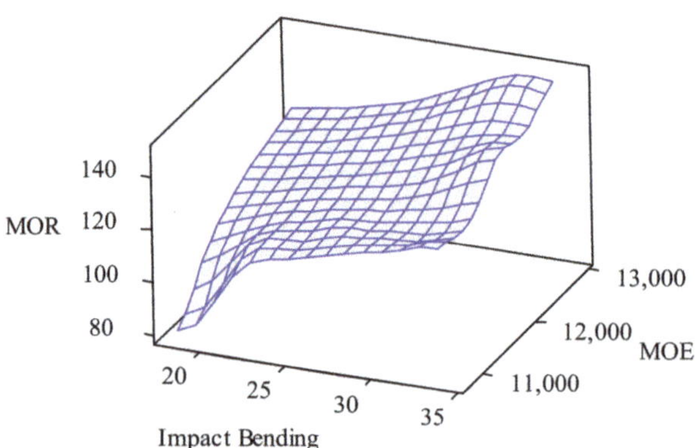

Figure 8. Contour and surface plots among MOR, MOE (mechanical properties), and impact bending in the seven treatments studied in the present project.

Figure 9. Cluster analysis based on all physical and mechanical properties, as well as crystallinity values of the seven treatments studied in the present project (HT = heat treatment at a determined temperature; NS = nanosilver impregnated).

4. Conclusions

The aim of this study was to investigate the physical and mechanical properties of thermally modified beech wood impregnated with silver nano-suspension and to examine their relationship with the crystallinity of cellulose. Specimens were impregnated with a 400 ppm nanosilver suspension (NS); at least, 90% of silver nano-particles ranged between 20 and 100 nano-meters. Heat treatment took place in a laboratory oven at three temperatures, namely 145, 165, and 185 °C. Physical properties and mechanical properties of treated wood demonstrated statistically insignificant fluctuations at low temperatures compared to control specimens. On the other hand, an increase of temperature to 185 °C had a significant effect on all properties. As a consequence of the thermal modification, part of amorphous cellulose was changed to crystalline cellulose. At low temperatures an increased crystallinity caused some of the properties to be improved. Change of amorphous cellulose to crystalline cellulose, as well as cross-linking in lignin, partially ameliorated the negative effects of thermal degradation at higher temperatures and therefore, compression parallel to grain and modulus of elasticity did not decrease significantly. Previous studies, reported increased thermal conductivity in solid wood and wood-based composites with addition of nano-metals and nano-minerals [14,19,20]. The increased thermal conductivity ultimately caused higher degradation in the main wood components and therefore, mechanical properties and pull-off strength decreased in specimens thermally modified at temperature higher than 165 °C [19,20]. In the present study, thermal conductivity coefficient was not measured however, the consequent effects of impregnating specimens with nanosilver suspension were observed in some of the properties. Impregnating specimens with silver nano-suspension prior to thermal modification enhanced the effects of thermal modification as a result of improved thermal conductivity.

Author Contributions: Data curation, S.B. and H.T.; Methodology, S.B.; Writing – original draft, H.T. and A.P.; Writing – review & editing, A.P.

Funding: This research received no external funding.

Acknowledgments: Senior authors appreciate the constant scientific support of Mr. Jack Norton (Retired, Horticulture & Forestry Science, Queensland Department of Agriculture, Forestry and Fisheries, Australia), as well as Alexander von Humboldt Stiftung, Germany.

Conflicts of Interest: The authors declare no conflict of interest.

References

1. Rowell, R.M. *Handbook of Wood Chemistry and Wood Composites*, 2nd ed.; CRC Press, Taylor and Francis Group: Boca Raton, FL, USA, 2012.
2. Gerardin, P. New alternatives for wood preservation based on thermal and chemical modification of wood—A review. *Ann. For. Sci.* **2016**, *73*, 559–570. [CrossRef]
3. Hill, C.A.S. *Wood Modification—Chemical, Thermal and Other Processes*; John Wiley and Sons Ltd.: West Sussex, UK, 2006.
4. Papadopoulos, A.N. Chemical modification of solid wood and wood raw materials for composites production with linear chain carboxylic acid anhydrides: A brief Review. *BioResources* **2010**, *5*, 499–506.
5. Mantanis, G.I. Chemical modification of wood by acetylation or furfurylation: A review of the present scaled-up technologies. *BioResources* **2017**, *12*, 4478–4489. [CrossRef]
6. Teng, T.; Arip, M.; Sudesh, K.; Lee, H. Conventional technology and nanotechnology in wood preservation: A review. *BioResources* **2018**, *13*, 9220–9252. [CrossRef]
7. Papadopoulos, A.N.; Bikiaris, D.N.; Mitropoulos, A.C.; Kyzas, G.Z. Nanomaterials and chemical modification technologies for enhanced wood properties: A review. *Nanomaterials* **2019**, *9*, 607. [CrossRef] [PubMed]
8. Boonstra, M.J.; Tjeerdsma, B. Chemical analysis of heat treated softwoods. *Holz als Roh und Werkstoff* **2006**, *64*, 204–211. [CrossRef]
9. Tjeerdsma, B.F.; Stevens, M.; Militz, H. *Durability Aspects of (Hydro) Thermal Treated Wood*; Document no. IRG/WP 00-4; International Research Group on Wood Preservation: Biarritz, France, 2000.
10. Boonstra, M.J.; van Acker, J.; Tjeerdsma, B.F.; Kegel, E.V. Strength properties of thermally modified softwoods and its relation to polymeric structural wood constituents. *Ann. For. Sci.* **2007**, *64*, 679–690. [CrossRef]

11. Tjeerdsma, B.F.; Boonstra, M.; Pizzi, A.; Tekely, P.; Militz, H. Characterization of thermal modified wood: Molecular reasons for wood performance improvement. CPMAS 13 CNMR characterization of thermally modified wood. *Holz Roh und Werkstoff* **1998**, *56*, 149–153. [CrossRef]

12. Tjeerdsma, B.F.; Militz, H. Chemical changes in hydrothermal treated wood: FTIR analysis of combined hydrothermal and dry heat-treated wood. *Holz als Roh und Werkstoff* **2005**, *63*, 102–111. [CrossRef]

13. Roco, M. Nanotechnology's Future. *Sci. Am.* **2006**, *13*, 427–445. [CrossRef]

14. Taghiyari, H.R.; Schimdt, O. Nanotachnology in wood based composite panels. *Int. J. Bioinorg. Hybrid Nanomater.* **2014**, *3*, 65–73.

15. Goffredo, G.B.; Accoroni, S.; Totti, T.; Romagnoli, T.; Valentini, L.; Munafò, P. Titanium dioxide based nanotreatments to inhibit microalgal fouling on building stone surfaces. *Build. Environ.* **2017**, *112*, 209–222. [CrossRef]

16. De Filpo, G.; Palermo, A.M.; Rachiele, F.; Nicoletta, F.P. Preventing fungal growth in wood by titanium dioxide nanoparticles. *Int. Biodeterior. Biodegrad.* **2014**, *85*, 217–222. [CrossRef]

17. Civardi, C.; Schwarze, F.; Wick, P. Micronized copper wood protection: An efficiency and potential health and risk assessment for copper based nanoparticles. *Environ. Pollut.* **2015**, *200*, 20–32. [CrossRef] [PubMed]

18. Moya, R.; Zuniga, A.; Berrocal, A.; Vega, J. Effect of silver nanoparticles synthesized with NPsAg-ethylene glycol on brown decay and white decay fungi of nine tropical woods. *J. Nanosci. Nanotechnol.* **2017**, *17*, 1–8. [CrossRef]

19. Taghiyari, H.R.; Enayati, A.; Gholamiyan, H. Effects of nano-silver impregnation on brittleness, physical and mechanical properties of heat-treated hardwoods. *Wood Sci. Technol.* **2012**, *47*, 467–480. [CrossRef]

20. Taghiyari, H.R.; Samandarpour, A. Effects of nanosilver-impregnation and heat treatment on coating pull-off adhesion strength on solid wood. *Drv. Ind.* **2015**, *66*, 321–327. [CrossRef]

21. Klemm, D.; Heublein, B.; Fink, H.P.; Bohn, A. Cellulose: Fascinating biopolymer and sustainable raw material. *Angew. Chem. Int. Ed.* **2005**, *44*, 3358–3393. [CrossRef]

22. Sharma, P.R.; Joshi, R.; Sharma, S.K.; Hsiao, B.S. A Simple Approach to Prepare Carboxycellulose Nanofibers from Untreated Biomass. *Biomacromolecules* **2017**, *18*, 2333–2342. [CrossRef]

23. Sharma, P.R.; Chattopadhyay, A.; Sharma, S.K.; Geng, L.; Amiralian, N.; Martin, D.; Hsiao, B.S. Nanocellulose from Spinifex as an Effective Adsorbent to Remove Cadmium(II) from Water. *ACS Sustain. Chem. Eng.* **2018**, *6*, 3279–3290. [CrossRef]

24. Sharma, P.R.; Chattopadhyay, A.; Sharma, S.K.; Hsiao, B.S. Efficient Removal of UO_2^{2+} from Water Using Carboxycellulose Nanofibers Prepared by the Nitro-Oxidation Method. *Ind. Eng. Chem. Res.* **2017**, *56*, 13885–13893. [CrossRef]

25. Sharma, P.R.; Zheng, B.; Sharma, S.K.; Zhan, C.; Wang, R.; Bhatia, S.R.; Hsiao, B.S. High Aspect Ratio Carboxycellulose Nanofibers Prepared by Nitro-Oxidation Method and Their Nanopaper Properties. *Appl. Nano Mater.* **2018**, *1*, 3969–3980. [CrossRef]

26. Sharma, P.R.; Sharma, S.K.; Antoine, R.; Hsiao, B.S. Efficient Removal of Arsenic Using Zinc Oxide Nanocrystal-Decorated Regenerated Microfibrillated Cellulose Scaffolds. *ACS Sustain. Chem. Eng.* **2019**, *7*, 6140–6151. [CrossRef]

27. Golmohammadi, H.; Morales-Narvaez, E.; Naghdi, T.; Merkoci, A. Nanocellulose in Sensing and Biosensing. *Chem. Mater.* **2017**, *29*, 5426–5446.

28. Sabo, R.; Yermakov, A.; Law, C.T.; Elhajjar, R. Nanocellulose-Enabled Electronics, Energy Harvesting Devices, Smart Materials and Sensors: A Review. *J. Renew. Mater.* **2016**, *4*, 297–312. [CrossRef]

29. Sharma, P.R.; Varma, A.A. Thermal stability of cellulose and their nanoparticles: Effect of incremental increases in carboxyl and aldehyde groups. *Carbohydr. Polym.* **2014**, *114*, 339–343. [CrossRef] [PubMed]

30. Sharma, P.R.; Varma, A.A. Functional nanoparticles obtained from cellulose: Engineering the shape and size of 6-carboxycellulose. *Chem. Commun.* **2013**, *49*, 8818–8820. [CrossRef]

31. Geng, L.; Peng, X.; Zhan, C.; Naderi, A.; Sharma, P.R.; Mao, Y.; Hsiao, B.S. Structure characterization of cellulose nanofiber hydrogel as functions of concentration and ionic strength. *Cellulose* **2017**, *24*, 5417–5429. [CrossRef]

32. Sharma, P.R.; Rajamohanan, P.R.; Varma, A.J. Supramolecular transitions in native cellulose-I during progressive oxidation reaction leading to quasi-spherical nanoparticles of 6-carboxycellulose. *Carbohydr. Polym.* **2014**, *113*, 615–662. [CrossRef]

33. Figueroa, M.; Bustos, C.; Dechent, P.; Reyes, L.; Cloutier, A.; Giuliano, M. Analysis of rheological and thermo-hygro-mechanical behaviour of stress-laminated timber bridge deck in variable environmental conditions Maderas. *Cienc. Tecnol.* **2012**, *14*, 303–319.

34. FitzPatrick, M.A. Characterization and Processig of Lignocellulosic Biomass in Ionic Liquids. Ph.D. Thesis, Queen's University, Kingston, ON, Canada, 2011.

35. Marson, G.A.; El Seoud, O.A. Cellulose dissolution in lithium chloride/*N*,*N*-dimethylacetamide solvent system: Relevance of kinetics of decrystallization to cellulose derivatization under homogeneouls solution conditions. *J. Polym. Sci. Part A Polym. Chem.* **1999**, *37*, 3738–3744. [CrossRef]

36. ASTM International. *ASTM D143—94. Standard Test Methods for Small Clear Specimens of Timber*; ASTM International: West Conshohocken, PA, USA, 2007.

37. Esteves, B.; Graca, J.; Pereira, H. Extractive composition and summative chemical analysis of thermally treated eucalypt wood. *Holzforschung* **2008**, *62*, 344–351. [CrossRef]

38. Phuong, L.X.; Shida, S.; Saito, Y. Effects of heat treatment on brittleness of Styrax tonkinensis wood. *J. Wood Sci.* **2007**, *53*, 181–186. [CrossRef]

39. Hatakeyama, T.; Nakamura, K.; Htakeyama, H. Studies on heat capacity of cellulose and lignin by differential scanning calorimetry. *Polymer* **1982**, *23*, 1801–1804. [CrossRef]

40. Borrega, M.; Kärenlampi, P.P. Hygroscopicity of Heat-Treated Norway Spruce (Picea abies) wood. *Eur. J. Wood Wood Prod.* **2010**, *68*, 233–235. [CrossRef]

41. Taghiyari, H.R.; Farajpour Bibalan, O. Effect of copper nanoparticles on permeability, physical, and mechanical properties of particleboard. *Eur. J. Wood Wood Prod.* **2013**, *71*, 69–77. [CrossRef]

42. Taghiyari, H.R.; Ghorbanali, M.; Tahir, P.M.D. Effects of the improvement in thermal conductivity coefficient by nano-wollastonite on physical and mechanical properties in medium-density fiberboard (MDF). *BioResources* **2014**, *9*, 4138–4149. [CrossRef]

Article

Preparation of Biomorphic Porous SiC Ceramics from Bamboo by Combining Sol–Gel Impregnation and Carbothermal Reduction

Ke-Chang Hung [1,†], Tung-Lin Wu [2,†], Jin-Wei Xu [1] and Jyh-Horng Wu [1,*]

[1] Department of Forestry, National Chung Hsing University, Taichung 402, Taiwan
[2] College of Technology and Master of Science in Computer Science, University of North America, Fairfax, VA 22033, USA
* Correspondence: eric@nchu.edu.tw
† These authors contributed equally to this work.

Received: 3 August 2019; Accepted: 31 August 2019; Published: 2 September 2019

Abstract: This study investigated the feasibility of using bamboo to prepare biomorphic porous silicon carbide (bio-SiC) ceramics through a combination of sol–gel impregnation and carbothermal reduction. The effects of sintering temperature, sintering duration, and sol–gel impregnation cycles on the crystalline phases and microstructure of bio-SiC were investigated. X-ray diffraction patterns revealed that when bamboo charcoal–SiO_2 composites (BcSiCs) were sintered at 1700 °C for more than 2 h, the resulting bio-SiC ceramics exhibited significant β-SiC diffraction peaks. In addition, when the composites were sintered at 1700 °C for 2 h, scanning electron microscopy micrographs of the resulting bio-SiC ceramic prepared using a single impregnation cycle showed the presence of SiC crystalline particles and nanowires in the cell wall and cell lumen of the carbon template, respectively. However, bio-SiC prepared using three and five repeated cycles of sol–gel impregnation exhibited a foam-like microstructure compared with that prepared using a single impregnation cycle. Moreover, high-resolution transmission electron microscopy and selected area electron diffraction revealed that the atomic plane of the nanowire of bio-SiC prepared from BcSiCs had a planar distance of 0.25 nm and was perpendicular to the (111) growth direction. Similar results were observed for the bio-SiC ceramics prepared from bamboo–SiO_2 composites (BSiCs). Accordingly, bio-SiC ceramics can be directly and successfully prepared from BSiCs, simplifying the manufacturing process of SiC ceramics.

Keywords: bamboo; carbothermal reduction; ceramic; silicon carbide; sol–gel process

1. Introduction

Over the past few decades, silicon carbide (SiC) ceramics have been extensively used in the structures of modern aviation vehicles, such as rocket nozzles, aeronautic jet engines, and aircraft brake materials, because such ceramics exhibit remarkable physical properties, including wear, corrosion, and thermal resistance [1–5]. In particular, SiC ceramics hold considerable promise for use as solar energy absorbers. Solar-energy-absorbing devices require ceramics with open porosity, excellent solar energy absorption performance, and high thermal conductivity. In general, SiC ceramics that exhibit larger grain sizes and are supplemented with α-SiC can produce relatively high thermal conductivity levels [6].

Reaction processing of SiC is a technique for producing advanced SiC-based structural ceramics. This technique involves a chemical reaction between carbon (C) and silicon (Si) during sintering [5]. In recent years, the use of wood as a carbon template for preparing SiC ceramics with a wood-like microstructure (wood-like SiC ceramics) has emerged as a relatively new research area [6–8]. Therefore, several techniques have been developed for fabricating porous wood-like SiC ceramics, including reactive infiltration with Si-containing melts [9,10], reactive silicon vapor infiltration [11–13], and SiO_2 sol–gel impregnation combined with carbothermal reduction [14–22]. In the process of SiO_2 sol–gel impregnation combined with carbothermal reduction, the charcoal is impregnated with SiO_2 sol at ambient temperature under reduced pressure. Subsequently, the water initiates the hydrolysis reaction of SiO_2 sol and then polycondensation reaction by dehydration or dealcoholation. The sols within charcoal are gelation as gels in 105 °C to prepare charcoal–SiO_2 composites [23], and then the charcoal–SiO_2 composites were subjected to carbothermal reduction reactions for fabricating SiC ceramics. Therefore, this technique has several advantages, for example, it is a low-cost approach, involves easy synthesis procedures, involves relatively low temperatures of synthesis, provides high-purity resultant products, and can retain the structure and morphology of starting carbonaceous materials [21].

Bamboo, a perennial woody plant belonging to the Gramineae family, is widely distributed across Asia [24–26] and exhibits higher growth rates than do other woody plants [27–29]. In Taiwan, bamboo is extensively used as a raw material for handicraft, furniture, and construction applications [30,31]. Bamboo materials are similar to wood, and they have a porous structure that renders them useful for realizing different penetration and treatment procedures [32]. Studies have focused on the crystalline-phase composition of wood-like SiC ceramics. The use of bamboo as an industrial raw material can improve economic efficiency. Accordingly, this study explored the feasibility of using bamboo to produce porous biomorphic SiC (bio-SiC) ceramics by combining sol–gel impregnation and carbothermal reduction. In addition, the study investigated the effects of sintering temperature, sintering duration, and repeated cycles of sol–gel impregnation on the crystalline phase and microstructure of the bio-SiC ceramics.

2. Materials and Methods

2.1. Experimental Materials and Procedure

This study obtained 4-year-old moso bamboo (*Phyllostachys pubescens* Mazel) from a local bamboo-processing factory (Nantou, Taiwan). The obtained bamboo was air-dried and then cut into rectangular specimens measuring approximately 50 mm (L) × 25 mm (T) × 5 mm (R). Each of the bamboo specimens was converted to a porous biocarbon template (bamboo charcoal) through carbonization in a nitrogen atmosphere at a temperature of 850 °C, heating rate of 5 °C/min, and holding time of 1 h. A SiO_2 precursor sol was formulated by mixing methyltrimethoxysilane (MTMOS) (Acros Chemical, Geel, Belgium) and methanol at a molar ratio of 0.13:1. In the sol–gel process, the bamboo and bamboo charcoal specimens were impregnated with the prepared sol under reduced pressure for 2 days. The impregnated specimens were then placed in an oven controlled to 105 °C for 24 h to age the gels. Repeated impregnation cycles (one, three, and five cycles) were applied, producing various bamboo charcoal–SiO_2 composites (BcSiCs) and bamboo–SiO_2 composites (BSiCs). Similar to the wood-SiO_2 composites reported in our previous study [33], the Fourier transform infrared (FTIR) absorption bands at 1270 (Si–CH_3, ν), 1104 (Si–O–Si, δ), and 778 cm^{-1} (Si–O–C, ν) were clearly observed in these BcSiCs and BSiCs (data not shown). This result confirms that silica mixtures are also formed in the bamboo-based composites. Consequently, the derived BcSiCs and BSiCs were subjected to carbothermal reduction reactions in nitrogen atmosphere (flow rate: 2.5 L/min) in a heater furnace (Nabertherm, LHT 04/17 SW, Lilienthal/Bremen, Germany) at 1500–1700 °C for 0.5–3 h to form porous bio-SiC ceramics. Figure 1 presents a summary of the processing scheme for the preparation of the porous bio-SiC ceramics.

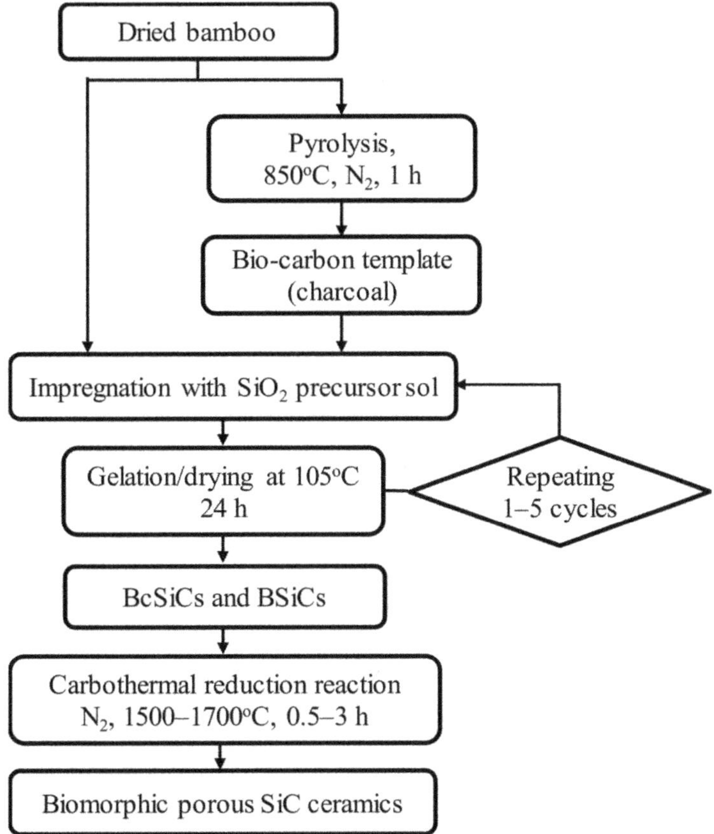

Figure 1. Processing scheme for manufacturing silicon carbide (SiC) ceramics with a biomorphic pore structure from bamboo by combining sol–gel impregnation and carbothermal reduction.

2.2. Characterization

The crystalline phases of bamboo charcoal and porous SiC ceramics were characterized through a powder X-ray diffraction (XRD) instrument (MAC Science, MXP18, Tokyo, Japan) operated in the 2θ range of 2–35° using CuKαl (λ = 1.5406 Å) radiation at 40 kV and 30 mA. The morphology and detailed structural features of the specimens were determined using a scanning electron microscopy (SEM) instrument (JEOL, JSM-7401F, Tokyo, Japan) and a high-resolution transmission electron microscopy (HRTEM) instrument (JEOL, JEM-2100F, Tokyo, Japan), respectively. Field-emission measurements for the nanowires of the synthesized SiC were performed in a vacuum chamber at a pressure of approximately 5.0×10^{-7} Torr. A thermal analyzer (PerkinElmer, Pyris 1, Shelton, CT, USA) was used to elucidate the synthesis and oxidation mechanisms of the samples. Thermogravimetric analysis (TGA) was conducted in an air atmosphere from 50 to 950 °C at a heating rate of 10 °C/min.

3. Results and Discussion

3.1. Effect of Sintering Temperature on the Properties of Porous Biomorphic Silicon Carbide (Bio-SiC) Ceramics

To understand the effect of sintering temperature on the properties of the porous bio-SiC ceramics fabricated in this study, the following test conditions were applied: impregnation of bamboo charcoal with a SiO₂ precursor sol (BcSiC) in a single cycle and then sintering at different temperatures for

2 h. Figure 2A illustrates the XRD patterns of bamboo charcoal and the BcSiCs sintered at 1500, 1600, and 1700 °C for 2 h. When the samples were sintered at temperatures below 1700 °C, two broad carbon diffraction peaks could be clearly observed at 2θ values of approximately 24° and 43°, which were attributed to the (002) and (101)/(100) planes of carbon, respectively. These results indicate that the samples had some unreacted carbon [3,21]. However, both specific carbon peaks almost completely disappeared after the samples were sintered at 1700 °C. Furthermore, when the samples were sintered at 1600 and 1700 °C, peaks associated with the α-SiC (2θ = 33.6° and 38.1°) and β-SiC phases (2θ = 35.3°, 60.0°, and 71.8°) were observed, and the intensity of the peaks increased with the sintering temperature. Similar observations were reported by Qian and Jin [21], who suggested that preparing β-SiC through carbothermal reduction usually results in a minor amount of α-SiC.

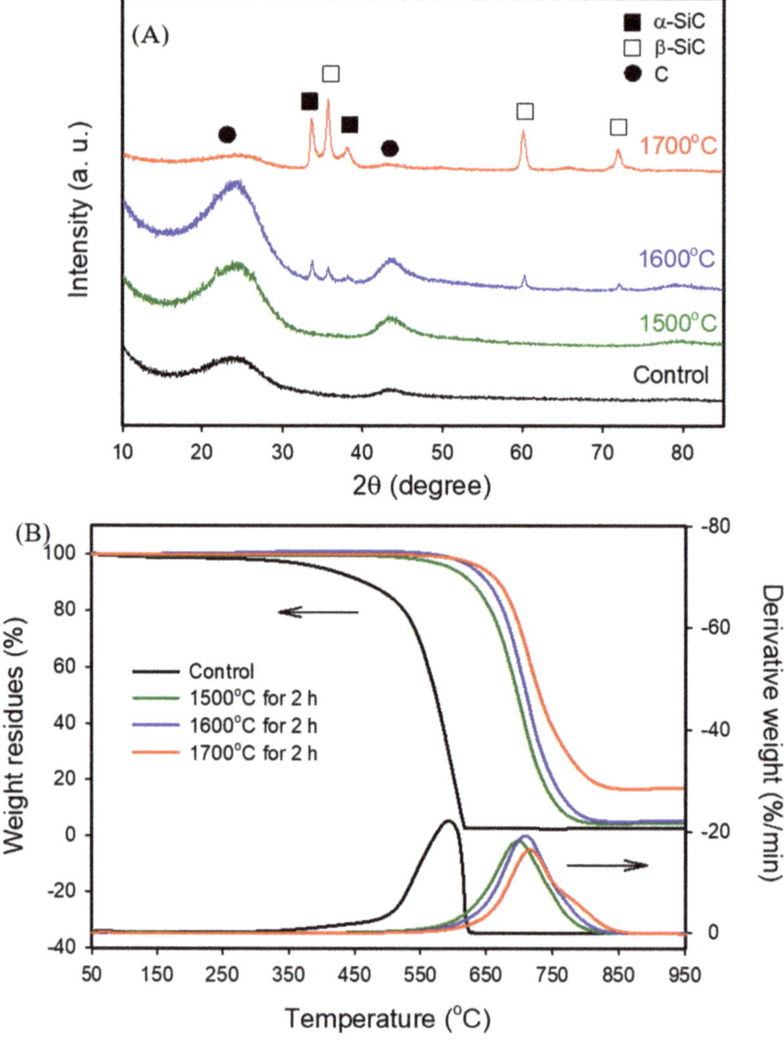

Figure 2. X-ray diffraction (XRD) patterns (**A**) and thermogravimetric analysis (TGA)/derivative thermogravimetry (DTG) curves (**B**) of BcSiCs sintered at different temperatures for 2 h.

Figure 2B presents the TGA curve of bamboo charcoal, indicating a considerable decrease in weight residue at sintering temperatures higher than 300 °C in an air atmosphere. A weight residue of only 2.7% was observed when the sintering temperature was up to 620 °C, this is because the carbon substrate was completely thermally degraded at approximately 600 °C [34]. By contrast, for the BcSiCs sintered at 1500, 1600, and 1700 °C, the thermal degradation onset temperatures were nearly 520, 600, and 610 °C, respectively. When the sintering temperature reached 950 °C, the weight residues of the three sintered BcSiCs were 4.7%, 5.3%, and 17.0%. Figure 2B also presents the derivative thermogravimetry (DTG) curves of the samples. As shown in the figure, the maximum thermal degradation temperature of bamboo charcoal was 590 °C, whereas the maximum thermal degradation temperatures of the BcSiCs sintered at 1500, 1600, and 1700 °C were 695, 700, and 710 °C respectively, signifying that the thermal stability of the BcSiCs could be improved by increasing the sintering temperature. The primary reason for this phenomenon is that SiC was generated through the carbothermal reduction of bamboo charcoal and SiO_2 during high-temperature sintering. A higher sintering temperature typically results in a more complete reaction and thus, a more thermally stable SiC ceramic. Figure 3 depicts the SEM micrographs of the BcSiCs sintered at 1500–1700 °C for 2 h. According to these micrographs, all specimens retained the inherently porous microstructure of bamboo, and the resulting bio-SiC ceramics replicated the original texture of the carbon template (Figure 3C–H). After the sintering process, SiC grains were prominently formed on the cell wall and roughened the cell wall surface. However, among the specimens, only the BcSiC specimen sintered at 1700 °C formed a large amount of SiC nanowires on the cell wall (carbon template) surface (Figure 3H), and the obtained SiC nanostructures comprised a nanowire core with extensional platelets. A similar phenomenon has been observed by Ding et al. [34] and Hata et al. [35]. Accordingly, 1700 °C was considered the optimum sintering temperature for the preparation of bio-SiC ceramics in this study.

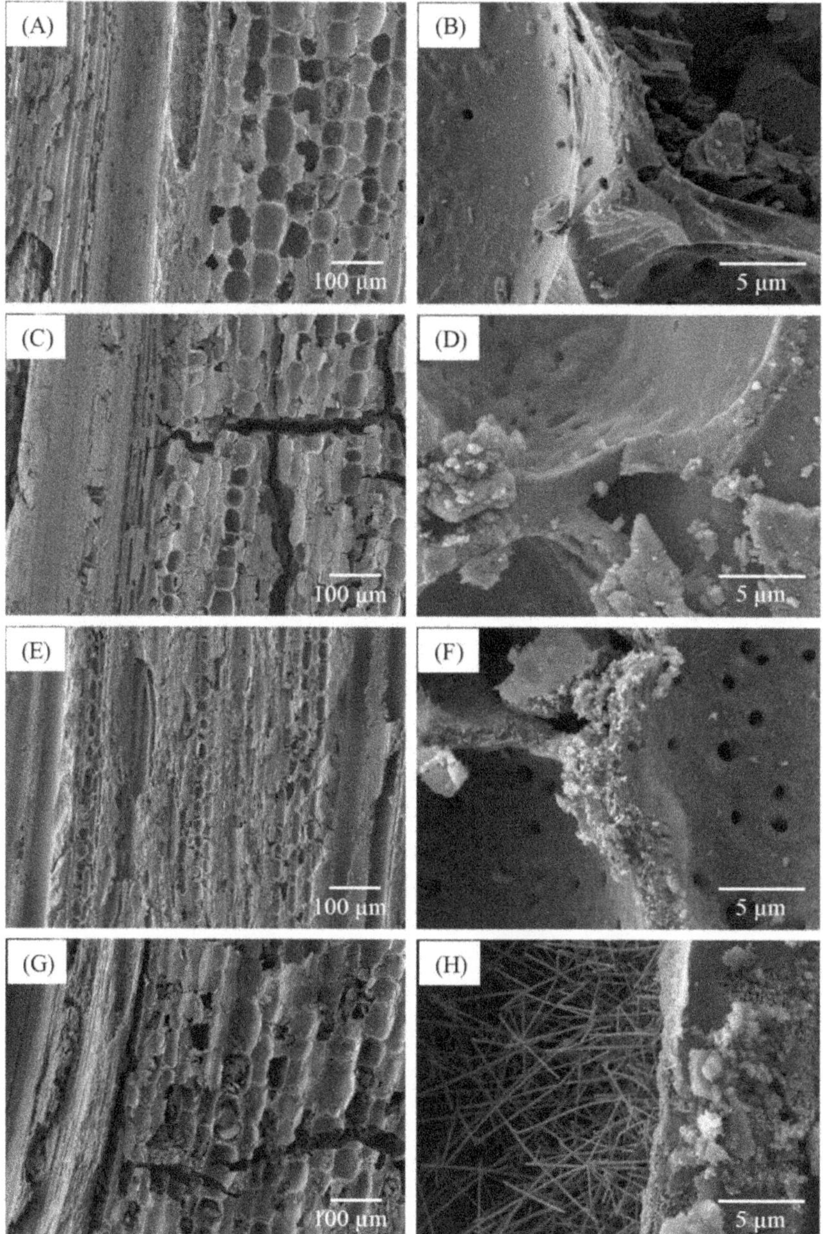

Figure 3. Scanning electron microscopy (SEM) micrographs of bamboo charcoal (**A,B**) and BcSiCs sintered at 1500 °C (**C,D**), 1600 °C (**E,F**), and 1700 °C (**G,H**) for 2 h.

3.2. Effect of Sintering Duration on the Properties of Porous Bio-SiC Ceramics

To understand the effect of sintering duration on the properties of porous bio-SiC ceramics, this study applied the following test conditions: impregnation of bamboo charcoal with the

SiO$_2$-precursor sol (in one cycle) and then the execution of sintering at 1700 °C for different durations. Figure 4 shows the XRD patterns of the BcSiCs sintered at 1700 °C for 0.5–3 h. For the specimen prepared at 1700 °C for 0.5 h, five distinct diffraction peaks of α- and β-SiC were observed along with two major peaks of carbon. However, as the sintering duration increased, the intensity of the SiC-associated peaks increased, whereas that of the carbon-associated peak decreased. This result is similar to that reported by Locs et al. [6], who demonstrated that prolonged sintering positively influenced SiC formation. After 2 h of sintering, the peak of carbon nearly disappeared, whereas the peaks of the SiC crystal phase were clearly observed. A similar result has also been reported by Locs et al. [6] and Qian and Jin [21], when a sufficient reaction time was allocated, a favorable reaction between carbon and SiO$_2$ was achieved. Accordingly, a well-crystallized bio-SiC ceramic could be produced through sintering at 1700 °C for more than 2 h.

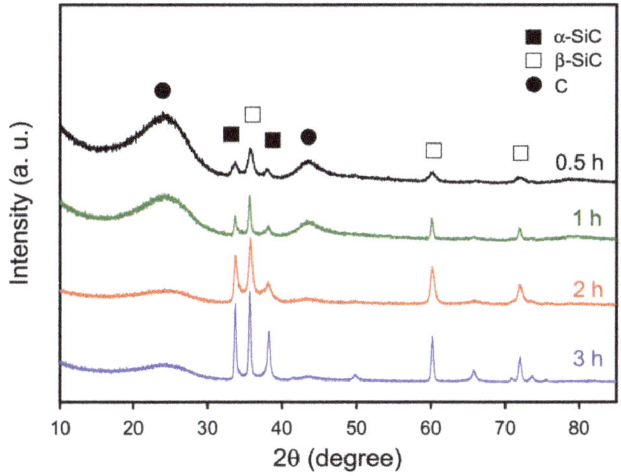

Figure 4. XRD patterns of BcSiCs sintered at 1700 °C for different durations.

In addition, comparison with previously reported XRD data for wood-derived porous SiC ceramics [1,3,6,15–18,21,22], this study shows that more α-SiC (hexagonal type) is formed in bamboo-derived porous SiC ceramics. Locs et al. [6] reported that the presence of different impurities (e.g., Ca, K, Na, etc.) in the biocarbon matrix could promote the formation of high-temperature SiC crystal phases (α-SiC). Generally, bamboo has higher alkaline extractives, ash, and silica contents as compared to wood [36,37]. Accordingly, the sintered BcSiCs form more α-SiC than wood-derived porous SiC ceramics, which can be explained as the bamboo charcoal is rich in alkaline metals and mineral substances.

Figure 5 depicts the SEM micrographs of the BcSiCs sintered at 1700 °C for different durations. Observing the microstructure of the specimen sintered for 0.5 h (Figure 5A) revealed that the cell wall surface was similar to the smooth surface of bamboo charcoal, but a small amount of SiC nanowires were formed in the cell lumen. However, when the sintering duration was extended to 1–2 h, the cell wall surface became rough and the cell lumen comprised a higher amount of SiC nanowires (Figure 5B,C). After 3 h of sintering, this study observed the formation of a large amount of SiC crystal particles and nanowires on the surface of the specimen (Figure 5D). Additionally, the SiC nanowires in this specimen were significantly shorter and wider than those in the other specimens.

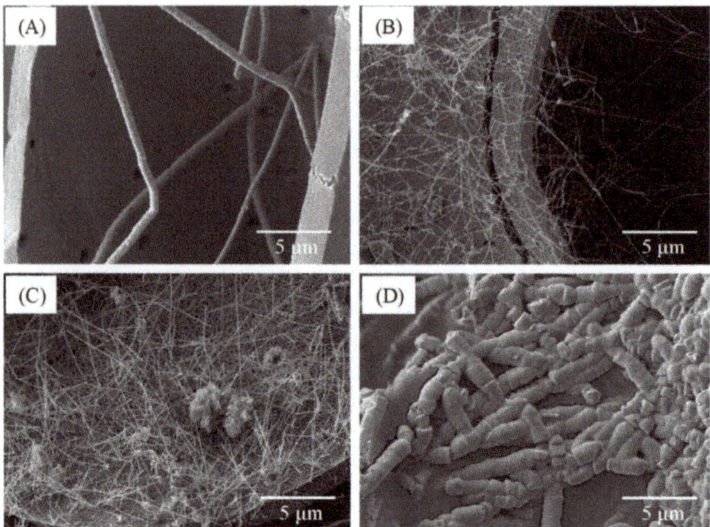

Figure 5. SEM micrographs of BcSiCs sintered at 1700 °C for 0.5 h (**A**), 1 h (**B**), 2 h (**C**), and 3 h (**D**).

3.3. Effect of Sol–Gel Impregnation Cycle on the Properties of Porous Bio-SiC Ceramics

To further investigate the effect of sol–gel impregnation cycle on the properties of the derived porous bio-SiC ceramics, bamboo charcoal specimens impregnated with the SiO_2 precursor sol in one, three, and five repeated cycles were analyzed. Figure 6 illustrates the XRD patterns of the bio-SiC ceramics that were prepared from the BcSiCs prepared using different sol–gel impregnation cycles. The XRD patterns did not show obvious differences among all bio-SiC ceramics. Peaks associated with β-SiC were observed at 2θ values of 35.3°, 60.0°, and 71.8°, along with three peaks associated with amorphous carbon. However, the intensity of the peaks associated with carbon decreased as the number of sol–gel impregnation cycles increased, and this finding is consistent with that reported by Qian et al. [18]. This phenomenon can be attributed to the fact that the amount of SiO_2 in the BcSiCs increased with the number of impregnation cycles, which resulted in a more efficient conversion of charcoal to SiC ceramics during the carbothermal reduction reaction.

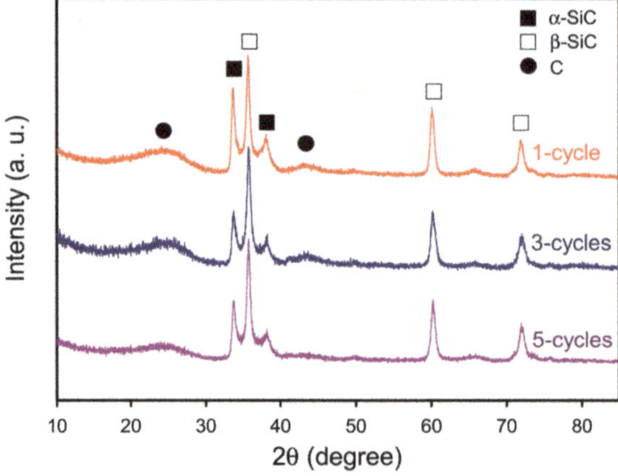

Figure 6. XRD patterns of sintered BcSiCs prepared using different repeated cycles of sol–gel impregnation process.

Figure 7 presents the microstructure of the bio-SiC ceramics prepared using three and five repeated cycles of sol–gel impregnation. The surface morphology of the cell walls of the bio-SiC ceramic prepared using three impregnation cycles (Figure 7A) was similar to that of the rough surface of the bio-SiC ceramic prepared using a single impregnation cycle (Figure 5C), except that SiC nanowires were not observed within the cell lumen. A foam-like cell wall was observed for the bio-SiC ceramic prepared using three impregnation cycles (Figure 7B). As reported by Qian and Jin [21], a possible reason for this observation is that CO was produced in addition to SiC during the reaction of carbon with SiO_2. Moreover, the microstructure of the bio-SiC ceramic prepared using five impregnation cycles (Figure 7C,D) was similar to that of the bio-SiC ceramic prepared using three impregnation cycles, however, some SiC nanowires were formed on the surface of the cell wall of the bio-SiC ceramic. These results reveal that the microstructures of the bio-SiC ceramics did not differ significantly when the number of repeated sol–gel impregnation cycles exceeded three. Accordingly, the bio-SiC ceramics have higher porosity when the number of repeated sol–gel impregnation cycles exceeded three. Pastore et al. [38] reported that carbon foams with low density and high open porosity exhibit significant microwave absorption capabilities. Therefore, the bio-SiC ceramics may have high potential for electromagnetic applications when the number of repeated sol–gel impregnation cycles exceeded three. Figure 8 depicts the HRTEM images and selected area electron diffraction (SAED) patterns of SiC nanowires of the bio-SiC ceramics prepared using five impregnation cycles. The nanowires had a diameter of nearly 100 nm and a single crystal structure (Figure 8A). The (111) plane, representing the atomic plane perpendicular to the growth direction, of β-SiC (Figure 8B) had a planar distance of 0.25 nm (Figure 8C), indicating that the preferred growth direction of the nanowires was perpendicular to the (111) plane. A similar result was reported by Ding et al. [34].

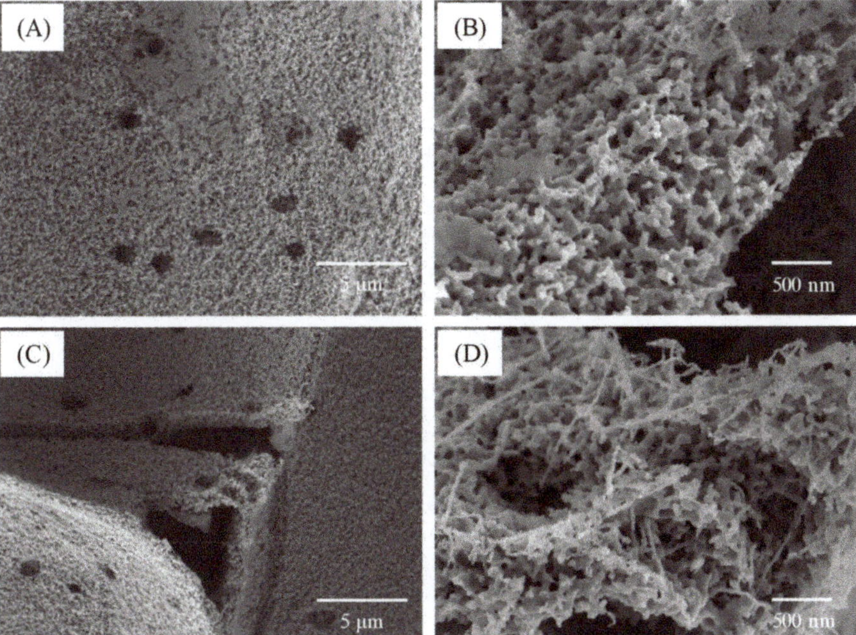

Figure 7. SEM micrographs of sintered BcSiCs prepared using three (**A,B**) and five repeated cycles (**C,D**) of sol–gel impregnation.

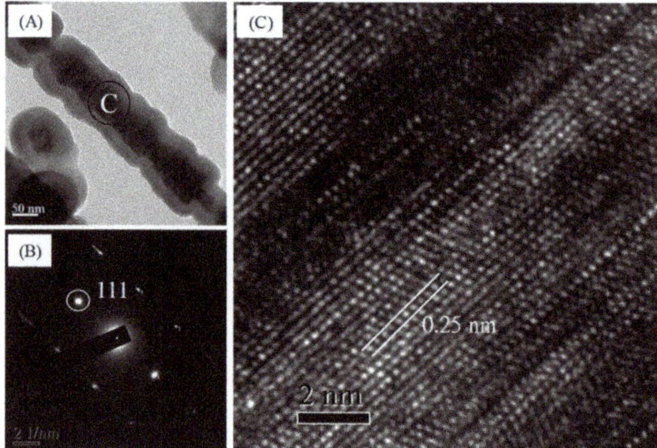

Figure 8. Transmission electron microscopy (TEM) image of a single β-SiC nanowire of sintered BcSiC prepared using five repeated cycles of sol–gel impregnation process (**A**), selected area electron diffraction (SAED) pattern (**B**), and high-resolution transmission electron microscopy (HRTEM) image (**C**) obtained for the β-SiC nanowire.

To further investigate the feasibility of directly producing bio-SiC ceramics from bamboo, BSiCs were prepared from air-dried bamboo by using MTMOS and a sol–gel process and then sintered at 1700 °C for 2 h in a nitrogen atmosphere. Figure 9 illustrates the XRD patterns of bio-SiC ceramics prepared from the BSiCs by using different sol–gel impregnation cycles. Similar to the XRD patterns of the sintered BcSiCs, β-SiC-associated peaks were observed at 2θ values of 35.3°, 60.0°, and 71.8°, in addition to peaks associated with carbon. The intensity of the peaks associated with carbon decreased as the number of repeated impregnation cycles increased. Figure 10 shows the HRTEM images and SAED patterns of a single SiC nanowire of a bio-SiC ceramic prepared from BSiCs using five impregnation cycles. The results in these images are similar to those obtained for the bio-SiC ceramic prepared from BcSiCs using five impregnation cycles. Accordingly, bio-SiC ceramics can be directly prepared from BSiCs, thus simplifying the manufacturing process of SiC ceramics. To the best of our knowledge, this is the first study to prepare biomorphic porous SiC ceramics directly from bamboo by combining sol–gel impregnation and carbothermal reduction.

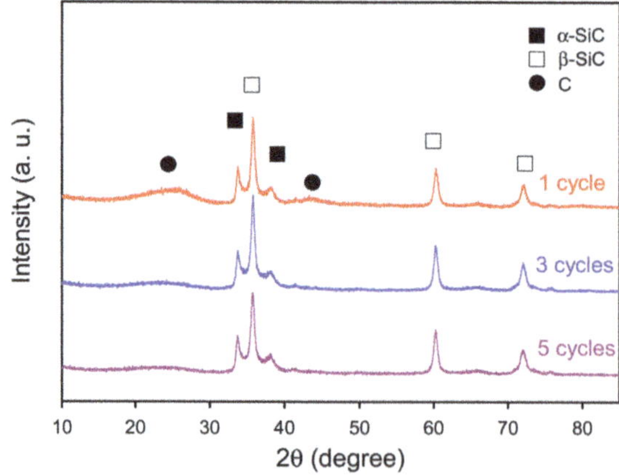

Figure 9. XRD patterns of sintered BSiCs prepared using different repeated cycles of sol–gel impregnation.

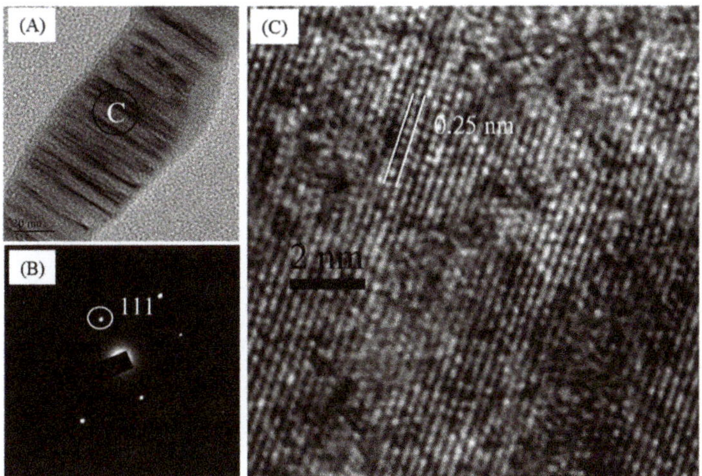

Figure 10. TEM image of a single β-SiC nanowire of sintered BSiC prepared using five repeated cycles of sol–gel impregnation (**A**), SAED pattern (**B**), and HRTEM image (**C**) observed for the β-SiC nanowire.

4. Conclusions

In this study, bio-SiC ceramics were prepared by combining sol–gel impregnation and carbothermal reduction, with MTMOS and bamboo serving as the precursors. The results revealed that a higher sintering temperature resulted in a more complete reaction and thus, more thermally stable SiC. The resulting SiC ceramics comprised β-SiC and a trace amount of α-SiC. In addition, the microstructures of the cell walls of bio-SiC ceramics prepared using three and five impregnation cycles were determined to be foam-like (porous SiC) structures, in contrast to the microstructure of a bio-SiC ceramic prepared using a single impregnation cycle. Moreover, bio-SiC ceramics could also be successfully prepared from BSiC in this study, and their XRD patterns and HRTEM images were similar to those obtained for bio-SiC ceramics prepared from BcSiCs. According to a review of the literature, this is the first study to synthesize bio-SiC ceramics directly from bamboo using sol–gel and carbothermal reduction.

Author Contributions: Conceptualization, K.-C.H., T.-L.W., and J.-H.W.; data curation, K.-C.H., T.-L.W., and J.-W.X.; investigation, K.-C.H., T.-L.W., and J.-W.X.; methodology, K.-C.H., T.-L.W., and J.-H.W.; resources, J.-H.W.; software, K.-C.H. and T.-L.W.; visualization, K.-C.H. and T.-L.W.; supervision, J.-H.W.; writing–original draft, K.-C.H., T.-L.W.; writing–review and editing, J.-H.W.

Funding: This work was financially supported by a research grant from the Ministry of Science and Technology, Taiwan (NSC 102-2628-B-005-006-MY3).

Conflicts of Interest: The authors declare that there is no conflict of interest regarding the publication of this paper.

References

1. Qian, J.M.; Wang, J.P.; Jin, Z.H. Preparation of biomorphic SiC ceramic by carbothermal reduction of oak wood charcoal. *Mat. Sci. Eng. A Struct.* **2004**, *371*, 229–235. [CrossRef]
2. Amaral-Labat, G.; Zollfrank, C.; Ortona, A.; Pusterla, S.; Pizzi, A.; Fierro, V.; Celzard, A. Structure and oxidation resistance of micro-cellular Si-SiC foams derived from natural resins. *Ceram. Int.* **2013**, *39*, 1841–1851. [CrossRef]
3. Gordic, M.; Bucevac, D.; Ruzic, J.; Gavrilovic, S.; Hercigonja, R.; Stankovic, M.; Mtovic, B. Biomimetic synthesis and properties of cellular SiC. *Ceram. Int.* **2014**, *40*, 3699–3705. [CrossRef]
4. Pizzi, A.; Zollfrank, C.; Li, X.; Cangemi, M.; Celzard, A. A SEM record of Proteins-derived microcellular silicon carbide foams. *J. Renew. Mat.* **2014**, *2*, 230–234. [CrossRef]

5.   Mao, W.G.; Chen, J.; Si, M.S.; Zhang, R.F.; Ma, Q.S.; Fang, D.N.; Chen, X. High temperature digital image correlation evaluation of in-situ failure mechanism: An experimental framework with application to C/SiC composites. *Mat. Sci. Eng. A Struct.* **2016**, *665*, 26–34. [CrossRef]

6.   Locs, J.; Berzina-Cimdina, L.; Zhurinsh, A.; Loca, D. Effect of processing on the microstructure and crystalline phase composition of wood derived porous SiC ceramics. *J. Eur. Ceram. Soc.* **2011**, *31*, 183–188. [CrossRef]

7.   Locs, J.; Berzina-Cimdina, L.; Zhurinsh, A.; Loca, D. Optimized vaccum/pressure sol impregnation processing of wood for the synthesis of porous, biomorphic SiC ceramics. *J. Eur. Ceram. Soc.* **2009**, *29*, 1513–1519. [CrossRef]

8.   Sun, D.; Hao, X.; Yu, X.; Chen, X.; Liu, M. Preparation and characterisation of carbon fibre-reinforced laminated woodceramics. *Wood Sci. Technol.* **2016**, *50*, 581–597. [CrossRef]

9.   Zollfrank, C.; Siebera, H. Microstructure and phase morphology of wood derived biomorphous SiSiC-ceramics. *J. Eur. Ceram. Soc.* **2004**, *24*, 495–506. [CrossRef]

10.   Esposito, L.; Sciti, D.; Pinacastelli, A.; Bellosi, A. Microstructure and properties of porous β-SiC template from soft woods. *J. Eur. Ceram. Soc.* **2004**, *24*, 533–540. [CrossRef]

11.   Vogli, E.; Mukerji, J.; Hoffmann, C.; Klandy, R.; Sieber, H.; Greil, P. Conversion of oak to cellular silicon carbide ceramic by gas-phase reaction with silicon monoxide. *J. Am. Ceram. Soc.* **2001**, *84*, 1236–1240. [CrossRef]

12.   Vogli, E.; Sieber, H.; Greil, P. Biomorphic SiC-ceramic perpetrated by Si-vapor phase infiltration of wood. *J. Eur. Ceram. Soc.* **2002**, *22*, 2663–2668. [CrossRef]

13.   Qian, J.M.; Wang, J.P.; Jin, Z.H. Preparation and properties of porous microcellular SiC ceramics by reactive infiltration of Si vapor into carbonized basswood. *Mater. Chem. Phys.* **2003**, *82*, 648–653. [CrossRef]

14.   Greil, P.; Vogli, E.; Fey, T.; Bezold, A.; Popovska, N.; Gerhard, H.; Siebera, H. Effect of microstructure on the fracture behaviour of biomorphous silicon carbide ceramics. *J. Eur. Ceram. Soc.* **2002**, *22*, 2697–2707. [CrossRef]

15.   Klinger, R.; Sell, J.; Zimmermann, T.; Herzog, A.; Vogt, U.; Graule, T.; Thurner, P.; Beckmann, F.; Muller, B. Wood-delivered porous ceramics via infiltration of SiO_2-sol and carbothermal reduction. *Holzforschung* **2003**, *57*, 440–446. [CrossRef]

16.   Furuno, T.; Fujitsawa, M. Carbonization of wood-silica composites and formation of silicon carbide in the cell wall. *Wood Fiber Sci.* **2004**, *36*, 269–277.

17.   Herzog, A.; Klingner, R.; Vogt, U.; Graule, T. Wood-derived porous SiC ceramics by sol infiltration and carbothermal reduction. *J. Am. Ceram. Soc.* **2004**, *87*, 784–793. [CrossRef]

18.   Qian, J.M.; Wang, J.P.; Qiao, G.J.; Jin, Z.H. Preparation of porous SiC ceramic with a woodlike microstructure by sol-gel and carbothermal reduction processing. *J. Eur. Ceram. Soc.* **2004**, *24*, 3251–3259. [CrossRef]

19.   Shin, Y.; Wang, C.; Exarhos, G.J. Synthesis of SiC ceramics by the carbothermal reduction of mineralized wood with silica. *Adv. Mater.* **2005**, *17*, 73–77. [CrossRef]

20.   Sieber, H. Biomimetic synthesis of ceramics and ceramic composites. *Mat. Sci. Eng. A Struct.* **2005**, *412*, 43–47. [CrossRef]

21.   Qian, J.M.; Jin, Z.H. Preparation and characterization of porous, biomorphic SiC ceramic with hybrid pore structure. *J. Eur. Ceram. Soc.* **2006**, *26*, 1311–1316. [CrossRef]

22.   Egelja, A.; Gulickovski, J.; Devečerski, A.; Ninić, M.; Radoslavljević-Mihaljović, A.; Matović, B. Preparation of biomorphic SiC ceramics. *Sci. Sinter.* **2008**, *40*, 141–145. [CrossRef]

23.   Sakka, S.; Miyafuji, H. *Handbook of Sol-Gel Science and Technology: Processing, Characterization and Applications. Volume III: Applications of Sol-Gel Technology*; Kluwer Academic Publishers: Boston, MA, USA, 2005.

24.   Scurlock, J.M.O.; Dayton, D.C.; Hames, B. Bamboo: An overlooked biomass resource? *Biomass Bioenerg.* **2000**, *19*, 229–244. [CrossRef]

25.   Liu, Z.; Fei, B.; Jiang, Z.; Cai, Z.; Liu, X. Important properties of bamboo pellets to be used as commercial solid fuel in China. *Wood Sci. Technol.* **2014**, *48*, 903–917. [CrossRef]

26.   Mori, Y.; Kuwano, Y.; Tomokiyo, S.; Kuroyanagi, N.; Odahara, K. Inhibitory effects of Moso bamboo (*Phyllostachys heterocycla* f. *pubescens*) extracts on phytopathogenic bacterial and fungal growth. *Wood Sci. Technol.* **2019**, *53*, 135–150. [CrossRef]

27.   Mi, Y.; Chen, X.; Guo, Q. Bamboo fiber-reinforced polypropylene composites: Crystallization and interfacial morphology. *J. Appl. Polym. Sci.* **1997**, *64*, 1267–1273. [CrossRef]

28. Lee, S.H.; Ohkita, T. Bamboo fiber (BF)-filled poly (butylenes succinate) bio-composite–Effect of BF-e-MA on the properties and crystallization kinetics. *Holzforschung* **2004**, *58*, 537–543. [CrossRef]

29. Shao, S.; Wen, G.; Jin, Z. Changes in chemical characteristics of bamboo (*Phyllostachys pubescens*) components during steam explosion. *Wood Sci. Technol.* **2008**, *42*, 439–451. [CrossRef]

30. Chang, S.T.; Wu, J.H. Stabilizing effect of chromated salt treatment on the green color of ma bamboo (*Dendrocalamus latiflorus*). *Holzforschung* **2000**, *54*, 327–330. [CrossRef]

31. Wu, T.L.; Chien, Y.C.; Chen, T.Y.; Wu, J.H. The influence of hot-press temperature and cooling rate on thermal and physicomechanical properties of bamboo particle-polylactic acid composites. *Holzforschung* **2013**, *67*, 325–331. [CrossRef]

32. Liu, H.; Jiang, Z.; Fei, B.; Hse, C.; Sun, Z. Tensile behavior and fracture mechanism of moso bamboo (*Phyllostachys pubescens*). *Holzforschung* **2015**, *69*, 47–52. [CrossRef]

33. Hung, K.C.; Wu, J.H. Characteristics and thermal decomposition kinetics of wood-SiO_2 composites derived by the sol-gel process. *Holzforschung* **2017**, *71*, 233–240. [CrossRef]

34. Ding, J.; Zhu, H.; Li, G.; Deng, C.; Li, J. Growth of SiC nanowires on wooden template surface using molten salt media. *Appl. Surf. Sci.* **2014**, *320*, 620–626. [CrossRef]

35. Hata, T.; Castro, V.; Fujisawa, M.; Imamura, Y.; Bonnamy, S.; Bronsveld, P.; Kikuchi, H. Formation of silicon carbide nanorods from wood-based carbons. *Fuller. Nanotub. Carbon Nano Struct.* **2005**, *13*, 107–113. [CrossRef]

36. Evbuomwan, B.O.; Abutu, A.S.; Ezeh, C.P. The effects of carbonization temperature on some physicochemical properties of bamboo based activated carbon by potassium hydroxide (KOH) activation. *Greener J. Phys. Sci.* **2013**, *3*, 187–191.

37. Huang, P.H.; Jhan, J.W.; Cheng, Y.M.; Cheng, H.H. Effects of carbonization parameters of moso-bamboo-based porous charcoal on capturing carbon dioxide. *Sci. World J.* **2014**, *2014*, 937867. [CrossRef] [PubMed]

38. Pastore, R.; Delfini, A.; Micheli, D.; Vricella, A.; Marchetti, M.; Santoni, F.; Piergentili, F. Carbon foam electromagnetic mm-wave absorption in reverberation chamber. *Carbon* **2019**, *144*, 63–71. [CrossRef]

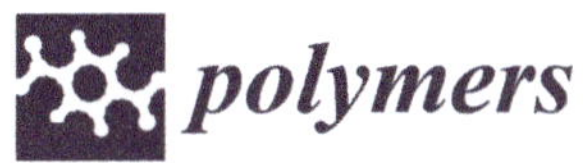 *polymers* 

Article

Preparation of a Fast Water-Based UV Cured Polyurethane-Acrylate Wood Coating and the Effect of Coating Amount on the Surface Properties of Oak (*Quercus alba* L.)

Jin Wang [1,2], Huagui Wu [3], Ru Liu [3], Ling Long [1,3,*], Jianfeng Xu [1,3], Minggui Chen [4] and Hongyun Qiu [3]

1   Research Institute of Forestry New Technology, Chinese Academy of Forestry, Haidian, Beijing 100091, China
2   Kingdecor (Zhejiang) Co., Ltd., Quzhou 324000, China
3   Research Institute of Wood Industry, Chinese Academy of Forestry, Haidian, Beijing 100091, China
4   Jiangsu Himonia Technology Co., Ltd., Jurong 212400, China
*   Correspondence: longling@caf.ac.cn; Tel.: +86-10-62889425

Received: 31 July 2019; Accepted: 27 August 2019; Published: 28 August 2019

Abstract: A fast water-based ultraviolet light (UV) curing polyurethane-acrylate (PUA) wood coating was prepared in the laboratory, and applied on oak (*Quercus alba* L.) at different coating amounts. The PUA wood coating can be fast cured within 22 min, which highly improved the drying speed compared to normal water-based wood coatings (often higher than 35 min). The coating amounts affected the coating properties after curing on oak. With the increase of coating amount, the adhesion, hardness and gloss value of surface increased to different extents. Meanwhile, the surface of sample became smooth gradually because the voids of the oak were filled. Thus, higher coating amount resulted in better coating properties. However, no significant increase of penetration depth was found. During curing, the hydroxyl groups of the wood reacted with the coating. The optimal parameter in this study was the coating amount of 120 g/m^2, where the adhesion reached 1 (with 0–5% cross-cut area of flaking along the edges), with the hardness of 2H and the gloss of 92.56°, which met the requirement of Chinese standard GB/T 18103–2013, and could be used for engineered wood flooring.

Keywords: water-based UV curing coating; coating amount; surface properties; polyurethane-acrylate; oak (*Quercus alba* L.)

---

1. Introduction

In recent years, with the improvement of environmental protection regulations and people's awareness, water-based wood coatings have become a research hotspot [1–5]. Water-based polyurethane acrylate (PUA) coatings are widely used owing to their wear-resistance, stain resistance, and ease of use [6]. They can form a variety of hydrogen bonds between molecular chains due to their functional acrylic groups and urethane bonds, which provides the coatings with weather resistance, solvent resistance, comprehensive mechanical properties and excellent appearance on wood products [7–10]. At the same time, the molecular structure and functionality can be adjusted to satisfy the different needs of furniture, cabinets, office and laboratory equipment. In addition, during film formation, little or no volatile organic compounds (VOCs) evaporate into the atmosphere, therefore they can be considered low cost and pollution-free.

However, a problem that has restricted the application of water-based coatings is their relatively slow drying rate compared with solvent-based coatings. It is known that the latent heat of water is much higher than that of organic solvents, such as benzene, xylene, etc. Besides, some properties of water-based coatings are not comparable to those of solvent-based coatings, like low hardness, and gloss. To expand the application of coatings to other areas, ultraviolet light (UV) curing technology is used in the manufacture of water-based coatings for fast curing [11–13]. UV cured coatings use high energy UV lights as sources, which can form coating films via self-polymerization [14]. Most UV-curable resins consist of multifunctional acrylate monomers and oligomers associated to an aromatic ketone that generates upon UV-exposure free radicals which will initiate the crosslinking polymerization [15]. Therefore, the PUA could meet these requirements. Some kinds of water-based UV curing PUA coatings have been investigated [15,16]. However, up to present, the water-based UV curing coatings have not really been applied on wooden products due to the limitations of the coating film properties.

It is well known that the properties of coatings are affected by many factors, such as the environment, degree of wetting between the coating and the substrate, chemical or physical effects, surface properties of the substrate, and so on [17–22]. Yang et al. [23] found that different environments could cause differences in the chemical composition on the surface of the PUA painting film and the surface morphology was thus different. Masson et al. [16] also found that basic coating performance features such as the drying speed were highly dependent on the temperature and film thickness.

Accordingly, many researchers have investigated the basic properties of the resin particles and the impact of paint in different environments [24–29]. However, the influence of the coating amount on the resulting surface performance of the coating has been less investigated, let alone for fast water-based UV curing PUA wood coatings. Coating amount, as a very important factor, affects the performance of cured coatings very much. At low coating amount, the coating thickness is thin and insufficient coverage will be obtained, while with an excessive amount of coating, drying defects such as wrinkles, and bubbling may occur. Therefore, in this paper, a kind of fast water-based UV curing coating was synthesized in the laboratory, and applied on oak for its wide application in furniture, flooring and other industries due to its resource advantages and good decorative properties in the solid wood market [30–32]. The effect of coating amount on the surface properties such as the surface morphology, chemical changes, permeability and surface energy was studied, with the aim to provide some basic data and reference for the application of the fast water-based UV curing wood coatings.

2. Materials and Methods

2.1. Materials

Oak (*Quercus alba* L.) wood which originated from Fujian Province in the southeast of China was purchased from a Beijing home building materials market. The wood was dried to a 8–10% moisture content prior to use. The oak was then sawn into slabs with dimensions of 630 mm (L) × 120 mm (T) × 20 mm (R). The slabs were sanded with 240 mesh sandpaper and the dust was removed to ensure that the substrate surface was clean. The reagents for the fast water-based UV curing PUA wood coating are listed in Table 1. All abbreviations concerning the reagents used are presented in parentheses in Table 1.

2.2. Synthesis of the Fast Water-Based UV Curing PUA Coating

The fast water-based UV curing PUA coating was prepared based on our previous study [33]. According to Table 1, the first step was to prepare a pre-polyacrylate emulsion. Monomers (24 g of AA, 4 g of ST, and 7 g of EA) were blended with 169 g of distilled water in a three-neck glass flask at 150 r/min. Afterwards, the initiator (8 g of APS) was added at 10 wt%. For the next 70–90 min, the rest of the monomers (96 g of AA, 16 g of ST, and 133 g of EA) was added dropwise into the reactor. Afterwards, the rest of the initiator (32 g of APS) and 36 g of KH550 were added dropwise into the reactor within 90–110 min. The reactor was heated up to 85 °C and held at this temperature for 4 h. The next step was to graft the pre-polyurethane onto the pre-polyacrylate chain. TDI (100 g), PEG-400 (200 g) and DBTD (8 g) were dissolved in *N*-methyl-2-pyrrolidone solvent and blended into a three-neck glass flask at 150 r/min. The reactor was heated to 40 °C. While stirring, HEA (66 g) was dripped into the reactor within 2 h. The pre-polyurethane was thus obtained. Afterwards, both the pre-polyacrylate and pre-polyurethane were mixed in the three-neck glass flask and blended at 150 r/min. The reaction mixture was heated up to 85 °C and kept at that temperature for 2 h. The pre-PUA emulsion was obtained after cooling and filtration. Defoamer agent (0.2%), photoinitiator (2%), and HDDA diluting agent (1.5%) were added into the emulsion at 100 r/min for 10 min. Then, the water-based UV curing PUA coating was obtained. After good dispersion, the coating was prepared and stored in a black bottle against light.

Table 1. Reagents for water-based UV curing PUA coating.

Reagent	Function	Manufacturer
Ammonium persulfate (APS)	Initiator	1
Acrylic acid (AA)	Monomer	1
Styrene (ST)	Monomer	1
Ethyl acrylate (EA)	Monomer	1
Hydroxyethyl acrylate (HEA)	Monomer	1
Toluene-2,4-diisocyanate (TDI)	Monomer	1
Polyethylene glycol (Mn: 400) (PEG-400)	Monomer	1
Dibutyltin dilaurate (DBTD)	Catalyst	1
N-methyl-2-pyrrolidone	Solvent	1
Hexanediol diacrylate (HDDA)	Diluting agent	1
γ-Aminopropyl triethoxysilane (KH550)	Silane	2
Polyether siloxane copolymer composition	Defoamer agent	3
2-Hydroxymethylphenylpropan-1-one	Photoinitiator	4

Notes: (**1**) Xilong Chemical Co.Ltd (Guangzhou.China); (**2**) Qufu Chengguang Chemical Co., Ltd (Shandong, China); (**3**) TEGO AIREX 901W; (**4**) Germany Evonik tego.

2.3. Coating on a Wood Sample

The coating amount was determined as a variable, and an uncoated wood slab was used as a control. The wet coating coating amount weights were 12, 60, and 120 g/m^2, respectively. The coating process was carried out by on an auto-coating rolling machine bracket with pressure. The coating speed was 80 mm/s. The drying temperature of the hot air dryer used in this test was determined to be 35 °C. The dryer mesh was subjected to a drying cycle of 2 min with 10 drying cycles, where the removal of water was achieved. Then, the coating it was cured by a UV curing machine. The photocuring energy was 250–350 mJ/cm^2 for 2 min.

2.4. Characterizations and Tests

The solid content of PUA was measured according to Chinese Standard GB/T 1725–2007 by drying the PUA coating at 125 °C for 1 h based on the residual percent of the coating. The viscosity was determined by a viscometer (NDJ-5S, Shanghai Pingxuan Scientific Instrument Co., Ltd., Shanghai, China). The average particle size of PUA was measured by a laser scattering particle distribution analyzer (MPT-Z, Malvern, London, UK).

The macro-structure was observed by an optical stereomicroscope (Leica S9 series, Jena, Germany). The microstructures of the above samples were observed by scanning electron microscope (SEM, JEOL JSM-6301F, Tokyo, Japan) with an acceleration voltage of 20 kV. The samples were sputter-coated with gold.

The dried coating thickness was measured by an ultrasonic velocity gauge (AT-140S, Guangzhou Amittari Design, Co. Ltd., Guangzhou, China). The throughout-dry state of the coating after 22 min of curing was tested following the ISO 9117:1990 standard with a plunger machine. The adhesion classification of coating films with wood was measured according to the ISO 2409:2013 standard method by a cross-cut test. The gloss values of the coatings were determined by the ISO 2813:2014 method using a geometry of 60°. Coating hardness was determined by ISO standard 15184:2012 by a pencil test.

Specimens including coated and uncoated oak were investigated. The chemical groups of the samples were examined by Fourier transform infrared spectroscopy (FTIR, BRUKER Vertex 70v, Hamburg, Germany) equipped with an attenuated total reflection (ATR) accessory. The surfaces were put in contact with the ZnSe crystal at a 45° angle of incidence.

The contact angle of pure oak with testing liquid (distilled water) and solution of 70 wt% PUA coating was measured according to ISO 15989:2004 by a contact angle meter (ISO, 2004; Krüss K11MK4, Hamburg, Germany). The contact angles of coated wood with different coating amounts with distilled water was also measured.

The surface morphologies were captured by atomic force microscopy (AFM) and collected by a scanning probe microscope in contact mode. The scanning area was 2 μm × 2 μm. The surface roughness was tested by a surface roughness tester using a stylographic method (SF200 Basic, Beijing Shidaishanfeng Technology Co. Ltd., Beijing, China). The values of the average profile error (Ra), maximum height (Rp), maximum valley depth (Rv), maximum profile height (Rt) and Maximum peak-valley height (Rz) were obtained.

Energy dispersed X-ray analysis (EDXA, Horiba 7021-H, Tokyo, Japan) was carried out in SEM analysis mapping mode. The PUA coated samples were cut vertically. The image and distribution of elemental N, mainly from the PUA resin, was captured digitally to allow further analysis of the penetration of PUA into each substrate.

3. Results and Discussion

3.1. Morphologies

Figure 1 shows images of the fast water-based UV curing PUA coating and the coated wood samples. It can be seen that the synthesized PUA coating was a stable and transparent liquid. Table 2 lists some parameters of the fast water-based UV curing PUA coating, such as the solid content, viscosity, and average particle size. It can be found that the PUA coating had very high solid content of 90 wt%, while the viscosity was not very high. The particle size of the PUA coating ranged from 80 nm to 500 nm (Figure 2) with an average value of 226 nm. Images of PUA coated oak are seen in Figure 1b–d. From the images, all wood samples were well coated by the PUA coating with a layer of coating film.

Table 2. Parameters of fast water-based UV curing PUA coating.

Parameter	Values
Solid content	90 wt%
Viscosity	50 mpa·s
Average particle size	226 nm

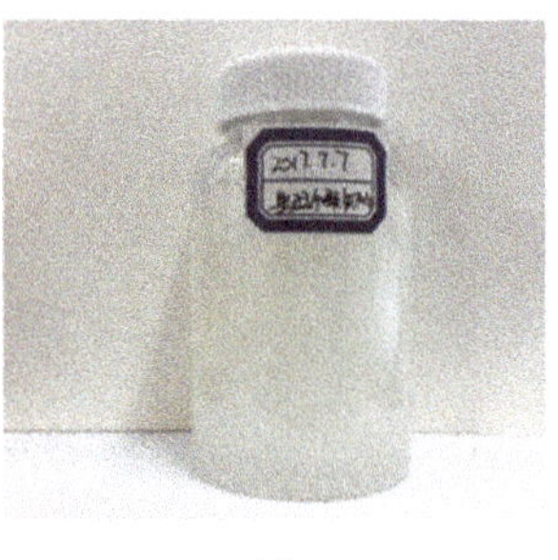

(a)   (b)

Figure 1. *Cont.*

(c)　　　　　　　　　　(d)

Figure 1. Images of PUA coating and its coated oak samples at different coating amounts. (**a**) PUA coating; (**b**) 12 g/m^2 PUA coated oak; (**c**) 60 g/m^2 PUA coated oak; (**d**) 120 g/m^2 PUA coated oak.

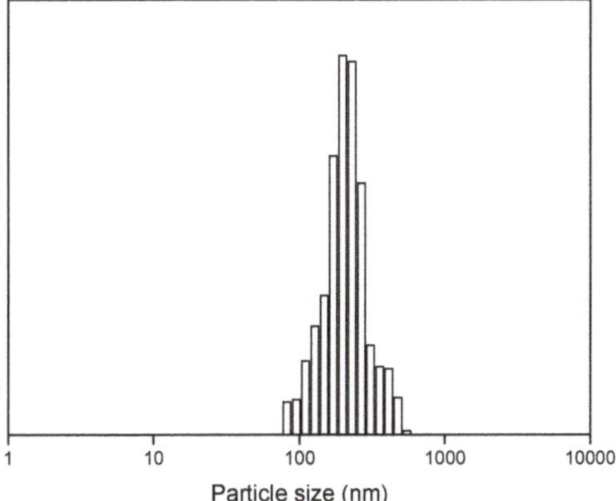

Particle size (nm)

Figure 2. Particle size distribution of the PUA coating.

Figure 3 shows a visual microscope picture of the oak surface with different coating amounts under an optical stereomicroscope. The wood samples were covered by the coating films. Besides, it can be seen that as the coating amount increased, the surface of the paint film became smoother. When the coating amount was 12 g/m^2, the oak vessels were partly occupied by paint, but most regions were still exposed. When the coating amount increased to 60 g/m^2, the vessels were substantially filled with few defects.

The surface morphologies of oak under different coating amounts tested by SEM are shown in Figure 4. The microstructures of the oak were clearly visible before painting. After the surface was coated with PUA coating, the microstructures were covered to varying degrees at different coating amounts. Within a certain range, as the coating amount increased, the surface smoothness of the sample gradually increased, which was consistent with the roughness values (discussed in the following part). In the case of 12 g/m^2 coated samples, the coating surface was incompletely covered, and some oak microstructures were exposed, namely, the surface of the painted film had a large number of grain protrusions, and the surface was rough. As the coating amount increased, the surface coverage of the oak increased, where the oak vessels could not be seen.

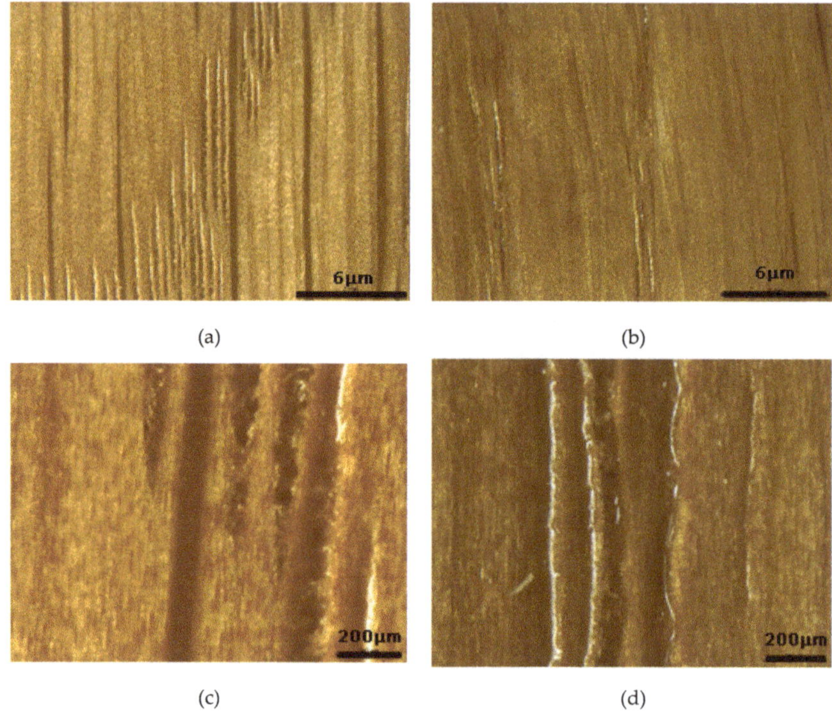

Figure 3. Optical microscope images of oak after coating at different coating amounts. (**a**,**c**) 12 g/m^2 PUA coated oak; (**b**,**d**) 60 g/m^2 PUA coated oak.

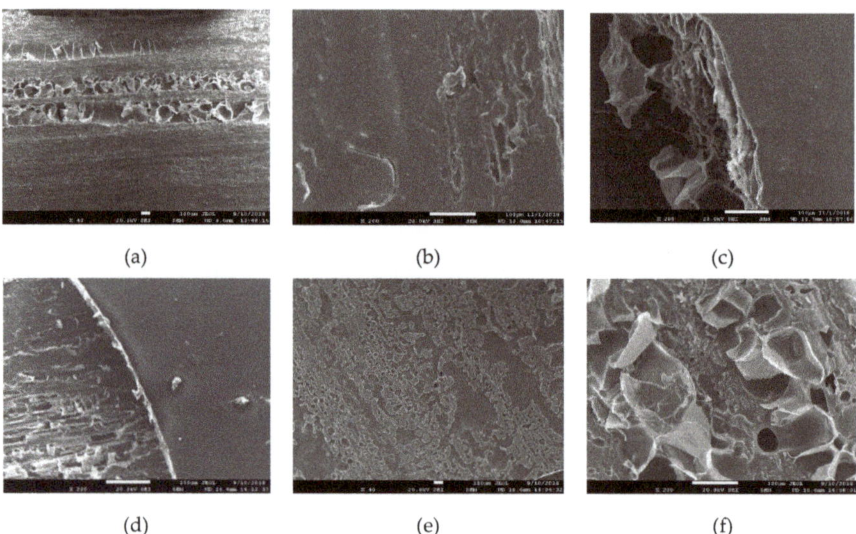

Figure 4. SEM images of oak before and after coating at different coating amount. (**a**) pure oak; (**b**) 12 g/m^2 PUA coated oak; (**c**) 60 g/m^2 PUA coated oak; (**d**) 120 g/m^2 PUA coated oak; (**e**,**f**) the back surface of coating peeling from the oak; (**a**,**e**) the magnification was ×40; (**b**–**d**,**f**) the magnification was ×200.

In order to further understand the adhesion between PUA coating and wood, the painted film was peeled off from the test piece, and the morphology of the back was observed (Figure 4e,f). It can be seen from the figure that the back of the paint film exhibited a microscopic topography similar to that of oak. However, some tyloses were easily found, which may be caused by the adhesion of coating in the vessels during drying of the water.

From the morphology results, it can be concluded that the PUA coating can form a layer of film that covers on the surface of the oak after 22 min of curing. Besides, we tested the drying state of all coatings according to ISO 9117:1990, and found no damage or markings on the surfaces. Therefore, the coatings were completely dried and be cured on oak by this process, which is much faster than the Chinese standard requirement for water-based coatings of 30–60 min (Table 3).

Table 3. Technical requirements for coatings and the related wooden products.

Technical requirements	Standards	Adhesion classification	Pencil hardness	Curing time (min)
Water-based coatings for woodenware for indoor decorating and refurbishing	GB/T 23999-2009	≤1	≥B	30–60
	EN 927:2006	≤1	–	–
UV curing coatings for woodenware	HG/T 3655–2012	≤2	≥H	–
Wooden furniture	GB/T 3324–2017	≤3	–	–
	CEN/TS 16209:2011	≤2	–	–
Solid wood floor	GB/T 15036–2018	≤3	≥H	–
	ISO 17959:2014	≤3	≥H	–
Engineered wood floor	GB/T 18103–2013	≤2	≥2H	–
	EN 14354–2005	≤2	–	–
Indoor wood-based door	LY/T 1923–2010	≤2	≥HB	–
Wood-based wall-board	LY/T 1697–2017	≤2	≥2B	–

Besides, the PUA coating has good adhesion to wood, because the back surface was similar to the morphology of wood. Therefore, the preparation and application of the fast water-based UV curing PUA wood coating was successful.

3.2. Basic Properties of PUA Coated Oak

The basic properties including dried coating thickness, adhesion, hardness, and gloss of the PUA coated oak samples are listed in Table 4. As the coating amount increased, the dried thickness, adhesion, hardness, and gloss of the sample were improved. By comparing with dried coating thickness values, it seemed that the dried coating thickness did not increase linearly. For example, at a coating amount of 120 g/m^2, the dried coating thickness was 74 μm, which was 21 times the dried coating thickness at 12 g/m^2. It may be that the surface of the sample was not completely covered at the coating amount of 12 g/m^2, and more PUA molecules were penetrated into the wood substrate, resulting in a low adhesion, hardness and surface gloss of the paint film. When the coating amount increased to 60 g/m^2, the coverage was improved, therefore, the adhesion classification, hardness, and surface gloss of the sample were improved. When the coating amount was 120 g/m^2, the sample was completely covered with the paint film, where the adhesion classification reached level 1 with the hardness of 2H and the gloss of 92.56°.

By comparing with the technical requirements for some related wooden products given by Chinese and international standards, listed in Table 3, it can be inferred that at 60 g/m^2, the adhesion classification and pencil hardness of the PUA coating met the requirements for water-based coatings and UV curing coatings. Besides, the curing time is much less than the requirement of water-based coatings of 30–60 min. For wooden products, all PUA coated samples can be used for wooden furniture and wood-based wallboard. However, for indoor wood-based doors and solid wood floors, the coating amount should increase to 60 g/m^2, while for applying on engineered wood floors, the coating amount must be further increased to 120 g/m^2.

Table 4. Adhesion, hardness, and gloss of oak coated at different coating amounts.

Sample	Dried coating thickness (μm)	Adhesion classification	Pencil hardness	Gloss (°)
12 g/m^2 PUA coated oak	3	2	B	28.7
60 g/m^2 PUA coated oak	34	1	H	83.2
120 g/m^2 PUA coated oak	74	1	2H	93.6

3.3. Chemical Analysis

Figure 5a shows the ATR-FTIR results for oak and pure PUA coating. The characteristic absorption bands attributed to the polyurethane-acrylate coating are reflected in the figure based on the literature [11,14,16]. The peaks observed at 1554 cm^{-1} and 818 cm^{-1} were assigned to –NH in and out of plane bending vibrations, which indicated that N element was in the coating. Therefore, the penetration depth of the coating can be determined according to the N element. In addition, bands at 1184 cm^{-1}, 1079 cm^{-1}, and 907 cm^{-1} were attributed to the C–O stretching vibrations, C–O stretching vibrations in ether bonds, and acid –OH out-of-plane bending vibrations, respectively. These indicated that carboxylic acid functions may be present in the coating, which can be reacted with the hydroxyl groups in the wood. Meanwhile, bands at 1726 cm^{-1} and 1672 cm^{-1} are characteristic peaks of carbonyl C=O and C=O stretching vibrations in unsaturated acids. Besides, other characteristic peaks of PUA can be observed, such as the 1251 cm^{-1} C–O stretching vibration in ester bonds, 1016 cm^{-1} C–OH stretching vibrations, 1385 cm^{-1} –OH in-plane bending vibrations, and 3390 cm^{-1} –OH stretching vibration.

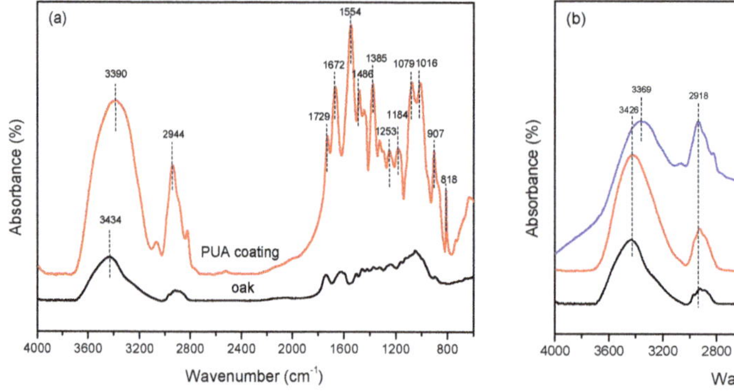
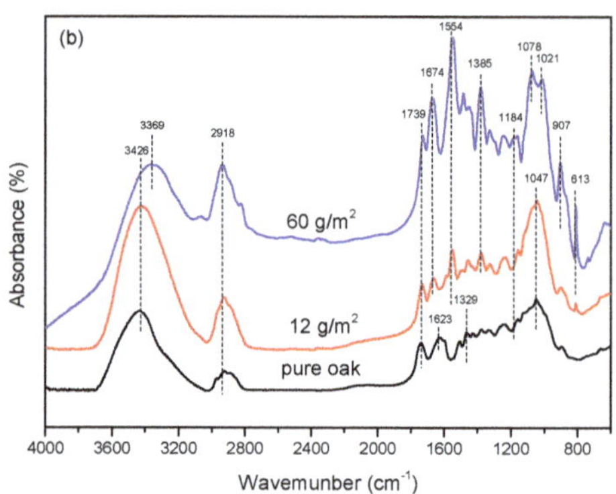

Figure 5. ATR-FTIR spectrum of samples. (**a**) Pure oak and PUA coating; (**b**) oak before and after coating with different coating amounts.

Figure 5b shows the FTIR results of pure oak and oak coated with PUA at amounts of 12 g/m^2 and 60 g/m^2, respectively. The comparison shows that the PUA characteristic peak gradually becomes obvious with the increase of the coating amount. However, some characteristic peaks shifted compared with the PUA spectrum shown in Figure 5a. Compared with the oak, 12 g/m^2 and 60 g/m^2 coated samples owned many characteristic peaks, such as the characteristic absorption peak of C=O stretching vibrations in unsaturated acids at 1674 cm^{-1}, the absorption peak of –NH$_2$ stretching vibrations at 1554 cm^{-1}, and the 1184 cm^{-1} C–O stretching vibration. These vibration absorption peaks indicated that the PUA coating covered the surface of the oak and exhibited the characteristic PUA absorption peaks. As the coating amount increased, in the FITR of oak coated with 60 g/m^2 PUA one can clearly observe that some chemical groups changed compared with that coated with 12 g/m^2 PUA, indicating that some chemical reactions took place. In details, the absorption peaks at 1047 cm^{-1} disappeared, while C–O stretching vibrations in ether bonds and C–OH stretching vibrations appeared at 1078 cm^{-1} and 1021 cm^{-1}. Additionally, the characteristic –OH out-of-plane bending vibration peaks at 907 cm^{-1}, –CH$_2$/CH$_3$ stretching vibration peaks at 2918 cm^{-1}, –OH internal bending vibration at 1385 cm^{-1}, –CH$_2$ out-of-plane bending vibrations at 613 cm^{-1}, and out-of-plane bending vibrations at 818c m^{-1} were obviously enhanced, showing the characteristics of PUA coating. Compared with pure PUA, the –OH absorption peak shifted to 3369 cm^{-1}, and the 1021 cm^{-1} C–OH stretching vibration absorption peak was weakened, while the stretching vibration at 1184 cm^{-1} for C–O was enhanced. This proved that the hydroxyl groups in the wood reacted with the coating by hydrogen bonds or by forming esters [34].

3.4. Contact Angle

Molecular interactions between the wood surface and the coating are also important in achieving good adhesion [35]. Figure 6 shows the image of contact angle of PUA droplets (70% solids) and water on the oak surface. The results are listed in Table 5. The coating is not easily spread on the oak compared to water because the polarity of PUA is lower than that of water. Although the wettability of PUA is not as good as that of water, the contact angle of PUA droplets on oak was lower than 90°, indicating the PUA can evenly spread on the oak surface. The contact angle between PUA coated oak and water were also tested. The results showed that with the increase of coating amount, the contact angle of water droplets on the surface increased because of the covering of a layer of coating film. Compared with oak coated at 12 g/m^2 PUA amount, the 60 g/m^2 possessed a larger contact angle with water, which could be explained by the incompletely covered surface at 12 g/m^2 PUA amount, where the water might penetrate into the wood sample.

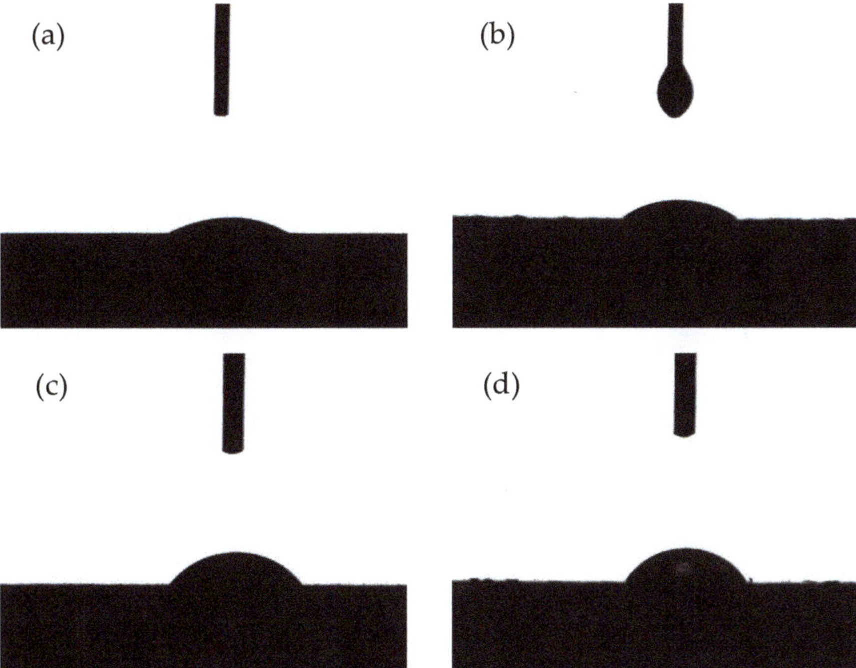

Figure 6. Contact angles of oak sample. (**a**) Pure oak with water; (**b**) pure oak with 70 wt% PUA droplet; (**c**) 12 g/m^2 PUA coated oak with water; (**d**) 60 g/m^2 PUA coated oak with water.

Table 5. Contact angles of PUA droplets (70% solids) and water on the oak surface.

Sample	Testing liquid	Contact angle (°)
Pure oak	Water	29.68
Pure oak	70 wt% PUA droplet	34.98
12g/m^2 PUA coated oak	Water	52.22
60g/m^2 PUA coated oak	Water	56.13

3.5. Surface Roughness

Table 6 shows the surface roughness of oak coated with different coating amounts. As the coating amount increased, Ra decreased continuously, and all the surface roughness values showed the same tendency. The Rp and the Rv decreased with the increase of the coating amount, indicating that the vessels and surface irregularities of the oak were continuously filled by PUA coating. Thus, the roughness values decreased, and the surface became smooth.

Table 6. Roughness of oak before and after coating at different coating amount.

Sample	Ra (μm)	Rp (μm)	Rv (μm)	Rt (μm)	Rz (μm)
Pure oak	3.456	5.930	12.521	55.110	18.450
12 g/m^2 PUA coated oak	3.026	3.451	10.000	45.975	13.451
60 g/m^2 PUA coated oak	2.422	1.651	6.924	48.010	8.577
120 g/m^2 PUA coated oak	0.310	0.516	0.400	2.834	0.916

The surface morphologies of the oak after coating at different amounts tested by AFM are shown in Figure 7. It can be seen from the figure that as the coating amount increased, the surface coverage of the oak increased, and the PUA coating can cover well on the oak under different coating amounts without obvious defects. Among them, when the coating amount was 12 g/m^2, the surface roughness value was 0.97 nm. As for the surface with a coating amount of 60 g/m^2, many small protrusions were observed, and the surface paint film was rough after drying, and the surface roughness value was 4.08 nm. Different from Table 5, it can be seen that the roughness of the coating increased with increasing coating amount, which might because that the paint surface was uneven, resulting in an increase in the surface roughness in a very small range.

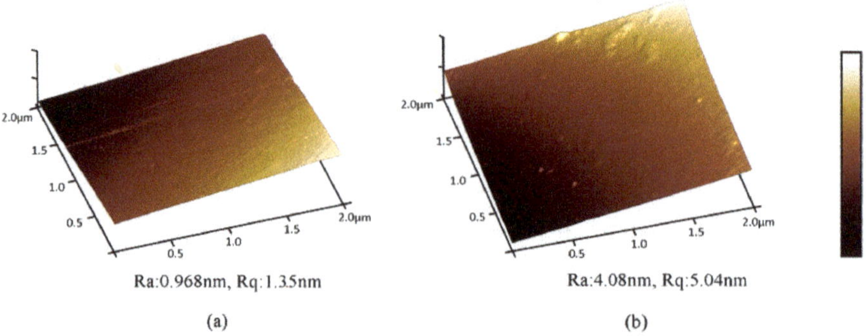

Figure 7. AFM images of oak after coating at different coating amount. (**a**) 12 g/m^2 PUA coated oak; (**b**) 60 g/m^2 PUA coated oak.

3.6. Penetration of PUA in Wood

In order to further analyze the permeability of different coatings in oak, the SEM-EDXA was performed on N-element (see red dots) for the coated samples. The results are shown in Figure 8.

As seen from the figure, the PUA was distributed in a large amount on the surface of the substrate, and a small amount was penetrated into the substrate. At the coating amount of 12 g/m², the PUA penetration depth on the oak surface was about 20 µm, and the small portion reached 35 µm. When the coating amount increase to 60 g/m², the voids in the oak penetration depth of 35 µm are substantially filled because of higher amount of PUA was applied. However, the penetration of the PUA coating in oak did not increase significantly, and 35 µm was the limit.

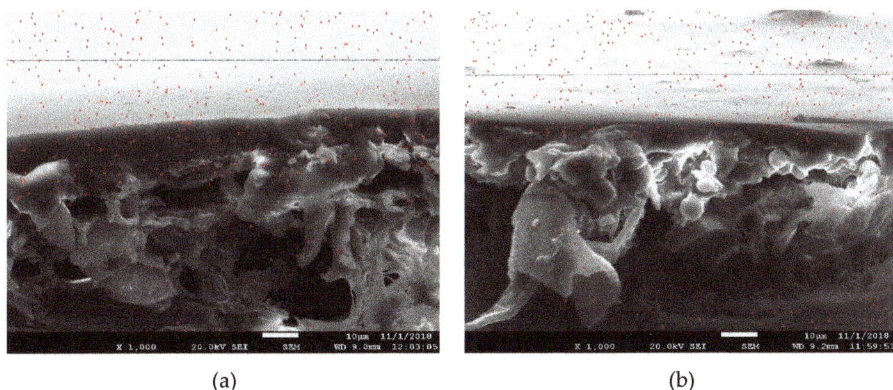

(a)   (b)

Figure 8. SEM-EDXA images for N distribution of oak after coating at different coating amount. (**a**) 12 g/m² PUA coated oak; (**b**) 60 g/m² PUA coated oak; the magnification was ×1000.

4. Conclusions

The preparation and application of the fast water-based UV curing PUA wood coating was successful, where it can be completely cured within 22 min. During curing, the hydroxyl groups in the oak reacted with the coating. It can be seen that as the coating amount increased, the number of hydroxyl groups decreased and the contact angle increased. At the same time, within a certain range, the adhesion, hardness and gloss value of oak surface increased to different extents. The coating coverage was improved and the surface gradually became smooth. The PUA can reach a depth of 35 µm from the wood surface, and with the increase of coating amount, the filling degree of PUA coating in wood increased.

Author Contributions: Investigation, J.W. and H.W.; writing—original draft preparation, R.L.; conceptualization, L.L.; writing—review and editing, J.X.; resources, M.C.; data curation, H.Q.

Funding: This study was financially supported by the National Key Research and Development Program of China (No. 2016YFD0600704).

References

1. Xu, H.; Qiu, F.; Wang, Y.; Wu, W.; Yang, D.; Guo, Q. UV-curable waterborne polyurethane-acrylate: Preparation, characterization and properties. *Prog. Org. Coat.* **2012**, *73*, 47–53. [CrossRef]
2. Qi, L.; Guo, L.; Teng, Q.; Xiao, W.; Du, D.; Li, X. Synthesis of waterborne polyurethane containing alkoxysilane side groups and the properties of the hybrid coating films. *Appl Surf. Sci.* **2016**, *377*, 66–74.
3. Chang, C.W.; Lu, K.T. Epoxy acrylate UV/PU dual-cured wood cotings. *J. Appl. Polym. Sci.* **2012**, *115*, 2197–2202. [CrossRef]
4. Meijer, M.D. Review on the durability of exterior wood coatings with reduced VOC-content. *Prog. Org. Coat.* **2001**, *43*, 217–225. [CrossRef]

5. Chang, W.Y.; Pan, Y.W.; Chuang, C.N.; Guo, J.J.; Chen, S.H.; Wang, C.K.; Hsieh, K.H. Fabrication and characterization of waterborne polyurethane (WPU) with aluminum trihydroxide (ATH) and mica as flame retardants. *J. Polym. Res.* **2015**, *22*, 243. [CrossRef]

6. Sun, Q.; Yu, H.; Liu, Y.; Li, J.; Lu, Y.; Hunt, J.F. Improvement of water resistance and dimensional stability of wood through titanium dioxide coating. *Holzforschung* **2010**, *64*, 757–761. [CrossRef]

7. Xu, H.; Qiu, F.; Wang, Y.; Yang, D.; Wu, W.; Chen, Z.; Zhu, J. Preparation, mechanical properties of waterborne polyurethane and crosslinked polyurethane-acrylate composite. *J. Appl. Polym. Sci.* **2012**, *124*, 958–968. [CrossRef]

8. Wang, Y.; Qiu, F.; Xu, B.; Xu, J.; Jiang, Y.; Yang, D.; Lia, P. Preparation, mechanical properties and surface morphologies of waterborne fluorinated polyurethane-acrylate. *Prog. Org. Coat.* **2013**, *76*, 876–883. [CrossRef]

9. Chou, P.L.; Chang, H.T.; Yeh, T.F.; Chang, S.T. Characterizing the conservation effect of clear coatings on photodegradation of wood. *Bioresource Technol.* **2008**, *99*, 1073–1079. [CrossRef]

10. Liu, F.; Liu, G. Enhancement of UV-aging resistance of UV-curable polyurethane acrylate coatings via incorporation of hindered amine light stabilizers-functionalized TiO_2-SiO_2 nanoparticles. *J. Polym. Res.* **2018**, *25*, 59. [CrossRef]

11. Decker, C.; Masson, F.; Schewalm, R. How to speed up the UV curing of water-based acrylic coatings. *JCT Res.* **2004**, *1*, 127–136. [CrossRef]

12. Dai, J.; Ma, S.; Liu, X.; Han, L.; Wu, Y.; Dai, X.; Zhu, J. Synthesis of bio-based unsaturated polyester resins and their application in waterborne UV-curable coatings. *Prog. Org. Coat.* **2015**, *78*, 49–54. [CrossRef]

13. Montazeri, M.; Eckelman, M.J. Life cycle assessment of UV-Curable bio-based wood flooring coatings. *J. Clean. Prod.* **2018**, *192*, 932–939. [CrossRef]

14. Xu, J.; Jiang, Y.; Zhang, T.; Dai, Y.; Yang, D.; Qiu, F.; Yu, Z.; Yang, P. Synthesis of UV-curing waterborne polyurethane-acrylate coating and its photopolymerization kinetics using FT-IR and photo-DSC methods. *Prog. Org. Coat.* **2018**, *122*, 10–18. [CrossRef]

15. Decker, C.; Masson, F.; Schwalm, R. Weathering resistance of waterbased UV-cured polyurethane-acrylate coatings. *Polym. Degrad. Stabil.* **2004**, *83*, 309–320. [CrossRef]

16. Masson, F.; Decker, C.; Jaworek, T.; Schwalm, R. UV-radiation curing of waterbased urethane-acrylate coatings. *Prog. Org. Coat.* **2000**, *39*, 115–126. [CrossRef]

17. Sonmez, A.; Budakci, M.; Bayram, M. Effect of wood moisture content on adhesion of varnish coatings. *Sci. Res. Essays.* **2010**, *12*, 1432–1437.

18. Brown, G.L. Formation of films from polymer dispersions. *J. Polym. Sci. A* **1956**, *102*, 423–434. [CrossRef]

19. Hwang, H.D.; Park, C.H.; Moon, J.I.; Kim, H.J.; Masubuchi, T. UV-curing behavior and physical properties of waterborne UV-curable polycarbonate-based polyurethane dispersion. *Prog. Org. Coat.* **2011**, *72*, 663–675. [CrossRef]

20. Hahn, K.; Ley, G.; Schuller, H.; Oberthür, R. On particle coalescence in latex films. *Colloid. Polym. Sci.* **1986**, *264*, 1092–1096. [CrossRef]

21. Tong, T.; Lu, Z.; Zhou, G.; Jia, W.; Wang, M. Effect of air velocity in dehumidification drying environment on one-component waterborne wood top coating drying process. *Dry. Technol.* **2016**, *34*, 7372015–7373937. [CrossRef]

22. Yang, X.; Liu, J.; Wu, Y.; Liu, J.; Cheng, F.; Jiao, X.; Lai, G. Fabrication of UV-curable solvent-free epoxy modified silicone resin coating with high transparency and low volume shrinkage. *Prog. Org. Coat.* **2019**, *129*, 96–100. [CrossRef]

23. Yang, F.; Zhu, L.; Han, D.; Li, W.; Chen, Y.; Wang, X.; Ning, L. Preparation and hydrophobicity failure behavior of two kinds of fluorine-containing acrylic polyurethane coatings. *Rsc Adv.* **2015**, *115*, 95230–95239. [CrossRef]

24. Kong, X.; Qu, J.; Zhu, Y.; Cao, S.; Chen, H. Adhesion behaviors and affecting factors of polymer latexes. *Paint. Coat. Ind.* **2010**, *40*, 37–40.

25. Chu, H.H.; Chang, C.Y.; Shen, B.H. An electrophoretic coating using a nanosilica modified polyacrylate resin. *J. Polym. Res.* **2018**, *25*, 44. [CrossRef]

26. Asif, A.; Shi, W.; Shen, X.; Nie, K. Physical and thermal properties of UV curable waterborne polyurethane dispersions incorporating hyperbranched aliphatic polyester of varying generation number. *Polymer* **2005**, *46*, 11066–11078. [CrossRef]

27. Asif, A.; Shi, W. UV curable waterborne polyurethane acrylate dispersions based on hyperbranched aliphatic polyester: Effect of molecular structure on physical and thermal properties. *Polym. Adv. Technol.* **2004**, *15*, 669–675. [CrossRef]
28. Asif, A.; Huang, C.Y.; Shi, W.F. Photopolymerization of waterborne polyurethane acrylate dispersions based on hyperbranched aliphatic polyester and properties of the cured films. *Colloid. Polym. Sci.* **2005**, *283*, 721–730. [CrossRef]
29. Asif, A.; Hu, L.; Shi, W. Synthesis, rheological, and thermal properties of waterborne hyperbranched polyurethane acrylate dispersions for UV curable coatings. *Colloid. Polym. Sci.* **2009**, *287*, 1041–1049. [CrossRef]
30. Phelps, J.E.; Workman, E.C., Jr. Vessel area studies in white oak (*Quercus alba* L.). *Wood Fiber Sci.* **1994**, *26*, 315–322.
31. Ozyhar, T.; Mohl, L.; Hering, S.; Hass, P.; Zeindler, L.; Ackermann, R.; Niemz, P. Orthotropic hygric and mechanical material properties of oak wood. *Wood Mater. Sci. Eng.* **2016**, *11*, 36–45. [CrossRef]
32. Catalin, C.; Cosmin, S.; Aurel, L.; Daniel, C.; Claudiu, R.I.; Alin, P.M.; Tibor, B.; Manuela, S.E.; Alexandru, P. Surface properties of thermally treated composite wood panels. *Appl. Surf. Sci.* **2018**, *438*, 114–126.
33. Xu, J.; Liu, R.; Wu, H.; Long, L.; Lin, P. A comparison of the performance of two kinds of waterborne coatings on bamboo and bamboo scrimber. *Coatings* **2019**, *9*, 161. [CrossRef]
34. Kuang, X.; Kuang, R.; Zheng, X.; Wang, Z. Mechanical properties and size stability of wheat straw and recycled LDPE composites coupled by waterborne coupling agents. *Carbohyd. Polym.* **2010**, *80*, 927–933. [CrossRef]
35. Landry, V.; Blanchet, P. Surface preparation of wood for application of waterborne coatings. *For. Prod. J.* **2012**, *62*, 39–45. [CrossRef]

Article

Thermal and Mechanical Behavior of Wood Plastic Composites by Addition of Graphene Nanoplatelets

Xingli Zhang [1],*, Jinglan Zhang [1] and Ruihong Wang [2]

[1]   College of Mechanical and Electrical Engineering, Northeast Forestry University, Harbin 150040, China
[2]   Key Laboratory of Functional Inorganic Material Chemistry, Heilongjiang University, Harbin 150010, China
*   Correspondence: zhang-xingli@nefu.edu.cn; Tel./Fax: +86-451-82192851

Received: 8 July 2019; Accepted: 15 August 2019; Published: 19 August 2019

Abstract: Wood plastic composites (WPCs) incorporating graphene nano-platelets (GNPs) were fabricated using hot-pressed technology to enhance thermal and mechanical behavior. The influences of thermal filler content and temperature on the thermal performance of the modified WPCs were investigated. The results showed that the thermal conductivity of the composites increased significantly with the increase of GNPs fillers, but decreased with the increase of temperature. Moreover, thermogravimetric analysis demonstrated that coupling GNPs resulted in better thermal stability of the WPCs. The limiting oxygen index test also showed that addition of GNPs caused good fire retardancy in WPCs. Incorporation of GNPs also led to an improvement in mechanical properties as compared to neat WPCs. Through a series of mechanical performance tests, it could be concluded that the flexural and tensile moduli of WPCs were improved with the increase of the content of fillers.

Keywords: wood plastic composite; graphene nano-platelets; thermal property; mechanical property

---

1. Introduction

Wood plastic composites (WPCs) are replacement products of wood made by dispersing wood flour into polymer matrix, and are widely utilized in construction, packaging, and furniture products [1]. WPC products show an excellent thermal insulation property, which may also limit their application, such as in heated floors. A possible solution to improve the thermal properties of WPCs is the introduction of additives. Nanocarbon materials are the main candidates for this purpose owing to their superior thermal conductivity [2–4]. Among the nanocarbon materials, graphene nanoplatelet (GNPs) fillers have attracted intensive research interests to obtain a dramatic improvement in the thermal properties of polymer composites. Compared with other nanocarbon materials, the high specific surface area of GNPs helps a great deal to result in an effective dispersion degree of the fillers. In addition, GNPs as a thermally conductive filler could maintain or even improve the mechanical properties of the composites because of their own excellent mechanical strength [5–7].

Recently, extensive research on enhancing the heat transfer of polymers by adding GNPs has been reported. The thermal properties of epoxy/GNPs composites were studied through thermal conductivity measurements, and the results showed that the introduction of 20 wt % chemically-functionalized GNPs enhanced thermal conductivity 29 times relative to that of pure epoxy [8]. The modeling results of Khan et al. showed that the significantly high improvement of the cross-plane thermal conductivity of GNP-based polymer is ascribed to good thermal transmission through GNP networks in the polymer matrix [9]. Zhou et al. reported that the high aspect ratio of GNPs is one of the key factors contributing to higher thermal conductivity of epoxy with relatively lower filler contents [10]. The above experimental studies mainly focused on the application of GNPs in polymers, but the attention given to the introduction of GNPs into WPCs is still scarce.

Herein, a competitive WPC reinforced with GNPs is reported. The resulting WPCs exhibited the highest thermal conductivity compared to the values of neat WPCs. The influences of filler content and temperature on the thermal and mechanical properties of composites were also investigated in detail.

2. Materials and Experiment

2.1. Materials Collection

Poplar wood fiber (WF) was obtained from Baiquan, Heilongjiang Province in China. Polyethylene (PE) (Fushun Petroleum Company, Fushun, China) was used as polyolefin matrix. Maleic anhydride grafted polyethylene (MAPE) (Rizhisheng New Materials Technology, Shanghai, China) was used as a compatibility agent to help the polymers bond together, and the weight content of MAPE in the total composite dry mass was never below 3%. The graphene nanoplatelets (GNPs) were purchased from Suzhou Rich Carbon Graphite Technology Company, with purity > 95%, thickness 1.0–1.77 nm, lamellar diameter 10–50 µm, surface area 360–450 m^2/g.

2.2. Sample Preparation

WF was dried by a convection oven at 103 °C for 24 h to obtain 3% moisture content. The dried WF, PE, MAPE and GNPs were then blended in a high-speed mixer for 10 min. The weight contents of raw materials are listed in Table 1 (control means the original wood plastic composites and WPC3 means the composites adding 3 wt % GNPs). A co-rotating twin screw extruder (Xawax Science Technology Company, Nanjing, China) was used in extruding the blends. The temperature of the extruder barrel was set between 140 and 180 °C with the screw rotating speed set at 30 rpm. The extruded melts were manufactured using a hot-press machine (BY114×8, Jianhu Machinery Factory, Nanjing, China) for 2 min. Finally, the WPC sheets were molded into test specimens sized 100 mm (L) × 100 mm (W) × 4 mm (H).

Table 1. Mass fraction of raw materials for wood-plastic composites.

Sample	WF (wt %)	PE (wt %)	MAPE (wt %)	GNPs (wt %)
Control	40 ± 1.5	57 ± 1.5	3 ± 0.1	0
WPC3	40 ± 2.0	54 ± 1.5	3 ± 0.2	3
WPC6	40 ± 1.4	51 ± 1.2	3 ± 0.1	6
WPC9	40 ± 1.7	48 ± 1.2	3 ± 0.1	9
WPC12	40 ± 2.0	45 ± 1.0	3 ± 0.2	12

2.3. Sample Characterization

2.3.1. Morphology Analysis

The microstructure of the GNPs was observed using a transmission electron microscope (TEM, 912 AB, Oberckochen, Munster, Germany) at an accelerating voltage of 100 kV. After sputtered with a thin layer of gold, the surface morphology of samples were taken using a scanning electron microscope (SEM, Quanta 200, FEI Company, Eindhoven, Netherlands) at an acceleration voltage of 12.5 kV.

2.3.2. Thermal Properties

The heat transfer performance of the modified WPCs sample was determined using a TC-3100 thermal measuring instrument (Xiatech Group, Xi'an, China) base using a transient hot wire method. The specimen dimensions were 40 mm × 40 mm × 4 mm, and the test temperatures were 20–80 °C. The accuracy of thermal conductivity measurements was about ±3%. Thermogravimetric analysis of the WPCs was measured by Perkin Elmer Pyris 6 TGA for analyzing the thermal decomposition temperature of composites. The heating condition was in the temperature range of 50–700 °C, and heating rate was 10 °C/min.

The limiting oxygen index (LOI) was measured according to ASTM D2863, and the apparatus used was a JF-3 oxygen index meter (Jiangning Analysis Instrument Company, Nanjing, China). The specimen dimensions used for the test were $130 \times 10 \times 3$ mm^3.

2.3.3. Mechanical Properties

Flexural and tensile properties of samples were measured according to ASTM D638-2004 standard, using a universal mechanical machine (Regear Instrument Cooperation, Shenzhen, China). Unnotched impact strength was measured according to ASTM D4812-2004 standard, using a JJ-20 impact tester (Intelligent Instrument Cooperation, Changchun, China) at a speed of 5 mm/min. For each mechanical analysis, six samples were tested.

Dynamic mechanical analysis (DMA) was performed in a dual cantilever mode using a dynamic mechanical analyzer (TA Instruments Inc, New Castle, DE, USA). The heating condition was in the temperature range of −10 to 120 °C, and heating rate was 3 °C/min.

3. Results and Discussion

3.1. Morphological Properties

This section shows the microstructure and the low-temperature fracture morphology of WPCs with an increase in GNPs content. As shown in the TEM images (Figure 1f), the general shape of GNPs is a semi-transparent film with twists and turns. In most cases, GNPs are in an overlapping state of multiple layers. The SEM images show that a stable heat conduction network chain is formed in the WPCs with the increase in GNP content, but in further increasing the packing material, the GNPs tended to form a higher structure, which lead to agglomeration in the PE matrix (see Figure 1d,e). Since the diffusion of phonons caused effective thermal conductivity in the GNPs, a good dispersion of GNPs in the PE matrix can steadily improve the thermal conductivity of the WPCs [11,12]. However, many GNPs overlap and agglomerates become defects in WPCs, which may disrupt the formation of network cross-linking and lead to a decrease in mechanical properties [13].

3.2. Thermal Properties

3.2.1. Effect of GNPs Content on the Thermal Conductivity

The WPCs samples with added GNPs exhibited a dramatic enhancement in thermal conductivity compared to neat WPCs, as shown in Figure 2. When the content of GNPs increased to 12 wt %, the thermal conductivity of WPCs increased by a factor of 258.9%. Compared to graphite [14], the addition of GNPs had a more significant enhancement effect on the heat transfer property of WPCs. Since the effective heat conduction in GNPs is attributed to phonon diffusion, the main method of modifying WPC heat conduction is also phonon heat conduction [15]. It can be found from SEM micrograph (Figure 1) that there are multiple interfaces between the GNPs fillers, WF, and PE matrix, which cause greater contact resistance and phonon scattering at interfaces. Due to the increase in the mass ratio of GNPs, the interfacial thermal contact resistance decreases, which minimizes the scattering of interfacial phonons and improves the thermal conductivity of the composites. Thus, better dispersion could lead to a decrease in the thermal contact resistance between fillers and polymer matrix, which are the key factors to exhibit higher thermal conductivity enhancement of nanocomposites.

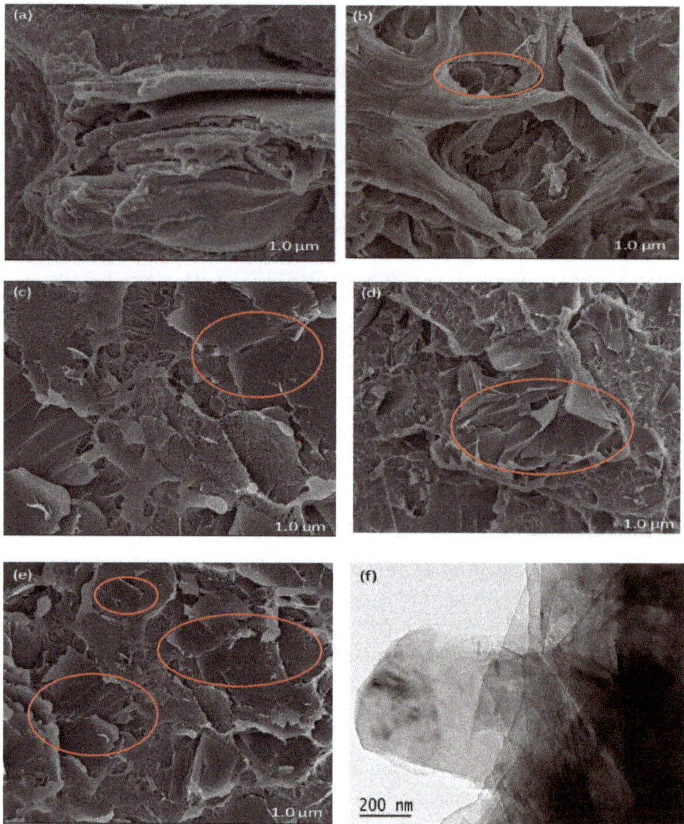

Figure 1. SEM and TEM micrographs of WPCs adding GNPs. (**a**) Control; (**b**) WPC3; (**c**) WPC6; (**d**) WPC9; (**e**) WPC12; (**f**) TEM image of GNPs.

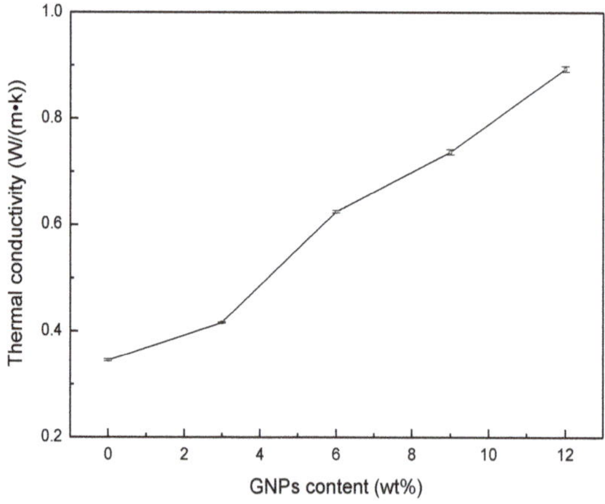

Figure 2. Thermal conductivity of the WPCs as a function of GNPs contents.

3.2.2. Effect of Temperature on the Thermal Conductivity

Figure 3 shows the thermal conductivity behavior of WPCs as a function of temperature. The thermal conductivities of WPC6 and WPC12 decreased gradually when the temperature increased from 20 °C to 80 °C. The thermal conductivity of GNPs and PE have been experimentally or theoretically assessed [16–18]. Many results show that temperature adversely affects the diffusion of phonons, so the thermal conductivity of WPCs produce a declining trend due to the predominant decrease in the thermal conductivity of GNPs and PE with temperature. In addition, the data show that when the GNPs content is higher, the thermal conductivity is less affected by the temperature change, indicating that a high content of GNPs is helpful to form a stable thermal conduction network in the polymer matrix. Therefore, GNP filler may effectively reduce the influence of temperature change on the thermal conductivity of a composite.

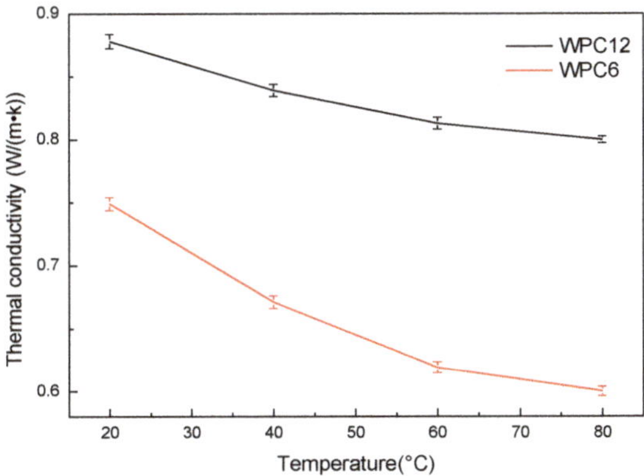

Figure 3. Thermal conductivity of the WPCs as a function of temperature.

3.2.3. Thermogravimetric Analysis (TG)

TGA/DTG profiles for test samples as a function of temperature are shown in Figure 4. TGA curves show that all samples exhibited similar thermal behavior. There are two times of weightlessness in the degradation process of samples. The first weight loss takes place at around 300–350 °C, which is attributed to the evaporation of wood components, such as hemicellulose, cellulose, and lignin. The second weight loss takes place at around 450–500 °C, which is caused by the degradation of the polymer matrix. Moreover, with the increase in GNPs content, the thermal degradation temperature of the composite gradually increased. The DTG curve also exhibits two derivative peaks, which are also related to the decomposition of wood fiber and PE, and the highest degradation temperature of the second peak is much higher than the first peak. Table 2 shows that the temperatures of 5%, 10% and 50% weight loss (T_5, T_{10}, T_{50}) increase with increasing GNPs content. It is believed that the presence of GNPs is beneficial to the enhancement of thermal stability of the composites.

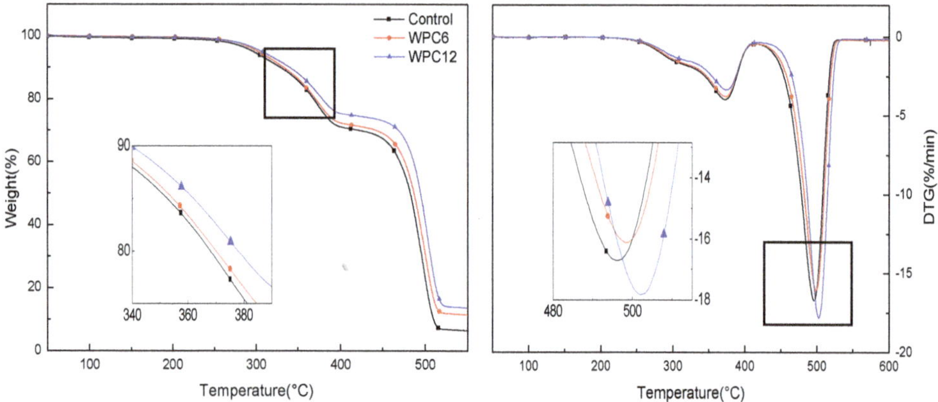

Figure 4. Mass loss curves and DTG curves of WPCs with added GNPs.

Table 2. Thermal decomposition parameters for WPCs with added GNPs.

Sample	T_5 (°C)	T_{10} (°C)	T_{50} (°C)	T_{max} (°C)
Control	296.85	329.00	482.53	373.53/496.33
WPC6	299.32	332.35	486.44	373.79/498.18
WPC12	299.56	339.12	494.07	374.14/502.14

3.2.4. The Limiting Oxygen Index (LOI) Test

As shown in Figure 5, the LOI values of WPCs with added GNPs are higher than those of neat WPCs, and increase with the GNP content. Previous reports demonstrated that graphene or its derivatives, as fire retardant nanofillers, could reduce the fire hazard of various polymers, as a protective intumescent carbonaceous char was formed on the surface of the materials [19,20]. Here the LOI test results confirm the conclusion that the addition of GNPs causes good fire retardancy in WPCs.

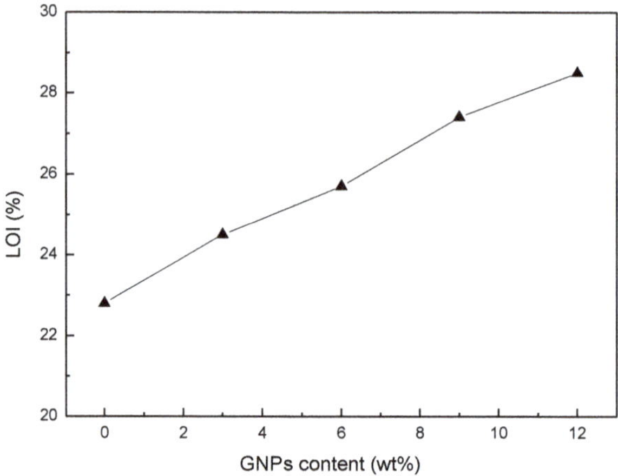

Figure 5. The LOI values of the WPCs with added GNPs.

3.2.5. Mechanical Properties

The mechanical properties of WPCs are significantly affected by GNPs content. As shown in Table 3, with increasing content of GNPs from 0 to 12 wt %, the flexural strength, flexural modulus and tensile modulus increase by 35.3%, 47.9% and 15.9%, respectively, but the tensile strength and impact strength of WPCs decrease slightly. It is well known that the adhesion and aggregation of nano-filler lead to defects in composites, as shown by SEM analysis (Figure 1), which is a major cause of instability in mechanics performance [21,22]. However, GNPs as filler have a beneficial impact on the flexural and tensile moduli of WPCs compared to other carbon nanomaterials. This may be explained by the excellent intrinsic mechanical resistance of GNPs. For flexural strength especially, the values decrease to an extent and then begin to increase gradually. The very large contacting interface between GNPs and PE matrix firstly cause a decrease of flexural strength. With increasing GNPs, their intrinsic mechanical properties play a leading role in improving the strength of WPCs.

Table 3. Mechanical properties WPCs adding GNPs.

WPC Type	Flexural Strength (MPa)	Flexural Modulus (GPa)	Tensile Strength (MPa)	Tensile Modulus (GPa)	Impact Strength (kJ/m^2)
Control	45.89 ± 0.62	1.96 ± 0.02	22.52 ± 0.41	1.44 ± 0.03	9.71 ± 0.56
WPC3	41.59 ± 0.65	2.32 ± 0.02	21.05 ± 0.47	1.45 ± 0.04	8.75 ± 0.43
WPC6	42.43 ± 0.71	2.50 ± 0.03	20.53 ± 0.42	1.56 ± 0.06	8.42 ± 0.46
WPC9	46.51 ± 0.57	2.89 ± 0.05	20.12 ± 0.32	1.65 ± 0.05	8.40 ± 0.51
WPC12	47.51 ± 0.35	2.90 ± 0.04	20.10 ± 0.41	1.67 ± 0.04	8.32 ± 0.23

The storage modulus (E′), loss modulus (E″), and tan δ determined using DMA are shown in Figure 6. The storage modulus (E′) tends to increase at first, then when the temperature exceeds 15 °C they exhibit a significant decreasing tendency. This result can be ascribed to the increase in molecular mobility of the PE chains at high temperatures [23]. In addition, the values of E′ also decrease with the increase in GNPs content, indicating that the presence of GNPs in the composite will increase the hindrance of the segmental motion of the polymer chains [24]. The loss modulus (E″) shows an opposite tendency compared to the storage modulus (in Figure 6b). The increase of the loss modulus may be due to the effects of interfacial interactions and entanglements [25]. The loss factor tan δ can be defined as the ratio of the loss modulus to the storage modulus, which is associated with the structural transformation in the composite [26]. In our study, three tan δ curves almost overlapped with the temperature increase, but the values were slightly different at the lower temperatures and tan δ decreased with an increase in GNP mass ratio. This phenomenon can be explained by the weak interfacial adhesion of GNPs in the PE matrix; however, the powerful mechanical properties of GNPs offset the shortage of interface between the polymer matrix and fillers, which is consistent with the mechanical test results.

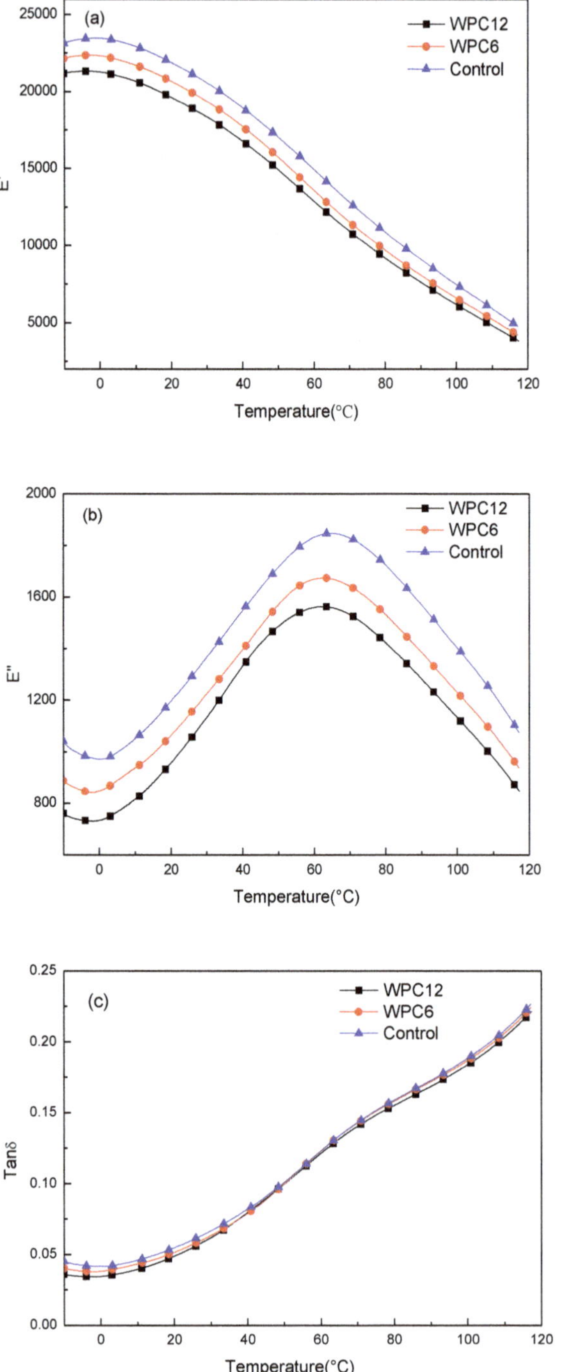

Figure 6. DMA curves of WPCs adding GNPs. (**a**) E′, (**b**) E″ and (**c**) tan δ.

4. Conclusions

In summary, the thermal transport and mechanical properties of WPCs reinforced by an amount of GNPs are investigated. SEM observation shows that GNPs with higher content present a multilayer overlapping state or agglomerations in the polymer matrix. Therefore, homogeneous dispersion of fillers within the matrix should be the key factor in maximizing the filler's effect. The thermal conductivity of the composite increases markedly with an increase in GNPs content, but decreases with increasing temperature. Additionally, TG analyses illustrate that WPCs modified by GNP fillers have better thermal stability than neat WPCs. The GNP fillers also have a positive effect on the mechanical properties of WPCs. The flexural strength of the composite decreases markedly at small amounts of GNPs, but begin to increase with more filler content. From the above study, it is evident that WPCs with added GNPs exhibited better improvement upon thermal conductivity, and avoid much of the decline in mechanical properties. This research also greatly extends the comprehensive utilization of WPCs.

Author Contributions: X.Z. conceived and designed the experiments; X.Z. and J.Z. performed the experiments and wrote the paper; R.W. and X.Z. analyzed the data.

Funding: This work has been supported by the National Natural Science Foundation of China (grant no. 51706039).

Acknowledgments: The authors are grateful to the support of Key Laboratory of Functional Inorganic Material Chemistry (Heilongjiang University), Ministry of Education.

Conflicts of Interest: The authors declare no conflict of interest.

References

1. Wang, Q.W.; Wang, W.H.; Song, Y.M. *Wood-Plastic Composites and Products*; Chemical Industry Press: Beijing, China, 2007.
2. Fang, H.M.; Bai, S.L.; Wong, P. Microstructure engineering of graphene towards highly thermal conductive composites. *Compos. Part A* **2018**, *112*, 216–238. [CrossRef]
3. Verdejo, R.; Bernal, M.M.; Romasanta, L.J. Graphene filled polymer nanocomposites. *J. Mater. Chem.* **2011**, *21*, 3301–3310. [CrossRef]
4. Birm, J.K.; Fei, Y.; Han, G.; Wang, Q.W.; Wu, Q.L. Mechanical and physical properties of core–shell structured wood plastic composites: Effect of shells with hybrid mineral and wood fillers. *Compos. Part B* **2013**, *45*, 1040–1048.
5. Teng, C.C.; Ma, C.C. Thermal conductivity and structure of non-covalent functionalized graphene/epoxy composites. *Carbon* **2011**, *49*, 5107–5111. [CrossRef]
6. Wejrzanowski, T.; Grybczuk, M.; Chmielewski, M. Thermal conductivity of metal-graphene composites. *Mater. Des.* **2016**, *99*, 163–173. [CrossRef]
7. Huang, C.; Qian, X.; Yang, R. Thermal conductivity of polymers and polymer nanocomposites. *Mater. Sci. Eng. R* **2018**, *132*, 1–22. [CrossRef]
8. Chu, K.; Wang, X.; Li, Y. Thermal properties of graphene/metal composites with aligned graphene. *Mater. Des.* **2018**, *140*, 85–94. [CrossRef]
9. Khan, M.F.S.; Alexander, A.B. Graphene-multilayer graphene nanocomposites as highly efficient thermal interface materials. *Nano Lett.* **2012**, *12*, 861–867.
10. Zhou, T.; Wang, X.; Cheng, P.; Wang, T.; Xiong, D.; Wang, X. Improving the thermal conductivity of epoxy resin by the addition of a mixture of graphite nanoplatelets and silicon carbide microparticles. *Express Polym. Lett.* **2013**, *7*, 585–594. [CrossRef]
11. Xiang, J.; Drzal, L.T. Thermal conductivity of exfoliated graphite nanoplatelet paper. *Carbon* **2011**, *49*, 773–778. [CrossRef]
12. Shahil, K.M.F.; Balandin, A.A. Thermal properties of graphene and multilayer graphene: Applications in thermal interface materials. *Solid State Commun.* **2012**, *152*, 1331–1340. [CrossRef]
13. Tang, B.; Hu, G.; Gao, H. Application of graphene as filler to improve thermal transport property of epoxy resin for thermal interface materials. *Int. J. Heat Mass Transf.* **2015**, *85*, 420–429. [CrossRef]

14. Zhang, X.; Hao, X.; Hao, J.; Wang, Q. Heat transfer and mechanical properties of wood-plastic composites filled with flake graphite. *Therm. Acta* **2018**, *664*, 26–31. [CrossRef]

15. Malekpour, H.; Chang, K.H.; Chen, J.C. Thermal conductivity of graphene laminate. *Nano Lett.* **2014**, *14*, 5155–5161. [CrossRef]

16. Gu, J.; Du, J.; Dang, J.; Geng, W.; Hu, S.; Zhang, Q. Thermal conductivities, mechanical and thermal properties of graphite nanoplatelets/polyphenylene sulfide composites. *RSC Adv.* **2014**, *4*, 22101–22105. [CrossRef]

17. Sayanthan, R.; Wang, X.; Jay, S.; John, W. Heat transfer performance enhancement of paraffin/expanded perlite phase change composites with graphene nano-platelets. *Energy Procedia* **2017**, *105*, 4866–4871.

18. Dilek, K.; Ismail, H.T.; Turhan, C. Thermal conductivity of particle filled polyethylene composite materials. *Compos. Sci. Technol.* **2003**, *63*, 113–117.

19. Guo, C.; Zhou, L.; Lv, J. Effects of Expandable Graphite and Modified Ammonium Polyphosphate on the Flame-Retardant and Mechanical Properties of Wood Flour-Polypropylene Composites. *Polym. Polym. Compos.* **2013**, *21*, 449–456. [CrossRef]

20. Bin, Y.; Shi, Y.; Yuan, B.; Qiu, S.; Xing, W.; Hu, W.; Song, L.; Loc, S.; Hu, Y. Enhanced thermal and flame retardant properties of flame-retardant-wrapped graphene/epoxy resin nanocomposites. *J. Mater. Chem. A* **2015**, *3*, 8034–8044.

21. Robertson, D.H.; Brenner, D.W.; Mintmire, J.W. Energetics of nanoscale graphitic tubules. *Phys. Rev. B* **1992**, *45*, 12592. [CrossRef]

22. Brenner, D.W.; Shenderova, O.A.; Harrison, J.A. A second-generation reactive empirical bond order (REBO) potential energy expression for hydrocarbons. *J. Phys. Condens. Matter* **2002**, *14*, 783. [CrossRef]

23. Yavari, F.; Fard, H.R.; Pashayi, K. Enhanced thermal conductivity in a nanostructured phase change composite due to low concentration graphene additives. *J. Phys. Chem. C* **2011**, *115*, 8753–8758. [CrossRef]

24. Papageorgiou, D.G.; Kinloch, I.A.; Young, R.J. Mechanical properties of graphene and graphene-based nanocomposites. *Prog. Mater. Sci.* **2017**, *90*, 75–127. [CrossRef]

25. Zhang, X.; Hao, X.; Hao, J. Thermal and mechanical properties of wood-plastic composites filled with multiwalled carbon nanotubes. *J. Appl. Polym. Sci.* **2018**, *135*, 46308. [CrossRef]

26. Rishi, A.M.; Kandlikar, S.G.; Gupta, A. Improved wettability of graphene nanoplatelets (GNP)/copper porous coatings for dramatic improvements in pool boiling heat transfer. *Int. J. Heat Mass Transf.* **2019**, *132*, 462–472. [CrossRef]

Article

Urea Formaldehyde Resin Resultant Plywood with Rapid Formaldehyde Release Modified by Tunnel-Structured Sepiolite

Xiaona Li [1,2], **Qiang Gao** [3], **Changlei Xia** [1,2], **Jianzhang Li** [2,3,*] **and Xiaoyan Zhou** [1,2,*]

1   College of Materials Science and Engineering, Nanjing Forestry University, Nanjing 210037, China
2   Jiangsu Key Open Laboratory of Wood Processing and Wood-Based Panel Technology, Nanjing 210037, China
3   MOE Key Laboratory of Wooden Material Science and Application, Beijing Forestry University, Beijing 100083, China
*   Correspondence: lijzh@bjfu.edu.cn (J.L.); zhouxiaoyan@njfu.edu.cn (X.Z.); Tel.: +86-10-6233-6912 (J.L.); +86-25-8542-8506 (X.Z.)

Received: 20 June 2019; Accepted: 23 July 2019; Published: 1 August 2019

Abstract: In order to reduce the cost of plywood and save edible resources (wheat flour), a cheap and resourceful clay, sepiolite, was used to modify urea formaldehyde (UF) resin. The performances of filler-filled UF resins were characterized by measuring the thermal behavior, cross section, and functional groups. Results showed that cured UF resin with SEP (sepiolite) formed a toughened fracture surface, and the wet shear strength of the resultant plywood was maximum improved by 31.4%. The tunnel structure of SEP was beneficial to the releasing of formaldehyde, as a result, the formaldehyde emission of the plywood bonded by UF resin with SEP declined by 43.7% compared to that without SEP. This study provided a new idea to reduce the formaldehyde emission, i.e., accelerating formaldehyde release before the product is put into use.

Keywords: wood adhesive; tunnel-structured; sepiolite; rapid formaldehyde release

---

1. Introduction

Urea formaldehyde (UF) resin and its modified products have been widely used in the plywood industry because of many advantages, such as colorlessness, fast curing, water solubility, and low cost [1–3]. However, UF resin and its bonded panels have a serious formaldehyde emission issue [4]. Researchers have focused on chemical modification and optimizing the synthetic process to reduce the formaldehyde emission, by means such as reducing the formaldehyde/urea (F/U) molar ratio [5], using melamine to replace part of the urea [6], or adding scavengers [7,8]. These methods successfully reduced the free formaldehyde content of UF resin to a low level and yielded environmentally friendly panels. In recent years, protein-based adhesives have been developed to replace the formaldehyde-based adhesives [9,10], however, their industrial application was limited by some disadvantages, such as low solid content, low dry bonding strength, and high cost [11,12]. Therefore, UF resin will still be the predominant adhesive in the manufacture of plywood for some time [13,14].

During the fabrication of plywood, 20–30% of wheat flour (WF) is introduced into UF resin to increase its viscosity for good workability and pre-press property and to prevent the UF resin from over penetration into wood [15]. About 2 million tons of WF are used in plywood production every year, which is an appalling waste of precious food resources. In order to reduce the cost of plywood and save edible resources, many types of materials have been studied as substitutes for WF, e.g., chestnut shell powder [8] or hydrolyzed corn cob powder [16]. However, compared with flour, the initial viscosity of other types of filler-based adhesives needs to be further improved. In recent years, inorganic clay minerals present a considerable advantage in the wood adhesive modification area owing to their low price and large output. Most studies have thus far focused on two-dimensional disk-like clays such as montmorillonite [17–19], identifying their exfoliation mechanism and properties. However, one-dimensional rod-like fillers (e.g. sepiolite [20] and attapulgite [21]) have recently gained attention because these fillers occupy less surface space than the disk-like ones, allowing fillers to be evenly distributed in polymeric matrices [22].

Sepiolite (SEP) is a clay mineral with the formula $Mg_4Si_6O_{15}(OH)_2 \cdot 6H_2O$. This clay mineral has a nanometer tunnels structure showing a micro-fibrous morphology with a particle size in the 2–10 μm length range [23]. The specific surface area of SEP is close to 320 m^2/g that is expressed in its admirable absorptive properties [24]. SEP is used to reinforce polymers because of the large interface area, which will result in a strong interaction between the polymer matrix and the nanofiller [25]. The studies using SEP as a filler of UF resin are relatively scarce. The unique tunnel structure of SEP can not only increase the adsorption area but also provide a channel for gas escape, and, therefore, we assumed that the unique tunnel structure of SEP will play a role in reducing formaldehyde emissions of UF resin resultant plywood.

In this study, different proportions of SEP were used to partially substitute WF as a UF resin filler in the plywood preparation process. Three-ply plywood was fabricated, and its wet shear strength and formaldehyde emission were measured. The effect of SEP addition on resin's functional groups, cross-sections of cured resins, and thermal behavior of resins were evaluated by Fourier-transform infrared spectroscopy (FTIR), scanning electron microscope (SEM), and thermogravimetric analysis (TGA), respectively.

2. Materials and Methods

2.1. Materials

Urea (industrial grade, 98%) was purchased from the Henan Zhongyuan Chemical Company, China. Formaldehyde (aqueous solution, industrial grade, 37%) was purchased from the Guangdong Xilong Chemical Factory, China. Sodium hydroxide (solid, analytical grade, 95%), formic acid (aqueous solution, analytical grade, 98%), and ammonium chloride (powder, analytical grade, 99%) were purchased from the Beijing Chemical Factory, China. WF was obtained from the Beijing Guchuan Flour Company, China. Poplar (*Populus tomentosa* Carr.) veneers having 8% moisture content were purchased from Hebei province, Langfang, China. SEP (moisture content ≤2%, 1 g/cm^3, 200–300 mesh) was purchased from the QiLiPing sepiolite factory, Nanyang, China.

2.2. Preparation of UF Resin

The UF resin was synthesized by formaldehyde and urea at an F/U molar ratio of 1:1 in the laboratory following the traditional "alkali–acid–alkali" three-step procedure as described previously in our research group [26,27]. The urea was added into formaldehyde solution three times at a mass ratio of 2.97:1.75:1. The formalin was placed in the reactor, adjusted to pH 8.0 with aqueous NaOH (30 wt %), and then the first amount of urea was added. The mixture was then heated to 90 °C for 1 h. The acidic reaction was brought by adding formic acid (20 wt %) to obtain a pH of 5.0 or so, and the condensation reactions were carried out until it reached a target viscosity. Then the mixture was adjusted to pH 8.0 by using NaOH, and the second amount of urea was added. After 30 min at 80 °C, the third amount of urea was added for a further stirring at 70 °C for 30 min. Then, the UF resin was cooled to room temperature, followed by adjusting the pH to 8.0.

Subsequently, 100.0 g of UF resin was added into a plastic cup followed by 0.6 g ammonium chloride; then SEP-based fillers (25 g) were added and stirred for 10 min for uniform mixing. The formulations of the fillers are shown in Table 1.

Table 1. The different formulations of sepiolite (SEP)-based fillers.

Resin No.	SEP (%)	WF (g)	SEP (g)
A1 (0% SEP + 100% WF)	0	25	0
A2 (20% SEP + 80% WF)	20	20	5
A3 (40% SEP + 60% WF)	40	15	10
A4 (60% SEP + 40% WF)	60	10	15
A5 (80% SEP + 20% WF)	80	5	20
A6 (100% SEP + 0% WF)	100	0	25

WF—wheat flour.

2.3. Preparation of Three-Ply Plywood

Poplar veneers with dimensions of 400 mm × 400 mm × 1.5 mm were used to prepare three-ply plywood. The resin was applied on double sides of a veneer at a spread rate of 300 g/m^2. The resin-coated veneer was then stacked between two uncoated veneers with the grain directions of two adjacent veneers perpendicular to each other. The stacked plywood was hot-pressed at 1.0 MPa and 120 °C for 6 min and then preserved at room temperature for 24 h, before evaluation of wet shear strength and formaldehyde emission.

2.4. Wet Shear Strength Measurement

The wet shear strength of plywood was measured in accordance with the National Standard of GB/T 17657-2013 (GB/T 17657, 2013). For each adhesive, eight replicates (25 mm × 100 mm) were soaked in 63 ± 2 °C water for 3 h and then cooled to room temperature for 10 min before measurement. The average value of the wet shear strength was calculated and recorded.

2.5. Formaldehyde Emission Measurement

The formaldehyde emission of plywood was determined using the desiccator method in accordance with the procedure described in China National Standard GB/T 17657-2013. The plywood was cut into dimensions of 50 mm × 150 mm. Ten specimens of each adhesive were directly put into a 9–11 L sealed desiccator at 20 ± 2 °C for 24 h. The emitted formaldehyde was absorbed by 300 mL deionized water in a container. The formaldehyde concentration in the sample solution was determined using acetyl acetone–ammonium acetate solution and the acetyl acetone method, with colorimetric detection at 412 nm. For purposes of comparison, ten specimens of each adhesive were placed indoor for 30 days and then the formaldehyde emission was measured according to the same method described above. Finally, the formaldehyde emission decline rate of the plywood was calculated using Equation (1):

$$\text{Decline rate} = (\text{initial} - \text{treatment})/\text{original} \tag{1}$$

where initial is the formaldehyde emission of the specimens tested directly, and treatment is the formaldehyde emission of the specimens after 30 days.

2.6. Fourier-Transform Infrared (FTIR) Spectroscopy

The FTIR spectroscopy (Nicolet FTIR spectrometer series 6700), in transmittance mode, was used for the characterization of the functional groups of the resins. The spectra were obtained in the spectral area 400–4000 cm^{-1} with a resolution of 4 cm^{-1} and 32 scans using potassium bromide (KBr) disc. Discs were prepared by mixing 1 mg of cured resins powder with 70 mg of KBr. Cured resins were prepared by drying the liquid resins in a convection oven at 120 °C for 2 h and then grinding them into a powder.

2.7. Thermal Stability of Cured Resins

Samples were placed in an oven at 120 °C until reaching a constant weight and then ground into powder. The stability of the cured adhesive was tested using a TGA instrument (TA Q50, WATERS Company, New Castle, DE, USA). About 5 mg of the ground powder was placed in a platinum tray and scanned from room temperature to 600 °C at a heating rate of 10 °C/min.

2.8. Scanning Electron Microscopy (SEM)

Samples were placed into a piece of aluminum foil and dried in an oven at 120 ± 2 °C until a constant weight was achieved. The sample fracture surface was sputter-coated with gold by using an ion sputter (Hitachi E-1010 Ion Sputter, Japan), and then a Hitachi S-3400N (Hitachi Science System, Ibaraki, Tokyo, Japan) scanning electron microscope was used to observe the fracture surface of the resultant adhesives.

2.9. Water Contact Angle Tests

The liquid Resin A5 and A1 in the same amount was uniformly coated on glass slides and then placed in an oven at 120 ± 2 °C for 2 h to form a cured resin film. The surface contact angles of the resin films were measured with a water contact angle (WCA) goniometer (DataPhysics Co. Ltd. Filderstadt, Germany) (OCA-20, DataPhysics Co. Ltd. Filderstadt, Germany). The measurements were performed via the sessile droplet method under standard parameters such as room temperature and drop volume (3 µL). At least three measurements of the angles on both film sides were recorded at an interval of 0.1 s for 180 s.

3. Results and Discussion

3.1. Wet Shear Strength Measurement

The wet shear strength of the plywood bonded by UF resin with different fillers is shown in Figure 1. All the values of wet shear strength of the plywood were above 0.7 MPa (labeled as the dash line) except the one of Resin A6. WF could absorb water and swell in the UF resin, which could prevent the UF resin molecule penetrating into the wood surface. Compared with WF, SEP does not swell in the resin, so that, using 100% SEP filler led to the resin's excessive penetration into the veneer. Therefore, the wet shear strength of plywood bonded by Resin A6 decreased by 44% to 0.57 MPa compared to the one with 100% WF (A1, 1.02 MPa). The plywood prepared by UF resin A5 showed the best wet shear strength (1.34 MPa), which improved by 31.4% compared to that of Resin A1. This could be due to the micro-fibrous morphology of SEP, which could make the UF resin molecules gather around SEP during the curing process, thus leading UF resin molecule distribution to be more homogeneous and forming a higher dense structure. Another possible reason was that the SEP with fibrous morphology could form an interpenetrated network with the UF resin system, which further improved the wet shear strength of the produced plywood.

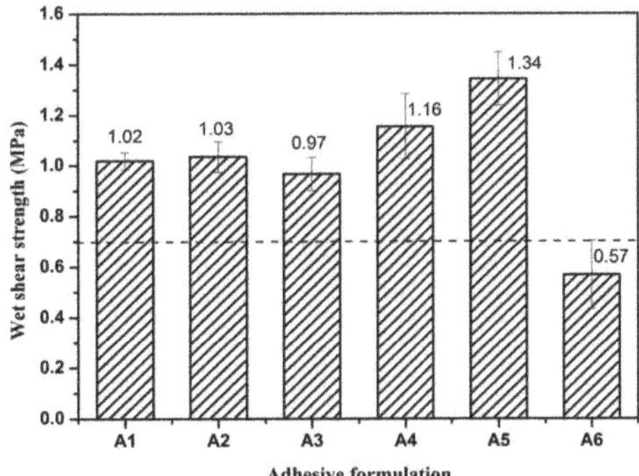

Figure 1. The wet shear strength of plywood bonded by urea formaldehyde (UF) resin with different fillers: Resin A1—0% SEP + 100% WF, Resin A2—20% SEP + 80% WF, Resin A3—40% SEP + 60% WF, Resin A4—60% SEP + 40% WF, Resin A5—80% SEP + 20% WF, Resin A6—100% SEP + 0%WF. Error bars represent standard deviations from eight replicates.

3.2. Formaldehyde Emission Measurement

In addition to the bonding strength, formaldehyde emission is another important property for the practical application of plywood bonded by UF resins. The formaldehyde emissions of plywood prepared by different UF adhesives are shown in Figure 2. The formaldehyde emission of plywood bonded by UF Resin A1 was 1.23 mg/L. When adding 20% SEP, the formaldehyde emission of plywood bonded by the developed Resin (A2) decreased by 36% to 0.79 mg/L. This was attributed to the free formaldehyde adsorption capacity of SEP. Because of the large specific surface area and special tunnel structure, the SEP could capture free formaldehyde into itself, which reduced the free formaldehyde content of the resin and formaldehyde emission of the resulting plywood. However, as the SEP addition further increased, the formaldehyde emission of plywood began to increase. For Resin A5, the formaldehyde emission increased by 22% to 1.50 mg/L compared to the plywood bonded by Resin A1 (1.23 mg/L). SEP adsorption characteristics are only a simple physical process, which indicates that the SEP adsorption capacity is limited. From the other perspective, this tunnel structure of SEP could accelerate the free formaldehyde release, which was a dominating factor for the plywood formaldehyde emission. As a result, formaldehyde emission was increased when the SEP addition exceeded 40%.

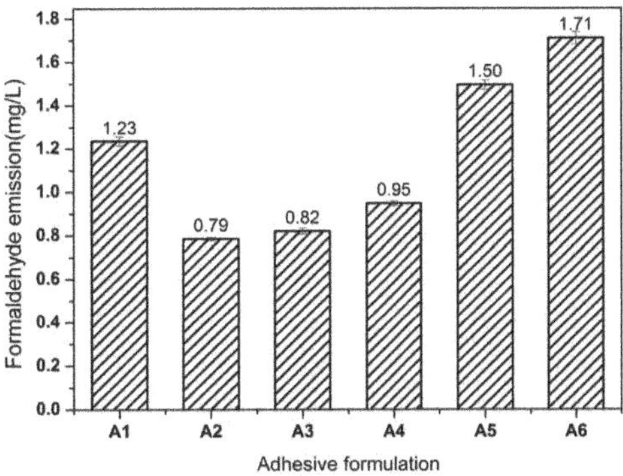

Figure 2. The formaldehyde emission of the plywood bonded by UF resin with different fillers: Resin A1—0% SEP + 100% WF, Resin A2—20% SEP + 80% WF, Resin A3—40% SEP + 60% WF, Resin A4—60% SEP + 40% WF, Resin A5—80% SEP + 20% WF, Resin A6—100% SEP + 0%WF. Error bars represent standard deviations from three replicates.

Based on the research of Ding et al. [28] and He et al. [29], the formaldehyde mainly emitted through vessels of the veneer. Therefore, the SEP tunnel structure of the inner layer is more conducive to the further release of formaldehyde of the resultant plywood over a long time. Furthermore, the formaldehyde emission of the resulting plywood specimens was retested after 30 days in this research. With the elongation of the storage time, the formaldehyde emission of the resin layer with tunnel structure should be faster than that without SEP.

The formaldehyde emission of the plywood after 30 days was evaluated and is shown in Table 2. The results showed that the declined rate of the samples increased as the SEP ratio increased in the filler, which indicated that the tunnel structure of SEP was beneficial to the releasing of formaldehyde. For UF Resin A5, the formaldehyde emission of the resulting plywood (1.50 mg/L) was 18% higher than that of UF Resin A1 (1.23 mg/L). After 30 days, the formaldehyde emission of the plywood bonded by UF Resin A1 and UF Resin A5 were 1.03 and 0.58 mg/L, reducing by 16.9% and 61.3%, respectively. Moreover, the final formaldehyde emission of plywood bonded by UF Resin A5 was 43.7% lower than that with Resin A1. In the plywood fabrication industry, the resulting plywood undergoes a hot stacking process for three to four weeks to cool down, which is beneficial to release formaldehyde, especially for the plywood bonded by UF resin with SEP. The schematic model of tunnel-structured SEP and the absorption and release method of formaldehyde are shown in Figure 3.

Table 2. The comparison of the plywood initial formaldehyde emission and the values after 30 days.

Samples	Initial Formaldehyde Emission (mg/L)	After 30 Days (mg/L)	Decline Rate (%)
A1	1.24	1.03	16.9
A2	0.79	0.65	17.7
A3	0.82	0.63	23.2
A4	0.95	0.69	27.4
A5	1.50	0.58	61.3
A6	1.71	1.44	15.8

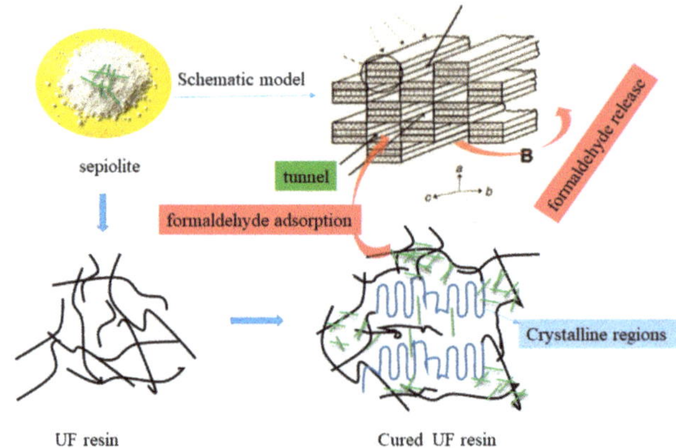

Figure 3. Schematic model of tunnel-structured SEP and the absorption and release method of formaldehyde.

3.3. FTIR Analysis

Figure 4 shows the FTIR spectra of different resins. The peak at 3349.05 cm^{-1} was assigned to the stretching vibration of N–H and O–H bonds in the primary amines and hydroxyl groups [30]. The flexing vibration of the C–H of methylene occurring at 2962.69 [31], 1601.63, and 1550.80 cm^{-1} were acylamino absorption bands; the peaks at 1388.08 and 1238.47 cm^{-1} were the deformation vibrations of the plane of methylene; the peak near 1135 cm^{-1} was the absorption band of ether linkage of

CH$_2$OCH$_2$ [32] and 1001.59 cm^{-1} was the absorption band of the methylol of CH$_2$O. Specifically, when SEP was added, the ether bond of the UF Resin A5 shifted toward a low wavenumber (from 1135.70 to 1131.32 cm^{-1}, red shift), which could be attributed to the hydrogen bonding interactions between the –OH on SEP and the ether bond of the UF resin [33].

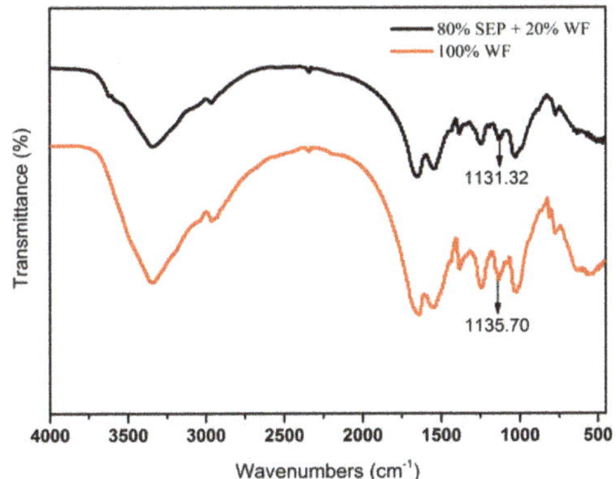

Figure 4. FTIR spectra of UF resin with different fillers: Resin A1—100% WF + 0% SEP; Resin A5—80% SEP + 20% WF.

3.4. TG and DTG Analysis

Figure 5 shows the TG and DTG curves of different resins. The thermal degradation process of UF resin could be divided into three stages. The first stage was at a temperature region of 20–120 °C, which was attributed to the evaporation of water [34]. The second stage started at 200 °C and ended at 260 °C, which could result from the loss of small molecules and the break of unstable chemical bonds, such as ether bond, methylene [35]. The mass loss at the second stage was 11.43% for Resin A5 and 19.00% for Resin A1, respectively. The third stage started at 260 °C and ended at 350 °C, indicating the resin structure degradation. The mass loss at the third stage was 55.56% for Resin A5 and 58.61% for Resin A1, respectively. The residue was 31.28% for Resin A5 and 19.82% for Resin A1, respectively.

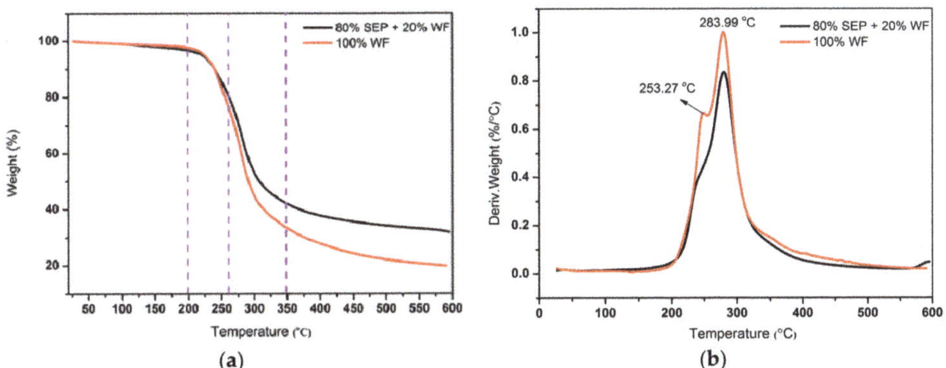

Figure 5. TG (**a**) and DTG (**b**) curves of UF resin with different fillers: Resin A1—100% WF + 0% SEP; Resin A5—80% SEP + 20% WF.

The TG curves showed that the degradation trend of the Resin A5 was similar to that of Resin A1. By contrast, the small mass loss and the higher residue amount of Resin A5 was mainly attributed to the high temperature resistance of SEP [36]. DTG results (Figure 5b) showed that Resin A5 and

A1 almost had the same maximum weight loss temperature at 283.99 °C. The TG and DTG results indicated that SEP only mildly affected the thermostability of the resultant resin.

3.5. SEM Analysis

The morphology of fillers (WF and SEP) and fracture surface micrographs of cured resins are presented in Figure 6. WF was composed of ellipsoid starch granules with a smooth surface. SEP was composed of irregular fiber-like aggregates. The fibrous morphology could increase the interface area of SEP with the polymer matrix, which would result in strong interaction and reinforced polymers. Because of the inherent brittleness of the UF resin, UF Resin A1 exhibited a neat brittle fracture characteristic (Figure 6c). On the contrary, the fracture surface of UF Resin A5 became rough and appeared wrinkled, indicating the toughness of the cured resin layer increased. The increased toughness may be due to the interpenetrated network formed by the fibrous SEP and the UF resin system. Thus, when subjected to external forces, the tough adhesive layer acted as a buffer, resulting in an uneven fracture surface with wrinkles. This also explained the reason why the wet shear strength of plywood prepared by UF Resin A5 was improved by 31.4% compared with that of UF resin without SEP.

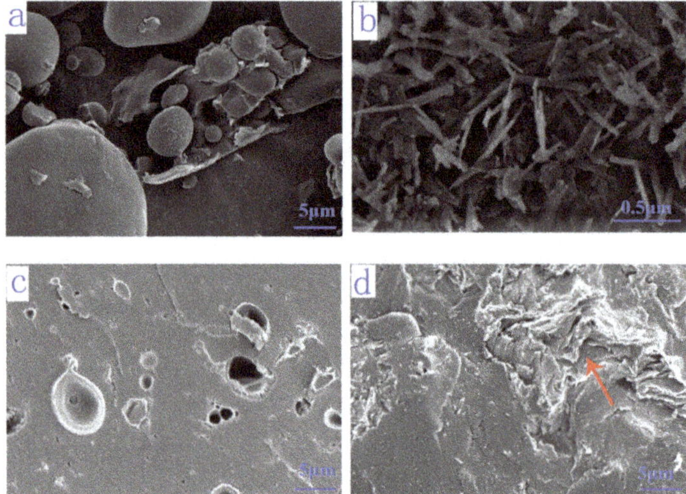

Figure 6. SEM images of WF (**a**), SEP (**b**), and cured UF resin with different fillers: Resin A1—100% WF + 0% SEP (**c**); Resin A5—80% SEP + 20% WF (**d**).

3.6. Water Contact Angle Analysis

The contact angle test results in 5 s of the cured Resin A5 and A1 film are presented in Figure 7. The water contact angles of Resin A1 and Resin A5 were 45.44° and 42.18°. Addition of SEP into UF resin decreased the contact angle, which was due to the natural hydrophilic properties of SEP [37]. From the wet shear strength analysis (Section 3.1), the plywood prepared by Resin A5 has the optimal strength, indicating that the hydrophilic property of SEP did not negatively affect the other properties of the UF resin.

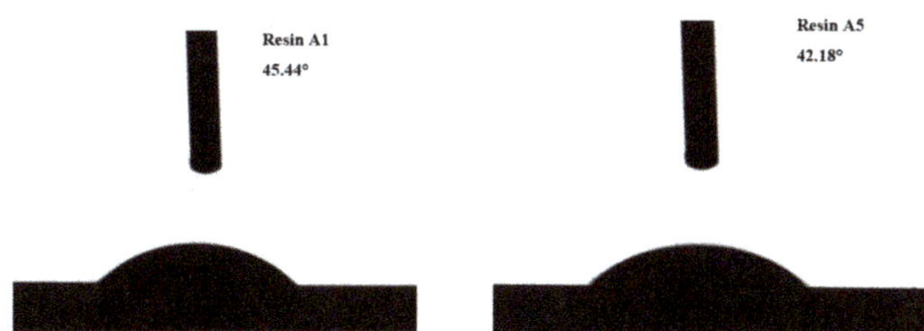

Figure 7. Water contact angles of Resin A1 and Resin A5.

4. Conclusions

When UF resin with 80% SEP + 20% WF filler (Resin A5) was used, the wet shear strength of the resulting plywood improved by 31.4%, which was attributed to the interpenetrated network formed by the fibrous SEP and the UF resin system, which resulted in a toughened fracture of the cured adhesive layer and further enhanced the bonding strength of the plywood.

Consistent with assumptions, the tunnel structure of SEP is beneficial to release formaldehyde, which was verified by the formaldehyde emission test after 30 days, i.e., the formaldehyde emission of the plywood bonded by UF Resin A5 declined by 43.7% compared with that of Resin A1.

Considering the wet shear strength, the formaldehyde emission, and cost of the plywood, the best strategy to modify the UF resin was the addition of 80% SEP + 20% WF filler.

Author Contributions: Formal analysis, X.L.; investigation, X.L.; methodology, X.L.; project administration, X.L.; supervision, Q.G., C.X., J.L. and X.Z.; writing—original draft, X.L.

Funding: This research received no external funding.

Acknowledgments: The authors are grateful for the financial support received from the China Postdoctoral Science Foundation (2018M642259) and the National Natural Science Foundation of China (31870549).

Conflicts of Interest: The authors declare no conflict of interest.

References

1. Jeong, B.; Park, B.D. Effect of molecular weight of urea-formaldehyde resins on their cure kinetics, interphase, penetration into wood, and adhesion in bonding wood. *Wood Sci. Technol.* **2019**, *53*, 665–685. [CrossRef]
2. Pizzi, A.; George, B.; Zanetti, M.; Meausoone, P.J. Rheometry of aging of colloidal melamine-urea-formaldehyde polycondensates. *J. Appl. Polym. Sci.* **2005**, *96*, 655–659. [CrossRef]
3. Jovanovic, V.; Samarzija-Jovanovic, S.; Petkovic, B.; Milicevic, Z.; Markovic, G.; Marinovic-Cincovic, M. Biocomposites based on cellulose and starch modified urea-formaldehyde resin: Hydrolytic, thermal, and radiation stability. *Polym. Compos.* **2019**, *40*, 1287–1294. [CrossRef]
4. Park, B.D.; Jeong, H.W. Hydrolytic stability and crystallinity of cured urea-formaldehyde resin adhesives with different formaldehyde/urea mole ratios. *Int. J. Adhes. Adhes.* **2011**, *31*, 524–529. [CrossRef]
5. Kim, M.G.; No, B.Y.; Lee, S.M.; Nieh, W.L. Examination of selected synthesis and room-temperature storage parameters for wood adhesive-type urea-formaldehyde resins by C-13-NMR spectroscopy. V. *J. Appl. Polym. Sci.* **2003**, *89*, 1896–1917. [CrossRef]
6. Hse, C.Y.; Fu, F.; Pan, H. Melamine-modified urea formaldehyde resin for bonding particleboards. *For. Prod. J.* **2008**, *58*, 56–61.
7. Kim, S.; Kim, H.J. Comparison of standard methods and gas chromatography method in determination of formaldehyde emission from MDF bonded with formaldehyde-based resins. *Bioresour. Technol.* **2005**, *96*, 1457–1464. [CrossRef]
8. Ayrilmis, N.; Kaymakci, A. Reduction of formaldehyde emission from light MDF panels by adding chestnut shell flour. *Holzforschung* **2012**, *66*, 443–446. [CrossRef]
9. Li, X.N.; Chen, M.S.; Zhang, J.Z.; Gao, Q.; Zhang, S.F.; Li, J.Z. Physico-Chemical Properties of Soybean Meal-Based Adhesives Reinforced by Ethylene Glycol Diglycidyl Ether and Modified Nanocrystalline Cellulose. *Polymers* **2017**, *9*, 463. [CrossRef]

10. Liu, X.R.; Wang, K.L.; Gu, W.D.; Li, F.; Li, J.Z.; Zhang, S.F. Reinforcement of interfacial and bonding strength of soybean meal-based adhesive via kenaf fiber-CaCO$_3$ anchored N-cyclohexyl-2-benzothiazole sulfenamide. *Compos. Part B Eng.* **2018**, *155*, 204–211. [CrossRef]

11. Zheng, P.T.; Chen, N.R.; Islam, S.M.M.; Ju, L.K.; Liu, J.; Zhou, J.H.; Chen, L.Y.; Zeng, H.B.; Lin, Q.J. Development of Self-Cross-Linked Soy Adhesive by Enzyme Complex from Aspergillus niger for Production of All-Biomass Composite Materials. *ACS Sustain. Chem. Eng.* **2019**, *7*, 3909–3916. [CrossRef]

12. Wu, Z.G.; Xi, X.D.; Lei, H.; Liang, J.K.; Liao, J.J.; Du, G.B. Study on Soy-Based Adhesives Enhanced by Phenol Formaldehyde Cross-Linker. *Polymers* **2019**, *11*, 365. [CrossRef]

13. Pizzi, A. Recent developments in eco-efficient bio-based adhesives for wood bonding: Opportunities and issues. *J. Adhes. Sci. Technol.* **2006**, *20*, 829–846. [CrossRef]

14. Mansouri, H.R.; Pizzi, A.; Leban, J.M. Improved water resistance of UF adhesives for plywood by small pMDI additions. *Holz Roh-Und Werkst.* **2006**, *64*, 218–220. [CrossRef]

15. Doosthoseini, K.; Zarea-Hosseinabadi, H. Using Na$^+$ MMT nanoclay as a secondary filler in plywood manufacturing. *J. Indian Acad. Wood Sci.* **2010**, *7*, 58–64. [CrossRef]

16. Zhu, X.F.; Xu, E.G.; Lin, R.H.; Wang, X.B.; Gao, Z.Z. Decreasing the Formaldehyde Emission in Urea-Formaldehyde Using Modified Starch by Strongly Acid Process. *J. Appl. Polym. Sci.* **2014**, *131*, 40202. [CrossRef]

17. Lei, H.; Du, G.; Pizzi, A.; Celzard, A.; Fang, Q. Influence of Nanoclay on Phenol-Formaldehyde and Phenol-Urea-Formaldehyde Resins for Wood Adhesives. *J. Adhes. Sci. Technol.* **2010**, *24*, 1567–1576. [CrossRef]

18. Li, Z.F.; Wang, J.; Li, C.M.; Gu, Z.B.; Cheng, L.; Hong, Y. Effects of montmorillonite addition on the performance of starch-based wood adhesive. *Carbohydr. Polym.* **2015**, *115*, 394–400. [CrossRef]

19. Kovacevic, Z.; Bischof, S.; Fan, M.Z. The influence of Spartium junceum L. fibres modified with montmorrilonite nanoclay on the thermal properties of PLA biocomposites. *Compos. Part B Eng.* **2015**, *78*, 122–130. [CrossRef]

20. Li, X.N.; Luo, J.L.; Gao, Q.; Li, J.Z. A sepiolite-based united cross-linked network in a soybean meal-based wood adhesive and its performance. *RSC Adv.* **2016**, *6*, 45158–45165. [CrossRef]

21. Li, H.Y.; Li, C.C.; Gao, Q.; Zhang, S.F.; Li, J.Z. Properties of soybean-flour-based adhesives enhanced by attapulgite and glycerol polyglycidyl ether. *Ind. Crops Prod.* **2014**, *59*, 35–40. [CrossRef]

22. Kim, J.S.; Choi, M.C.; Jeong, K.M.; Kim, G.H.; Ha, C.S. Enhanced interaction in the polyimide/sepiolite hybrid films via acid activating and polydopamine coating of sepiolite. *Polym. Adv. Technol.* **2018**, *29*, 1404–1413. [CrossRef]

23. Ruiz-Hitzky, E. Molecular access to intracrystalline tunnels of sepiolite. *J. Mater. Chem.* **2001**, *11*, 86–91. [CrossRef]

24. Falco, G.; Giulieri, F.; Volle, N.; Pagnotta, S.; Sbirrazzuoli, N.; Disdier, E.P.; Mija, A. Self-organization of sepiolite fibbers in a biobased thermoset. *Compos. Sci. Technol.* **2019**, *171*, 226–233. [CrossRef]

25. Chivrac, F.; Pollet, E.; Schmutz, M.; Averous, L. Starch nano-biocomposites based on needle-like sepiolite clays. *Carbohydr. Polym.* **2010**, *80*, 145–153. [CrossRef]

26. Zhang, J.Z.; Wang, X.M.; Zhang, S.F.; Gao, Q.; Li, J.Z. Effects of melamine addition stage on the performance and curing behavior of melamine-urea-formaldehyde (MUF) resin. *Bioresources* **2013**, *8*, 5500–5514. [CrossRef]

27. Zhang, J. Synthesis, Structure, and Properties of Modified Urea-Formaldehyde Resins Based on Multiple Copolycondensation. Ph.D. Thesis, Beijing Forestry University, Beijing, China, 2015; pp. 34–45.

28. Ding, W.B.; Li, W.Y.; Gao, Q.; Han, C.R.; Zhang, S.F.; Li, J.Z. The Effects of Sealing Treatment and Wood Species on Formaldehyde Emission of Plywood. *Bioresources* **2013**, *8*, 2568–2582. [CrossRef]

29. He, Z.K.; Zhang, Y.P.; Wei, W.J. Formaldehyde and VOC emissions at different manufacturing stages of wood-based panels. *Build. Environ.* **2012**, *47*, 197–204. [CrossRef]

30. Ye, J.; Qiu, T.; Wang, H.Q.; Guo, L.H.; Li, X.Y. Study of glycidyl ether as a new kind of modifier for urea-formaldehyde wood adhesives. *J. Appl. Polym. Sci.* **2013**, *128*, 4086–4094. [CrossRef]

31. Tong, X.M.; Zhang, M.; Wang, M.S.; Fu, Y. Effects of surface modification of self-healing poly(melamine-urea-formaldehyde) microcapsules on the properties of unsaturated polyester composites. *J. Appl. Polym. Sci.* **2013**, *127*, 3954–3961. [CrossRef]

32. Kandelbauer, A.; Despres, A.; Pizzi, A.; Taudes, I. Testing by Fourier transform infrared species variation during melamine-urea-formaldehyde resin preparation. *J. Appl. Polym. Sci.* **2007**, *106*, 2192–2197. [CrossRef]

33. Zhao, Y.T.; He, M.; Zhao, L.; Wang, S.Q.; Li, Y.P.; Gan, L.; Li, M.M.; Xu, L.; Chang, P.R.; Anderson, D.P.; et al. Epichlorohydrin-cross-linked ydroxyethyl cellulose/soy protein isolate composite films as biocompatible and biodegradable implants for tissue engineering. *ACS Appl. Mater. Interfaces* **2016**, *8*, 2781–2795. [CrossRef]
34. Roumeli, E.; Papadopoulou, E.; Pavlidou, E.; Vourlias, G.; Bikiaris, D.; Paraskevopoulos, K.M.; Chrissafis, K. Synthesis, characterization and thermal analysis of urea-formaldehyde/nanoSiO$_2$ resins. *Thermochim. Acta* **2012**, *527*, 33–39. [CrossRef]
35. Siimer, K.; Kaljuvee, T.; Pehk, T.; Lasn, I. Thermal behaviour of melamine-modified urea-formaldehyde resins. *J. Therm. Anal. Calorim.* **2010**, *99*, 755–762. [CrossRef]
36. Turhan, Y.; Doğan, M.; Alkan, M. Characterization and some properties of poly (vinyl chloride)/sepiolite nanocomposites. *Adv. Polym. Technol.* **2013**, *32*, E65–E82. [CrossRef]
37. Can, M.F.; Avdan, L.; Bedeloglu, A.C. Properties of biodegradable PVA/sepiolite-based nanocomposite fiber mats. *Polym. Compos.* **2015**, *36*, 2334–2342. [CrossRef]

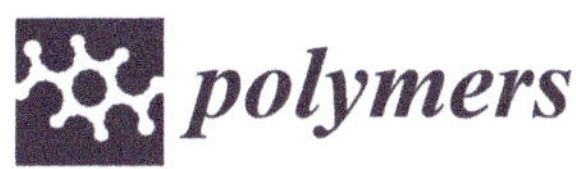 

Article

Long-Term Creep Behavior Prediction of Sol-Gel Derived SiO$_2$- and TiO$_2$-Wood Composites Using the Stepped Isostress Method

Ke-Chang Hung [1], Tung-Lin Wu [2] and Jyh-Horng Wu [1,*

[1] Department of Forestry, National Chung Hsing University, Taichung 402, Taiwan
[2] College of Technology and Master of Science in Computer Science, University of North America, Fairfax, VA 22033, USA
* Correspondence: eric@nchu.edu.tw

Received: 8 July 2019; Accepted: 19 July 2019; Published: 20 July 2019

Abstract: In this study, methyltrimethoxysilane (MTMOS), methyltriethoxysilane (MTEOS), tetraethoxysilane (TEOS), and titanium(IV) isopropoxide (TTIP) were used as precursor sols to prepare wood-inorganic composites (WICs) by a sol-gel process, and subsequently, the long-term creep behavior of these composites was estimated by application of the stepped isostress method (SSM). The results revealed that the flexural modulus of wood and WICs were in the range of 9.8–10.5 GPa, and there were no significant differences among them. However, the flexural strength of the WICs (93–103 MPa) was stronger than that of wood (86 MPa). Additionally, based on the SSM processes, smooth master curves were obtained from different SSM testing parameters, and they fit well with the experimental data. These results demonstrated that the SSM was a useful approach to evaluate the long-term creep behavior of wood and WICs. According to the Eyring equation, the activation volume of the WICs prepared from MTMOS (0.825 nm^3) and TEOS (0.657 nm^3) was less than that of the untreated wood (0.832 nm^3). Furthermore, the WICs exhibited better performance on the creep resistance than that of wood, except for the WIC$_{MTEOS}$. The reduction of time-dependent modulus for the WIC prepared from MTMOS was 26% at 50 years, which is the least among all WICs tested. These findings clearly indicate that treatment with suitable metal alkoxides could improve the creep resistance of wood.

Keywords: activation volume; creep behavior; sol-gel process; stepped isostress method; wood-inorganic composites

---

1. Introduction

Wood and wood-based composites have been widely used in everyday human lives. However, those materials have some disadvantage properties (e.g., dimensional instability, susceptibility to biological degradation, and flammability), thus limiting their exterior application and long-term utilization [1,2]. Over the past few decades, various wood modifications, including heat treatment, esterification, and inorganic modification by sol-gel technology, have been employed to improve their properties and to enhance their quality [3–7]. Among these, the sol-gel derived wood-inorganic composite (WIC) approaches have received a lot of attention over the last few years [8–12]. These WICs have been proven to be effective in improving the flame retardancy, thermal stability, UV stability, and fungal resistance compared to wood [10,13–17]. However, the creep property of WICs was rarely investigated.

Creep is one of the fundamental properties of materials limiting their long-term application as excessive deformation or reduced stiffness occurs over an extended period of time [18]. Therefore, the evaluation of creep behavior is essential in engineering applications. At realistic service-life durations, conventional creep tests are time-consuming [19]. To reduce the duration of a creep test, an accelerated creep test is required to obtain a master curve that is based on the superposition principle, consisting of the duration time, exposure temperature, and applied stress. Developed from the time-temperature superposition principle (TTSP), the stepped isothermal method (SIM) can be used in the same manner as time-equivalence for a single sample to predict the long-term creep property of viscoelastic materials from stepped increments of temperature [20–23]. In the past few years, the stepped isostress method (SSM), which can evaluate the creep behavior of a single sample by a stepped increase of the stress level [24–27], was shown to be more beneficial in assessing the creep behavior of low-thermal-conductivity materials (e.g., wood and wood composites) compared to the SIM [28]. A previous study demonstrated that wood treated with methyltrimethoxysilane (MTMOS) could effectively enhance its creep resistance [29]. However, to the best of our knowledge, the effect of different metal alkoxides on the creep behavior of WICs has not been studied using the SSM. Therefore, in the present study, MTMOS, methyltriethoxysilane (MTEOS), tetraethoxysilane (TEOS), and titanium(IV) isopropoxide (TTIP) were used as precursor sols to prepare various sol-gel derived WICs, and the long-term creep behavior of all the WICs was predicted using the SSM.

2. Experimental

2.1. Materials

Japanese cedar (*Cryptomeria japonica* (L. f.) D. Don) sapwood (20–30 years old) supplied by the experimental forest of the National Taiwan University was used in this study. The MTMOS, MTEOS, TEOS, and TTIP were purchased from Acros Chemical (Geel, Belgium). All other chemicals and solvents used were of the highest quality available.

2.2. Preparation of Wood-Inorganic Composites

The oven-dried wood specimens, with dimensions of 3 mm (R) × 12 mm (T) × 58 mm (L), selected for this study were free of defects and exhibited a modulus of elasticity (MOE) of approximately 10.0 GPa to reduce material variability. Before the investigation, the samples were Soxhlet-extracted using a 1:2 (*v/v*) mixture of ethanol and toluene for 24 h and then washed with distilled water. The extracted wood samples were placed in an oven at 105 °C for 12 h, and their masses recorded. The WIC_{TTIP} was made directly from the oven-dried wood, while the wood used to make WIC_{MTMOS}, WIC_{MTEOS}, and WIC_{TEOS} was conditioned at 20 °C/65% RH for one week before preparation. The precursor sol was formulated with the desired metal alkoxide (i.e., MTMOS, MTEOS, TEOS, or TTIP), solvent (methanol or 2-propanol), and acetic acid at a molar ratio of 0.12/1/0.08 for preparing the WICs. The oven-dried or conditioned specimens were impregnated with the precursor sol for three days under reduced pressure. The impregnated specimens were then aged at 50 °C for 24 h and 105 °C for another 24 h [16,30]. The oven-dried weights of WICs were recorded to determine the weight percent gain (WPG).

2.3. Determination of Composite Properties

The density and flexural tests were carried out according to the ASTM [31,32] standards, respectively. The three-point static bending test with a loading rate of 1.28 mm/min and a span of 48 mm was used to determine the modulus of rupture (MOR) and the MOE of the specimens. Five specimens of each WIC were tested. All specimens were conditioned at 20 °C/65% RH for two weeks prior to testing.

2.4. Accelerated and Experimental Creep Tests

The universal testing machine (Shimadzu AG-10kNX, Tokyo, Japan) was used to implement the short-term SSM of wood and WICs for assessing the extended creep behavior. The creep strain at a reference stress is provided by Equation (1) based on the SSM:

$$\varepsilon(\sigma_r, t) = \varepsilon(\sigma, t/\alpha_\sigma) \tag{1}$$

where ε is the creep strain as a function of stress and time, σ_r is the reference stress, σ is the elevated stress, and α_σ is the shift factor. The SSM creep tests were conducted at isostresses between 30 and 80% of the average breaking load (ABL). Additionally, various SSM testing parameters were carried out to investigate the differences between the SSM creep tests. The stepped stress increments were 5%, 7.5%, 10%, and 12.5% ABL, and the dwell times were 2, 3, or 5 h.

All SSM tests were performed at 20 °C, which is below the glass transition temperature (T_g), thus, the Eyring model (Equation (2)) is used to calculate the activation volume in this study [29]. This model was applied to estimate the shift factor (α_σ), which shows the following express rate with the stress level [25,28]:

$$\log \alpha_\sigma = \log(\frac{\dot{\varepsilon}}{\dot{\varepsilon}_r}) = \frac{V^*}{2.303kT}(\sigma - \sigma_{ref}) \tag{2}$$

where $\dot{\varepsilon}$ and $\dot{\varepsilon}_r$ are respectively the creep rate at the elevated stress (σ) and reference stress (σ_{ref}), V^* is the activation volume, k is Boltzmann's constant (1.38×10^{-23} J/K), and T is the absolute temperature.

On the other hand, a full-scale experimental creep test was performed as a basis for comparison to validate the master curves derived from the accelerated creep tests. Three specimens of each WIC were tested for creep at an applied stress of 30% ABL, and a linear variable differential transducer (LVDT) was used to measure and record the mid-span deflection values of the samples for a period of 120 days. All the samples during the experimental creep tests were held at 20 °C/65% RH.

2.5. Analysis of Variance

The software Statistical Package for the Social Science (SPSS 12.0) (Chicago, IL, USA) for the Windows program was used to perform statistical analysis. The significance of difference was calculated by Scheffe's test, and p values < 0.05 were considered to be significant.

3. Results and Discussion

3.1. Flexural Properties of Wood-Inorganic Composites

The density and flexural properties of wood and WICs with WPG of 20% are shown in Table 1. The density of all five specimens ranged from 426 to 535 kg/m^3. In addition, the MOE of all the WICs is in the range of 9.8–10.5 GPa, which is not significantly different from the untreated wood (10.0 GPa). According to Pasquini et al. [33], the stiffness of wood is found to depend mainly on the crystallinity of the cellulose. Therefore, the influence of sol-gel treatment on the crystallinity of the wood was the limit. However, the MOR of all the WICs increased to 93–103 MPa and is significantly higher than that of the wood (86 MPa) for all WICs except for the WIC$_{MTEOS}$. The possible reason for the increase in MOR is the reaction of the inorganic compound with the wood via the sol-gel process to deposit or coat the cell wall, cell lumen, and intercellular space of the wood. However, further comparing the specific flexural properties in Table 1, the results showed that the sMOR, of all WICs, was not significantly different from that of untreated wood, while the sMOE of WIC$_{TEOS}$ and WIC$_{TTIP}$ was lower than that of untreated wood.

Table 1. Density and flexural properties of wood and wood-inorganic composites.

Specimen	WPG (%)	Density (kg/m^3)	Flexural Properties		Specific Flexural Properties	
			MOE (GPa)	MOR (MPa)	sMOE (GPa)	sMOR (MPa)
Wood	-	426 ± 35^b	10.0 ± 1.1^a	86 ± 3^b	24.6 ± 2.6^a	205 ± 7^{ab}
WIC$_{MTMOS}$	19.7 ± 0.7^a	453 ± 18^b	10.5 ± 1.7^a	103 ± 12^a	23.1 ± 2.5^{ab}	226 ± 26^a
WIC$_{MTEOS}$	19.9 ± 1.1^a	482 ± 29^{ab}	10.1 ± 0.4^a	93 ± 3^{ab}	20.9 ± 0.8^{ab}	193 ± 6^{ab}
WIC$_{TEOS}$	21.0 ± 1.5^a	521 ± 44^a	9.8 ± 0.4^a	103 ± 6^a	18.8 ± 0.7^b	198 ± 12^{ab}
WIC$_{TTIP}$	20.0 ± 1.0^a	535 ± 12^a	10.1 ± 0.4^a	101 ± 4^a	18.9 ± 0.7^b	188 ± 7^b

Values are the means ± SD ($n = 5$). Different superscript letters (a and b) within a column indicate significant difference at $p < 0.05$.

3.2. Accelerated Creep via SSM

Generally, the SSM requires the following four adjustment steps to produce the master curve: vertical shifting, rescaling, eliminating, and horizontal shifting. In this section, for ease of understanding, the SSM creep curve of WIC$_{MTEOS}$ at reference stress of 30% ABL with a 5% stepwise jump stress and a 3 h dwelling time was chosen as an example to outline the SSM and experimental creep tests. Figure 1A shows that an immediate strain jump between each load step was clearly observed in the SSM creep curve. However, there was no creep strain at each jump since the composites are elastic under instantaneous strain. Based on this, a vertical shifting is required to eliminate the elastic component in the recorded strain. This shifting links the end of the current loading curve to the start of the next loading curve at each load step, resulting in the continuous creep strain curve as shown in Figure 1B. The second step rescaling accounts for the deformation and damage from the stress and strain history of previous steps, and it was conducted by the modified method of Yeo and Hsuan [34]. As shown in Figure 1C, a series of independent creep curves were shifted along the logarithmic time axis to the reference stress level (30% ABL). Subsequently, the time before the onset time at the primary creep region was eliminated from each curve (Figure 1D), which is influenced by the history of the creep strain and the stress level. As a result of rescaling and eliminating, the master curve construction should be horizontally shifted along the time axis of the individual creep curves according to the shift factor $\log(\alpha_\sigma)$, where the magnitude of this shift factor is a function of the stress level. After this adjustment step, the final smooth master curve of WIC$_{MTEOS}$ was produced (Figure 1E). Figure 1F shows that the SSM-fitted curve closely matches with the experimental data. The effects of the test parameters (stress increment and dwelling time) on the SSM master curves for WIC$_{MTEOS}$ are presented in Figure 2. Clearly, the test conditions did not affect the master curve for a given WIC$_{MTEOS}$ sample (Figure 2A), and these curves showed a high correlation to the long-term experimental creep behavior (Figure 2B). Similarly, the master curves of all the WICs had a similar trend with their long-term experimental creep data (Figure 3), and each individual master curve (SSM test result) could be used to predict its long-term creep behavior. Accordingly, SSM is a useful method for evaluating and comparing the creep behavior of newly developed materials.

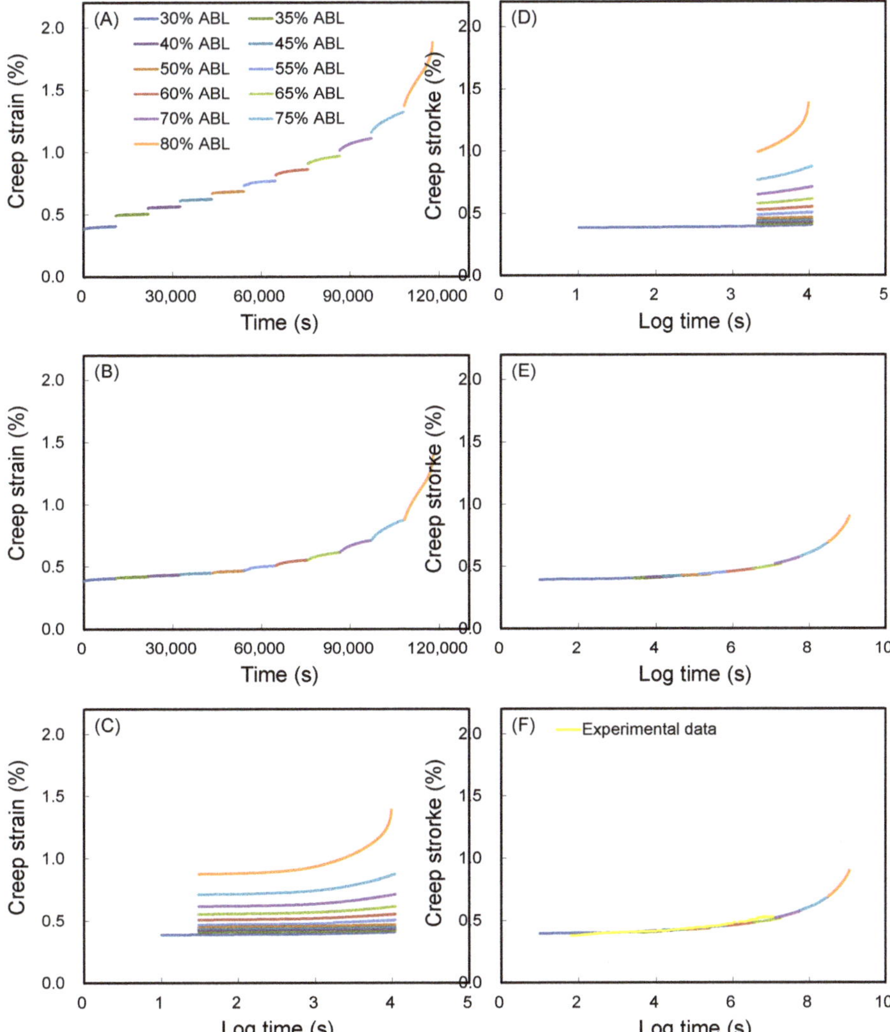

Figure 1. (**A**) The SSM creep data of the WIC$_{MTEOS}$ (reference stress: 30% ABL; interval stress: 5% ABL; dwelling time: 3 h). The handing of the SSM test data for WIC$_{MTEOS}$: (**B**) vertical shifting, (**C**) rescaled creep curves, (**D**) eliminating the period before the onset time of each stress step, and (**E**) horizontal shifting. (**F**) Experimental data and master curve.

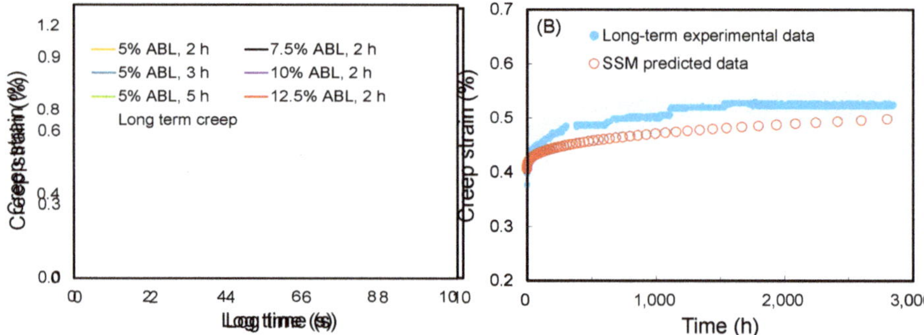

Figure 2. (**A**) Master curves of WIC$_{MTEOS}$ from different SSM testing parameters at a logarithmic time scale. (**B**) SSM-predicted creep curve and experimental creep data of WIC$_{MTEOS}$ at a normal time scale. Experimental creep data are displayed as the mean (blue line) ± SD (light blue ribbon) (n = 3).

Additionally, a linear regression was used to determine the slope of the plot of the shift factor versus the stress level. Figure 4 shows the relationship between the shift factor and stress level, as validated by the values of the coefficient of determination (R^2) being > 0.90. This result revealed that the superposition method used in the SSM approach was a valid approach to produce the creep master curve and that the same creep mechanism was active for each SSM test with different test parameters. The activation volume (V^*) was calculated according to Eyring theory as Equation (2). The V^* of all the WICs was in a range of 0.657–0.948 nm^3. Of these, the WIC$_{MTMOS}$ (0.825 nm^3) and WIC$_{TEOS}$ (0.657 nm^3) were lower than that of the untreated wood (0.832 nm^3).

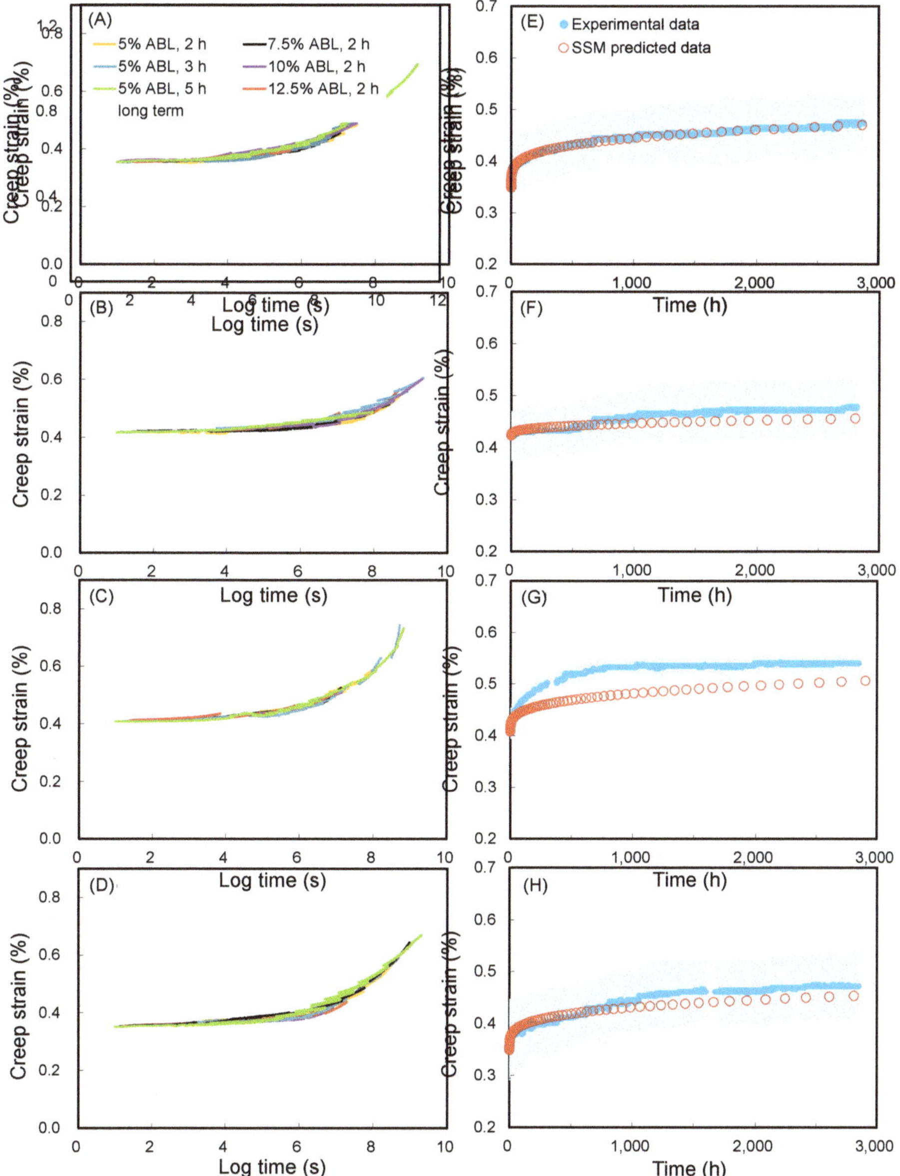

Figure 3. Master curves of (**A**) wood, (**B**) WIC$_{MTMOS}$, (**C**) WIC$_{TEOS}$, and (**D**) WIC$_{TTIP}$, using different SSM testing parameters at a logarithmic time scale. The SSM-predicted creep curve and experimental creep data of (**E**) wood, (**F**) WIC$_{MTMOS}$, (**G**) WIC$_{TEOS}$, and (**H**) WIC$_{TTIP}$ at a normal time scale. Experimental creep data are displayed as the mean (blue line) ± SD (light blue ribbon) ($n = 3$).

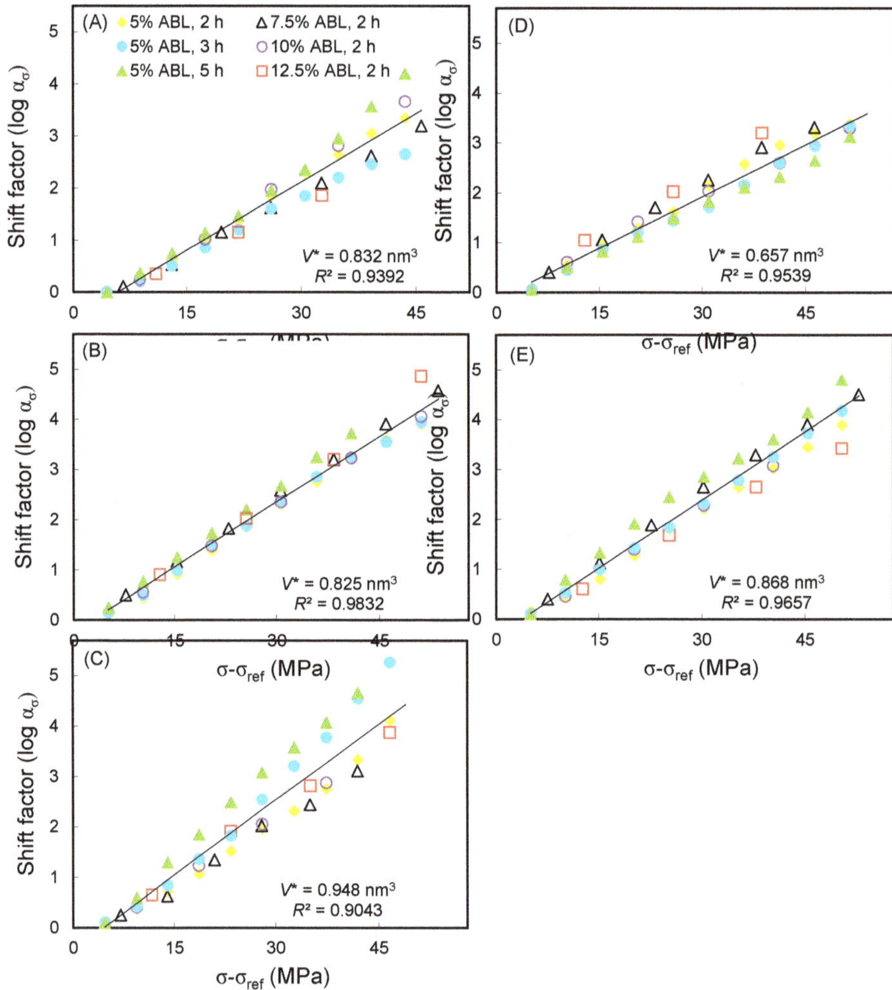

Figure 4. Typical Eyring plots of (**A**) wood, and WICs prepared from (**B**) MTMOS, (**C**) MTEOS, (**D**) TEOS, and (**E**) TTIP at a reference load of 30% ABL.

3.3. SSM-Predicted Creep Curves

The SSM-predicted compliance master curves of wood and WICs prepared from MTMOS, MTEOS, TEOS, and TTIP are presented in Figure 5. These results show that all the WICs had a lower creep compliance during the creep duration, except for WIC$_{MTEOS}$. In addition, the creep master curves were fit to a Findley power law equation with three parameters, which is described by the following Equation (3):

$$S(t) = S_0 + at^b \tag{3}$$

where $S(t)$ is the time-dependent compliance value, S_0 is the instantaneous elastic compliance value, a and b are constant values, and t is the elapsed time. The fitted parameters of the Findley power law model are shown in Table 2. It was seen that the model fits the SSM master curves of the WICs very well, all giving R^2 values of greater than 0.99.

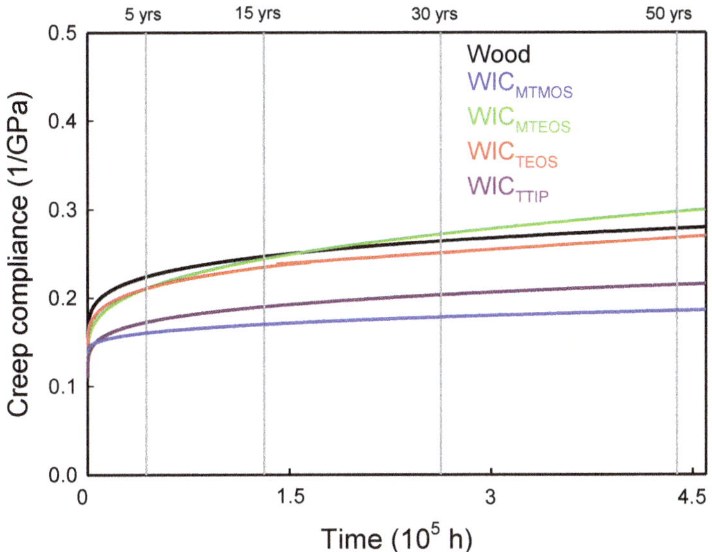

Figure 5. SSM-predicted creep data of wood and WICs prepared from MTMOS, MTEOS, TEOS, and TTIP.

Table 2. SSM-predicted creep compliances of wood and wood-inorganic composites.

Specimen	S_0 (GPa^{-1})	a	b	R^2	$S(t)$ (GPa^{-1}) Time (Years)				Modulus Reduction (%) Time (Years)			
					5	15	30	50	5	15	30	50
Wood	0.134	0.0097	0.208	0.9982	0.22	0.25	0.26	0.28	40	45	49	52
WIC_MTMOS	0.138	0.0008	0.317	0.9932	0.16	0.17	0.18	0.19	15	20	23	26
WIC_MTEOS	0.134	0.0023	0.328	0.9911	0.21	0.24	0.27	0.30	36	45	51	55
WIC_TEOS	0.142	0.0025	0.264	0.9925	0.21	0.24	0.25	0.27	34	41	45	49
WIC_TTIP	0.109	0.0061	0.220	0.9942	0.17	0.19	0.20	0.21	37	43	46	49

$S(t) = S_0 + at^b$, where $S(t)$ is the time-dependent compliance value, S_0 is the instantaneous elastic compliance value, and a and b are constant values.

The instantaneous elastic compliances (S_0) and the predicted time-dependent compliances ($S(t)$) of untreated wood and all WICs over the 5–50-year periods are listed in Table 2. The WIC_TTIP has the lowest S_0 value (0.109 GPa^{-1}), while the S_0 value of untreated wood and all other WICs are in the range of 0.134–0.142 GPa^{-1}. For the predicted compliance, the compliance values of the WICs were less than that of wood over a 50-year period, except for the WIC_MTEOS. Among all WICs, WIC_MTMOS exhibited the lowest compliance values of 0.16, 0.17, 0.18, and 0.19 GPa^{-1} at 5, 15, 30, and 50 years, respectively. Miyafuji and Saka [35] pointed out that SiO_2 with an even distribution in the cell walls was more effective in improving wood properties. Therefore, a possible reason for the above phenomenon is that the bulking effect of the MTMOS was higher than that of the other three precursor sols. In other words, during the sol-gel process, the MTMOS is mostly deposited in cell walls of wood rather than in cell lumens, which also results in a lower density of WIC_MTMOS under the same WPG as shown in Table 1.

Furthermore, the modulus reduction was calculated to evaluate the creep resistance of a sample under long-term conditions, and is given by Equation (4):

$$\text{Modulus reduction (\%)} = \left[1 - \frac{S_0}{S(t)}\right] \times 100 \tag{4}$$

As listed in Table 2, the modulus of the untreated wood would decrease by 52% over 50 years. However, the modulus reduction of WICs decreased in a range of 26–49% over a 50-year period, except for the WIC$_{MTEOS}$ (55%). Of these, the least modulus reduction was found for WIC$_{MTMOS}$ (26%). Accordingly, these results demonstrated that the creep resistance of the wood would be improved with MTMOS, TEOS, and TTIP treatment.

3.4. Accelerated Creep of Timber via SSM

To further understand and validate the suitability of the SSM approach for timber, the extended creep behavior was estimated from the Japanese cedar timber samples with dimensions of 12 mm (R) × 50 mm (T) × 230 mm (L) using the SSM method at a reference stress of 30% ABL. As shown in Figure 6, the test parameters did not affect the master curve for a given timber sample. The SSM-predicted creep curves match very well with the experimental creep behavior. The difference between the creep curve predicted by the SSM and the experimental creep of timber is very small and similar to the wood samples described above. Accordingly, the SSM can undoubtedly be applied for constructing the master curve of timber.

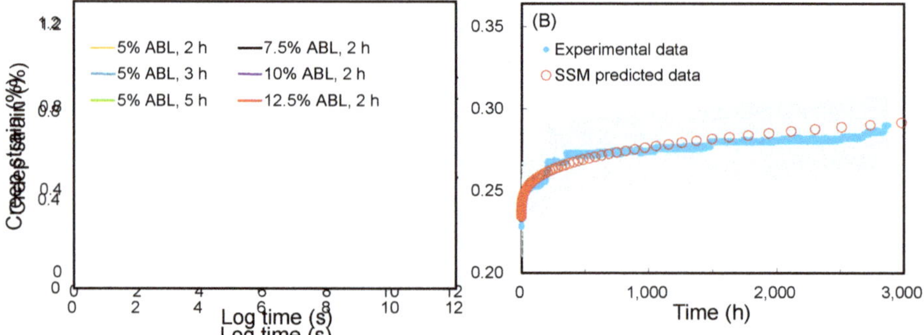

Figure 6. (**A**) Master curves of timber from different SSM testing parameters at a logarithmic time scale. (**B**) SSM-predicted creep curve and experimental creep data of timber at a normal time scale. Experimental creep data are displayed as the mean (blue line) ± SD (light blue ribbon) ($n = 3$).

4. Conclusions

The extended creep behavior of various WICs, including WIC$_{MTMOS}$, WIC$_{MTEOS}$, WIC$_{TEOS}$, and WIC$_{TTIP}$, were estimated using the stepped isostress method (SSM). The results of the SSM showed that it was suitably applied for constructing the master curves of wood and WICs. Accordingly, the WICs had lower creep compliance than the untreated wood during the creep duration, except for the WIC$_{MTEOS}$. The reduction in time-dependent modulus of the untreated wood was 52% at 50 years, but after MTMOS, TEOS, and TTIP treatment, the reduction value decreased. Among WICs tested, WIC$_{MTMOS}$ showed the least modulus reduction (26%) over a 50-year period. The activation volumes were 0.825, 0.948, 0.657, and 0.868 nm^3 for WIC$_{MTMOS}$, WIC$_{MTEOS}$, WIC$_{TEOS}$, and WIC$_{TTIP}$, respectively. Overall, this study provided a reliable method for predicting the extended creep behavior of wood and various WICs.

Author Contributions: K.-C.H. and J.-H.W. conceived and designed the experiments; K.-C.H., T.-L.W. and J.-H.W. performed the experiments; K.-C.H. and J.-H.W. analyzed the data; J.-H.W. contributed reagents/materials/analysis tools; K.-C.H. and J.-H.W. wrote the paper.

Funding: This work was financially supported by a research grant from the Ministry of Science and Technology, Taiwan (MOST 105-2622-B-005-002-CC3).

Conflicts of Interest: The authors declare no conflicts of interest.

References

1. Saka, S.; Ueno, T. Several SiO_2 wood-inorganic composites and their fire-resisting properties. *Wood Sci. Technol.* **1997**, *31*, 457–466. [CrossRef]
2. Li, Y.F.; Dong, X.Y.; Liu, Y.X.; Li, J.; Wang, F.H. Improvement of decay resistance of wood via combination treatment on wood cell wall: Swell-bonding with maleic anhydride and graft copolymerization with glycidyl methacrylate and methyl methacrylate. *Int. Biodeterior. Biodegrad.* **2011**, *67*, 1087–1094. [CrossRef]
3. Hill, C.A.S. *Wood Modification: Chemical, Thermal and Other Process*; John Wiley & Sons: Chichester, UK, 2006.
4. Gao, J.; Kim, J.S.; Terziev, N.; Daniel, G. Decay resistance of softwoods and hardwoods thermally modified by the Thermovouto type thermos-vacuum process to brown rot and white rot fungi. *Holzforschung* **2016**, *70*, 877–884. [CrossRef]
5. Moghaddam, M.; Walinder, M.E.P.; Claesson, P.M.; Swerin, A. Wettability and swelling of acetylated and furfurylated wood analyzed by multicycle Wilhelmy plate method. *Holzforschung* **2016**, *70*, 69–77. [CrossRef]
6. Zhang, X.; Lei, H.; Zhu, L.; Zhu, X.; Qian, M.; Yadavalli, G.; Wu, J.; Chen, S. Thermal behavior and kinetic study for catalytic co-pyrolysis of biomass with plastics. *Bioresour. Technol.* **2016**, *220*, 233–238. [CrossRef] [PubMed]
7. Beck, G.; Strohbusch, S.; Larnøy, E.; Militz, H.; Hill, C. Accessibility of hydroxyl groups in anhydride modified wood as measured by deuterium exchange and saponification. *Holzforschung* **2017**, *72*, 17–23. [CrossRef]
8. Shabir Mahr, M.; Hübert, T.; Schartel, B.; Bahr, H.; Sabel, M.; Militz, H. Fire retardancy effects in single and double layered sol-gel derived TiO_2 and SiO_2-wood composites. *J. Sol-Gel Sci. Technol.* **2012**, *64*, 452–464. [CrossRef]
9. Shabir Mahr, M.; Hübert, T.; Stephan, I.; Bücker, M.; Militz, H. Reducing copper leaching from treated wood by sol-gel derived TiO_2 and SiO_2 depositions. *Holzforschung* **2013**, *67*, 429–435. [CrossRef]
10. Wang, X.; Liu, J.; Chai, Y. Thermal, mechanical, and moisture absorption properties of wood-TiO_2 composites prepared by a sol-gel process. *Bioresources* **2012**, *7*, 893–901. [CrossRef]
11. Pries, M.; Mai, C. Fire resistance of wood treated with a cationic silica sol. *Eur. J. Wood Prod.* **2013**, *71*, 237–244. [CrossRef]
12. Gholamiyan, H.; Tarmian, A.; Ranjbar, Z.; Abdulkhani, A.; Azadfallah, M.; Mai, C. Silane nanofilm formation by sol-gel processes for promoting adhesion of waterborne and solventborne coatings to wood surface. *Holzforschung* **2016**, *70*, 429–437. [CrossRef]
13. Kartal, S.N.; Yoshimura, T.; Imamura, Y. Decay and termite resistance of boron-treated and chemically modified wood by in situ co-polymerization of allyl glycidyl ether (AGE) with methyl methacrylate (MMA). *Int. Biodeterior. Biodegrad.* **2004**, *53*, 111–117. [CrossRef]
14. Tshabalala, M.A.; Libert, R.; Schaller, C.M. Photostability and moisture uptake properties of wood veneers coated with a combination of thin sol-gel films and light stabilizers. *Holzforschung* **2011**, *65*, 215–220. [CrossRef]
15. Qin, C.; Zang, W. Antibacterial properties of titanium alkoxide/poplar wood composite prepared by sol-gel process. *Mater. Lett.* **2012**, *89*, 101–103. [CrossRef]
16. Hung, K.C.; Wu, J.H. Characteristics and thermal decomposition kinetics of wood-SiO_2 composites derived by the sol-gel process. *Holzforschung* **2017**, *71*, 233–240. [CrossRef]
17. Hung, K.C.; Wu, J.H. Comparison of physical and thermal properties of various wood-inorganic composites (WICs) derived by the sol-gel process. *Holzforschung* **2018**, *72*, 379–386. [CrossRef]
18. Tajvidi, M.; Simon, L.C. High-temperature creep behavior of wheat straw isotactic/impact-modified polypropylene composites. *J. Thermoplast. Compos. Mater.* **2015**, *28*, 1406–1422. [CrossRef]
19. Tanks, J.D.; Rader, K.E.; Sharp, S.R. Accelerated creep and creep-rupture testing of transverse unidirectional carbon/epoxy lamina based on the stepped isostress method. *Compos. Struct.* **2017**, *159*, 455–462. [CrossRef]
20. Jones, C.J.F.P.; Clarke, D. The residual strength of geosynthetic reinforcement subjected to accelerated creep testing and simulated seismic events. *Geotext. Geomembr.* **2007**, *25*, 155–169. [CrossRef]
21. Alwis, K.G.N.C.; Burgoyne, C.J. Accelerated creep testing for aramid fibres using the stepped isothermal method. *J. Mater. Sci.* **2008**, *43*, 4789–4800. [CrossRef]
22. Yeo, S.S.; Hsuan, Y.G. Evaluation of creep behavior of high density polyethylene and polyethylene-terephthalate geogrids. *Geotext. Geomembr.* **2010**, *28*, 409–421. [CrossRef]

23. Achereiner, F.; Engelsing, K.; Bastian, M.; Heidemeyer, P. Accelerated creep testing of polymers using the stepped isothermal method. *Polym. Test.* **2013**, *32*, 447–454. [CrossRef]

24. Hadid, M.; Rechak, S.; Tati, A. Long-term bending creep behavior prediction of injection molded composite using stress-time correspondence principle. *Mater. Sci. Eng. A Struct.* **2004**, *385*, 54–58. [CrossRef]

25. Giannopoulos, I.P.; Burgoyne, C.J. Prediction of the long-term behaviour of high modulus fibres using the stepped isostress method (SSM). *J. Mater. Sci.* **2011**, *46*, 7660–7671. [CrossRef]

26. Giannopoulos, I.P.; Burgoyne, C.J. Accelerated and real-time creep and creep-rupture results for aramid fibers. *J. Appl. Polym. Sci.* **2012**, *125*, 3856–3870. [CrossRef]

27. Huang, C.W.; Yang, T.C.; Wu, T.L.; Hung, K.C.; Wu, J.H. Effects of maleated polypropylene content on the extended creep behavior of wood polypropylene composites using the stepped isothermal method and the stepped isostress method. *Wood Sci. Technol.* **2018**, *52*, 1313–1330. [CrossRef]

28. Hadid, M.; Guerira, B.; Bahri, M.; Zouani, A. Assessment of the stepped isostress method in the prediction of long term creep of thermoplastics. *Polym. Test.* **2014**, *34*, 113–119. [CrossRef]

29. Hung, K.C.; Wu, J.H. Effect of SiO_2 content on the extended creep behavior of SiO_2-based wood-inorganic composites derived via the sol-gel process using the stepped isostress method. *Polymers* **2018**, *10*, 409. [CrossRef] [PubMed]

30. Miyafuji, H.; Kokaji, H.; Saka, S. Photostable wood-inorganic composites prepared by the sol-gel process with UV absorbent. *J. Wood Sci.* **2004**, *50*, 130–135. [CrossRef]

31. ASTM. *Standard Test Methods for Evaluating Properties of Wood-Based Fiber and Particle Panel Materials*; ASTM D1037-06a; ASTM International: West Conshohocken, PA, USA, 2006.

32. ASTM. *Standard Test Methods for Flexural Properties of Unreinforced and Reinforced Plastics and Electrical Insulating Materials*; ASTM D790-09; ASTM International: West Conshohocken, PA, USA, 2009.

33. Pasquini, D.; Teixeira, E.M.; Curvelo, A.A.S.; Belgacem, M.N.; Dufresne, A. Surface esterifi cation of cellulose fibres: Processing and characterisation of low-density polyethylene/cellulose fibres composites. *Compos. Sci. Technol.* **2008**, *68*, 193–201. [CrossRef]

34. Yeo, S.S.; Hsuan, Y.G. *Service Life Prediction of Polymeric Materials: Global Perspectives*; Springer: New York, NY, USA, 2009.

35. Miyafuji, H.; Saka, S. Topochemistry of SiO_2 wood-inorganic composites for enhancing water-repellency. *Mater. Sci. Res. Int.* **1999**, *5*, 270–275. [CrossRef]

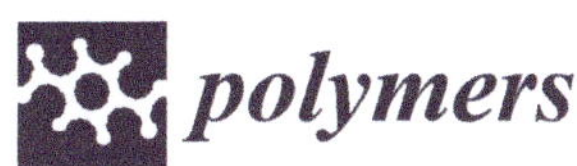 

Article

Environmentally-Friendly High-Density Polyethylene-Bonded Plywood Panels

Pavlo Bekhta [1,*] and Ján Sedliačik [2]

[1] Department of Wood-Based Composites, Cellulose, and Paper, Ukrainian National Forestry University, 79057 Lviv, Ukraine
[2] Department of Furniture and Wood Products, Technical University in Zvolen, 960 53 Zvolen, Slovakia
* Correspondence: bekhta@nltu.edu.ua; Tel.: +38-032-2384499

Received: 11 June 2019; Accepted: 3 July 2019; Published: 8 July 2019

Abstract: Thermoplastic films exhibit good potential to be used as adhesives for the production of veneer-based composites. This work presents the first effort to develop and evaluate composites based on alder veneers and high-density polyethylene (HDPE) film. The effects of hot-pressing temperature (140, 160, and 180 °C), hot-pressing pressure (0.8, 1.2, and 1.6 MPa), hot-pressing time (1, 2, 3, and 5 min), and type of adhesives on the physical and mechanical properties of alder plywood panels were investigated. The effects of these variables on the core-layer temperature during the hot pressing of multiplywood panels using various adhesives were also studied. Three types of adhesives were used: urea–formaldehyde (UF), phenol–formaldehyde (PF), and HDPE film. UF and PF adhesives were used for the comparison. The findings of this work indicate that formaldehyde-free HDPE film adhesive gave values of mechanical properties of alder plywood panels that are comparable to those obtained with traditional UF and PF adhesives, even though the adhesive dosage and pressing pressure were lower than when UF and PF adhesives were used. The obtained bonding strength values of HDPE-bonded alder plywood panels ranged from 0.74 to 2.38 MPa and met the European Standard EN 314-2 for Class 1 plywood. The optimum conditions for the bonding of HDPE plywood were 160 °C, 0.8 MPa, and 3 min.

Keywords: alder plywood; high-density polyethylene film; bending strength; modulus of elasticity in bending; shear strength; thickness swelling; water absorption

---

1. Introduction

The wood-based composites sector plays an important role in national economies in many countries. Plywood is widely used for different applications, such as construction, furniture manufacturing, means of transportation, packaging, decorative purposes, and many others. In comparison with conventional solid wood products, plywood has various advantages: increased dimensional stability, uniformity and higher mechanical strength, reduced processing cost, availability in larger sizes, better appearance, and biological benefits. On the other hand, one of the main disadvantages of plywood products is using a large amount of adhesive during its manufacture, which can be up to 20% of its total mass [1]. This disadvantage decreases the plywood product's ecological balance and makes it less favorable than solid wood, especially when considering resins derived from petrochemical resources. Global production of plywood reached 157 mln m^3 in 2017 [2]. To produce such an amount of plywood, approximately 15 mln tons of resin are used. Synthetic thermosetting resins based on phenol, urea, formaldehyde, and isocyanates are usually used.

Urea–formaldehyde (UF) resins are incombustible, provide good bonding strength, resistance to fluctuations of temperature, light, and corrosion, have a small curing time, simple manufacturing technology, and low production costs. However, they also have significant disadvantages, such as fragility, low water resistance, and significant emissions of formaldehyde. Phenol–formaldehyde (PF) resins can improve the bonding strength and water resistance, but they require a longer curing time, higher curing temperatures, higher production costs, and also emit phenol and formaldehyde [3]. The formaldehyde can irritate the eyes, respiratory and nervous systems, and possibly lead to cancer and leukemia. Therefore, formaldehyde was reclassified in 2004 by the International Agency for Research on Cancer (IARC) as "carcinogenic to humans (Group 1)" [4], compelling companies to reduce formaldehyde emission to lower levels.

Significant efforts have been made to reduce formaldehyde emissions from wood-based panels by the addition of various additives to the thermosetting resins or by the protection of the product with veneer, varnish, paint, and other coatings [5]. One of the possible directions is the creation of wood composites based on environmentally-friendly products, where thermoplastics (polyethylene, polypropylene, poly(vinyl chloride), and their copolymers) are used as adhesives. Already, there is a positive experience in the creation and use of wood composites based on thermoplastics [6–19]. Waste polyethylene can be used in the manufacture of oriented strand board (OSB) panels, resulting in the enhancement of thickness swelling, humidity, dimensional stability, water absorption, and screw withdrawal resistance [6]. Laminated veneer lumber (LVL) was manufactured using high-density polyethylene (HDPE) as a binding agent [7]. The properties of the composite boards were quite similar to or even better than those found in LVL made using thermosetting resin. The thermoplastic polymers were successfully used for coating of birch plywood [8–10]. Formaldehyde-free wood–plastic plywood has been successfully produced using thermoplastic polymers as wood adhesive [11–19].

The various thermoplastic polymers such as HDPE [7,12–15,17,19–21], polystyrene [16,22], polypropylene [18,21], or poly(vinyl chloride) [23,24] in different forms, such as textile fiber waste (polyurethane, polyamide-6) [21], recycled plastic shopping bags [7,11], or film [12–15,17–20,23–25] were used for veneer bonding.

The use of thermoplastic film as an adhesive for the bonding of veneer, apart from the fact that the plastic film is formaldehyde-free, has several other advantages compared with using liquid adhesives. Dry adhesive film is simpler to apply than wet adhesives; all of the untidy and unpleasant mixing and spreading operations in wet gluing are wholly removed from the plywood factory by the use of dry adhesive film. The dry adhesive film contains in each square meter of surface precisely the same quantity of adhesive, equal quality, uniform composition, exactly the same bond strength, and the same standard thickness [26].

Unfortunately, thermoplastic polymers are often hydrophobic, which leads to severe problems in the adhesion, causing poor mechanical properties [27]. Therefore, it is very important to promote the adhesion between a hydrophilic wood and a hydrophobic thermoplastic polymer, which can be done by using coupling agents [28,29], various surface treatments [23,24,30], thermal, or chemical modification of wood [25,27,31]. Some researchers used modified HDPE or polypropylene to manufacture plywood [32–34]. However, most of these approaches result in an obvious increase in cost and complexity of the preparation process. An alternative way of enhancing the physical and mechanical properties of plastic-bonded plywood is the modification of fabrication conditions, having an obvious advantage of low cost and easy processing [19].

Most of the mentioned studies used poplar, rarely eucalyptus or oak for bonding with thermoplastic polymers. No literature is available on using alder wood veneer. Alder is one of the most promising under-utilized wood species in Europe. Due to its workability and properties, alder can be considered as a suitable material for plywood manufacturing. Currently, producers of plywood in Ukraine often replace the traditional birch raw material with the alder raw material. This work presents the first effort to develop and evaluate composites based on alder veneers and HDPE film.

2. Materials and Methods

2.1. Materials

Rotary-cut alder wood veneer (*Alnus glutinosa* Goertn.) with dimensions of 300 mm × 300 mm × 1.6 mm and an average moisture content of 6% was used to make plywood panels. To minimize the influence of wood structure defects on the results of the experiment, the veneer sheets were selected and evaluated for the production of plywood panels. The veneer sheets were visually checked, and sheets without shocks, cracks, curling, and colors of more or less uniform thickness were selected. Observation on the wood appearance did not show any visible defects.

HDPE film with a thickness of 0.14 mm, density of 0.93 g/cm^3 and melting point of 135 °C was used for the bonding of plywood samples. The plastic film was cut into the same dimensions as the veneers. UF and PF resins were also used for the comparison. UF and PF adhesives were prepared according to the manufacturer's instructions. For the preparation of UF adhesive, 5% hardener (ammonium nitrate) and 15% filler (wheat flour) were used.

2.2. Manufacturing of Plywood Samples

Three-layer plywood samples were prepared. Instead of traditional UF and PF adhesives, a HDPE film was used as an adhesive for manufacturing the plywood samples. One sheet of HDPE film was incorporated between two veneer sheets, which were laid with the directions of the fiber perpendicular to each other. The laying of the dry HDPE film was very simple, and the design of the package is shown in Figure 1.

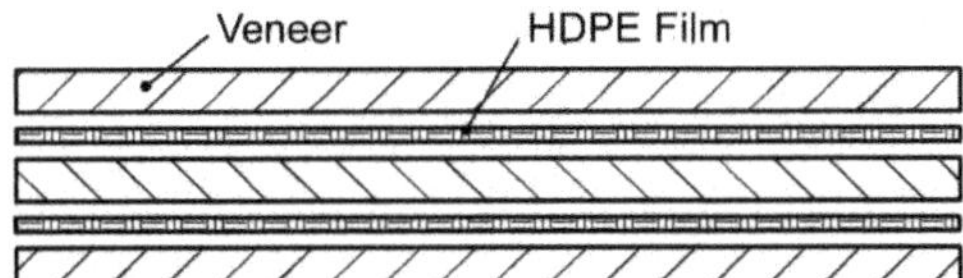

Figure 1. Structure of three-layer high-density polyethylene HDPE-bonded plywood panels.

The influence of hot-pressing pressure (0.8, 1.2, and 1.6 MPa), hot-pressing temperature (140, 160, and 180 °C), and hot-pressing time (1, 2, 3, and 5 min) on the properties of plywood were evaluated. The pressing temperature depends on the adhesive. The melting temperature of the HDPE film was 135 °C, implying the value of the lower limit of hot-pressing temperature. This temperature must be over 135 °C, thus making HDPE flow and penetrate the alder veneers. By contrast, conventional plywood made from the commercial UF and PF resins is generally hot-pressed at approximately 105 and 145 °C, respectively. Therefore, a range of 140 to 180 °C was considered for hot pressing plywood panels using HDPE film as adhesive. The pressing conditions for manufacturing of plywood samples are given in Table 1. The physical and mechanical properties of HDPE-bonded plywood were compared with UF and PF plywood, and relevant plywood standards.

Table 1. Manufacturing conditions of plywood.

Test No.	Manufacturing Conditions				
	Adhesive Type	Adhesive Spread Rate (g/m^2)	Pressing Pressure (MPa)	Pressing Temperature (°C)	Pressing Time (min)
1	HDPE	130	0.8	140, 160, 180	1, 2, 3, 5
2	HDPE	130	0.8, 1.2, 1.6	160	3
4	UF	160	1.8	105	3
5	PF	160	1.8	145	3

For the comparison, the plywood samples using UF and PF adhesives were produced according to the regimens usually used in practice. UF plywood samples were manufactured at the hot-pressing conditions: pressure of 1.8 MPa, temperature of 105 °C, time of 3 min, and an adhesive dosage of 160 g/m^2. PF plywood samples were produced at a pressure of 1.8 MPa, temperature of 145 °C, time of 3 min, and adhesive dosage of 160 g/m^2. The adhesive mixture was applied to the surface of the veneer by hand using a roller, and the open assembly time was about 5 min. The dosage of HDPE film at a thickness of 0.14 mm equals 130 g/m^2, which was 19% less than in the case of using UF and PF adhesive.

After hot pressing, the plywood samples were subjected to a cold-press stage that was performed at room temperature for 5 min, which was used to reduce the distortion and stress of the plywood. Three replicate panels were manufactured for all the conditions and control.

After bonding, the plywood panels were air conditioned for 5 days. After air conditioning in a standard climate (T = 20 ± 2 °C, RH = 65 ± 5%), standard samples were taken from each panel to determine the appropriate physical and mechanical properties: 20 samples per shear strength test, 6 samples per MOR/MOE test, and 6 samples per dimensional changes test.

2.3. Physical and Mechanical Properties

Thickness, density, bending strength (MOR), and modulus of elasticity (MOE) in bending, shear strength, water absorption, and thickness swelling of plywood samples were determined according to the standards [35–39]. For the shear strength test, one half of the samples were tested in dry conditions and the other half in wet conditions after soaking in water at 20 ± 3 °C for 24 h. Mechanical properties of the samples, MOR and MOE in bending were carried out in parallel ($\parallel$) and perpendicular ($\perp$) directions, depending on the surface layer. Physical properties of the samples, water absorption (WA) and thickness swelling (TS), were conducted in accordance with EN-317 [39]. Before testing, the weight and thickness of each sample were measured. Conditioned samples of each type of plywood panel were fully immersed in distilled water at room temperature for 2, 24, 48, and 72 h. The samples were removed from the water, patted dry, and then measured again. The samples were weighed to the nearest 0.01 g and measured to the nearest 0.001 mm immediately.

The compression ratio (CR) of plywood panels was calculated as shown below:

$$CR = (T_V - T_P)/T_V \times 100(\%) \tag{1}$$

where CR is the compression ratio of plywood panels, T_V is the total thickness of all veneers and HDPE films (mm), and T_P is the thickness of the panel (mm).

Furthermore, the measurement of the core temperature that can be achieved inside the veneer package under given pressing conditions of plywood samples was undertaken. Temperature changes were measured using thermocouples connected to an PT-0102K digital multichannel device.

In addition, micromorphological properties were evaluated by microscopic imaging.

2.4. Statistical Analysis

Statistical analysis was conducted using SPSS software program version 22 (IBM Corp., Armonk, NY, USA). Analysis of variance (ANOVA) was performed on the data to determine significant differences at the 95% level of confidence. Duncan's multiple range test was used to determine the significant difference between and among the groups.

3. Results and Discussion

3.1. Statistical Analysis

The influence of different factors on physical and mechanical properties of plywood was analyzed using ANOVA analysis. The results are summarized in Tables 2 and 3. The observations made in this study and the results of the statistical analysis indicated that both mechanical and physical properties were significantly influenced by the various parameters.

Table 2. ANOVAs of the variable parameters on the mechanical and physical properties of plywood panels.

Source	Thickness		Density		MOR (∥)		MOR (⊥)		MOE (∥)		MOE (⊥)		Shear Strength	
	F	P	F	P	F	P	F	P	F	P	F	P	F	P
T	23.419	0.000 *	29.001	0.000 *	470.672	0.000 *	10.349	0.000 *	60.390	0.000 *	10.726	0.000 *	196.558	0.000 *
Tp	8.907	0.001 *	13.653	0.000 *	544.381	0.000 *	20.852	0.000 *	60.400	0.000 *	34.705	0.000 *	411.222	0.000 *
$T \times Tp$	15.852	0.000 *	28.080	0.000 *	311.957	0.000 *	7.087	0.000 *	34.411	0.000 *	11.866	0.000 *	98.263	0.000 *
P	314.877	0.000 *	54.816	0.000 *	3.973	0.041 *	7.858	0.007 *	18.195	0.000 *	50.216	0.000 *	5.665	0.007 *
A	417.616	0.000 *	33.237	0.000 *	5.540	0.006 *	25.482	0.000 *	22.496	0.000 *	27.623	0.000 *	22.974	0.000 *

Note: T = temperature of pressing; Tp = time of pressing; P = pressure of pressing; A = type of adhesive; F = F value; MOR (∥) = bending strength in parallel direction; MOR (⊥) = bending strength in perpendicular direction; MOE (∥) = modulus of elasticity in bending in parallel direction; MOE (⊥) = modulus of elasticity in bending in perpendicular direction. * Significant difference at the 5% level ($P \leq 0.05$).

Table 3. ANOVAs of the variable parameters on the physical properties of plywood panels.

Source	TS (2 h)		TS (24 h)		TS (48 h)		TS (72 h)		WA (2 h)		WA (24 h)		WA (48 h)		WA (72 h)	
	F	P	F	P	F	P	F	P	F	P	F	P	F	P	F	P
T	1.256	0.296 ns	1.378	0.264 ns	1.268	0.292 ns	1.580	0.219 ns	2.049	0.142 ns	2.691	0.080 ns	0.631	0.537 ns	8.609	0.001 *
Tp	3.023	0.060 ns	2.065	0.140 ns	0.436	0.650 ns	0.221	0.802 ns	1.298	0.284 ns	3.459	0.041 *	1.426	0.252 ns	3.955	0.027 *
$T \times Tp$	2.627	0.063 ns	1.230	0.311 ns	2.542	0.070 ns	1.258	0.302 ns	1.421	0.251 ns	2.392	0.083 ns	0.971	0.416 ns	2.927	0.045 *
P	3.649	0.051 ns	10.007	0.002 *	14.203	0.000 *	14.974	0.000 *	0.389	0.684 ns	0.978	0.399 ns	2.670	0.102 ns	9.711	0.002 *
A	15.427	0.000 *	75.756	0.000 *	60.828	0.000 *	76.271	0.000 *	7.921	0.001 *	19.885	0.000 *	13.428	0.000 *	3.561	0.035 *

Note: T = temperature of pressing; Tp = time of pressing; P = pressure of pressing; A = type of adhesive; F = F value. * Significant difference at the 5% level ($P \leq 0.05$); ns = not significant.

3.2. Thickness and Density of Plywood Panels

The aim of the veneer thickness measurement was to find effects of the different conditions of pressing on the tolerance of a pressed plywood panel. In the case of using HDPE film as adhesive in the production of plywood, it is very important to choose the pressing parameters so that the thickness of finished plywood is within acceptable limits, to avoid unnecessary losses of wood raw material. The average values of the thickness, density, and moisture content of plywood samples, as well as the compression ratio, are given in Table 4.

Table 4. Thickness and density of plywood panels.

Type of Adhesive	Pressing Conditions			Moisture Content (%)	Compression Ratio CR (%)	Thickness (mm)	Density (kg/m^3)
	Pressure (MPa)	Temperature (°C)	Time (min)				
UF	1.8	105	3	5.0	8.5	4.23 (0.04) *	601.7 (12.42)
PF	1.8	145	3	3.5	13.5	3.99 (0.05)	619.1 (9.67)
HDPE	0.8	140	1	4.1	4.2	4.70 (0.01)	536.8 (23.05)
			2	3.8	4.8	4.69 (0.04)	548.5 (7.79)
			3	3.7	5.2	4.62 (0.04)	567.2 (8.02)
		160	1	3.8	5.3	4.55 (0.06)	573.4 (9.04)
			2	3.5	7.1	4.58 (0.01)	592.2 (4.09)
			3	3.1	6.1	4.63 (0.02)	553.2 (5.02)
		180	1	3.4	4.5	4.65 (0.02)	555.1 (5.73)
			2	3.2	5.7	4.67 (0.01)	567.0 (10.34)
			3	3.2	5.7	4.55 (0.03)	588.6 (4.48)
	1.2	160	3	2.8	3.0	4.57 (0.01)	553.2 (9.06)
	1.6	160	3	2.9	8.1	4.37 (0.02)	595.2 (9.32)

* Values in parenthesis are standard deviations. UF: urea–formaldehyde. PF: phenol–formaldehyde.

ANOVA analysis showed that the temperature, pressure, and time of the hot pressing, as well as the type of adhesive used, significantly affects the thickness and density of the HDPE-bonded plywood panels. It can be seen that the average thickness of HDPE-bonded plywood panels made at different hot-pressing temperatures, pressures, and times is not smaller but even exceeds the thickness of control plywood using UF and PF adhesives (Table 4), which is essential for the industrial application of this technology. The smallest thickness (3.99 mm) and the highest density (619.1 kg/m^3) had plywood samples made using PF adhesive. The largest thickness (4.70 mm) and the smallest density (536.8 kg/m^3) had plywood samples made using HDPE film. Primarily, this can be explained by the high hot-pressing pressures of plywood samples and the increased adhesive dosage in the case of using UF and PF adhesives.

As the time of the hot pressing increases from 1 to 3 min, the thickness of the plywood samples decreases at temperatures of 140 and 180 °C, and increases at a temperature of 160 °C (Figure 2). This is because at a higher hot-pressing temperature and longer time of pressing, the wood becomes more plastic and more easily compressed. The difference in thickness values of plywood samples pressed for 1 and 2 min was insignificant ($P \leq 0.05$) based on Duncan's test. Of course, if the thickness of the plywood samples decreases with the increase in temperature and time of pressing, then it is natural that the density of such samples, in contrast, increases (Figure 2). It was found that differences in density values of the panels pressed at temperatures of 160 and 180 °C for 2 and 3 min were insignificant ($P \leq 0.05$) based on Duncan's test. Nevertheless, according to the F values (Table 2), it can be seen that, in the ranking from highest to lowest, the hot-pressing temperature has the greatest influence on the thickness and density of the plywood samples, after that an interaction of hot-pressing temperature and hot-pressing time and finally hot-pressing time.

The hot-pressing pressure also significantly influences the thickness and density of plywood samples. With an increase in pressing pressure from 0.8 to 1.6 MPa, the thickness of the plywood samples decreases by 6.0%, and the density of samples increases by 7.6% (Figure 2). Between the pressures of 0.8 and 1.2 MPa there is no significant difference ($P \leq 0.05$) based on Duncan's test in the effect on the density and thickness of the plywood samples.

Nevertheless, plywood panels containing HDPE film were pressed at a lower pressure than the control panels (Table 1). In this case, the average compression ratio of plywood made using HDPE film was smaller—5.4, 3.04, and 8.05%, respectively, for 0.8, 1.2, and 1.6 MPa compared with the compression ratio of 8.5 and 13.5% for control UF and PF plywood, respectively. Moreover, plywood containing HDPE film was manufactured with 19% less adhesive spread than the adhesive spread used for the control panels. In addition, its compression ratio will be smaller because less moisture was brought with the adhesive into the veneer package, and such package, in turn, is less densified (wood is deformed more heavily).

The European Standard EN 315 [40] specifies tolerances of unsanded plywood panels for a nominal thickness of 4 mm as −0.4 mm (min) and +0.8 mm (max); i.e., the thickness of the finished unsanded plywood panels should be in the range 3.6–4.8 mm. In this study, the values of plywood thicknesses were 4.23 ± 0.04 mm and 3.99 ± 0.05 mm for panels made using UF and PF adhesives, respectively, and 4.37 (± 0.02)–4.70 (± 0.01) mm for panels made using HDPE film (Table 4), and they did not go beyond tolerances for unsanded panels in accordance with this standard.

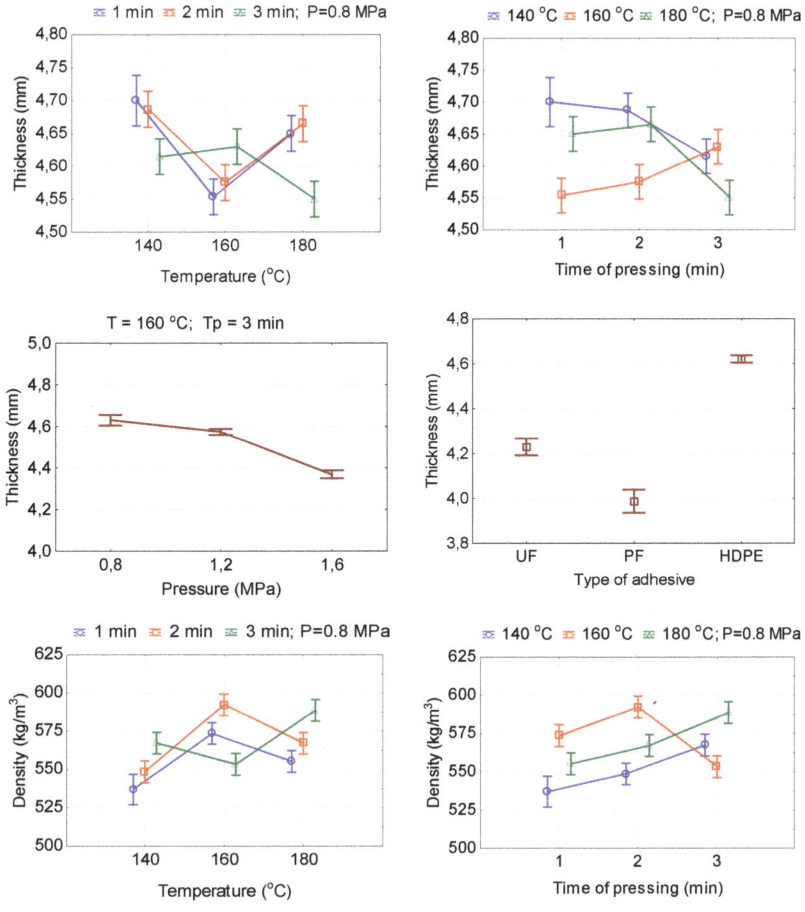

Figure 2. *Cont.*

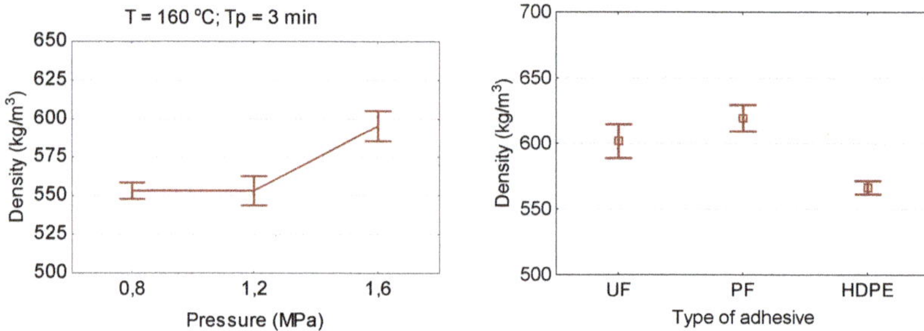

Figure 2. Relationship between the thickness and density of plywood and hot-pressing parameters.

3.3. Shear Strength

Average values of the dry and wet shear strength of plywood panels are presented in Table 5. The quality of the bonding of the thermoplastic adhesive and the wood surface depends on the processing details of the adhesive bonding, such as the porosity of the wood surface, the viscosity of the molten adhesive, the applied pressure, and the processing duration [41].

Figure 3 shows the effects of pressing temperature and time on the shear strength of plywood samples. The obtained strength ranged from 0.74 to 2.38 MPa, practically all of which met the European Standard EN 314-2 [38] for Class 1 (dry conditions) plywood. In this study, the bonding strength mean values obtained from the samples of HDPE-bonded plywood panels were above the limit value (1.0 MPa) indicated in the European Standard EN 314-2 [38] standard. HDPE-bonded plywood panels produced using 140 °C and 1–2 min and 160 °C and 1 min did not meet the European Standard EN 314-2 [38].

Table 5. Shear strength values of plywood panels pressed using different types of adhesive and pressing conditions.

Type of Adhesive	Pressing Conditions			Shear Strength (MPa)	
	Pressure (MPa)	**Temperature (°C)**	**Time (min)**	**Dry Test**	**Wet Test**
UF	1.8	105	3	2.99 (0.15) *	2.05 (0.14)
PF	1.8	145	3	2.95 (0.10)	2.79 (0.23)
HDPE	0.8	140	1	1.72 (0.14)	0.74 (0.12)
			2	1.73 (0.14)	0.99 (0.07)
			3	2.44 (0.09)	1.71 (0.12)
			5	2.01 (0.13)	1.08 (0.05)
		160	1	1.60 (0.11)	0.79 (0.05)
			2	1.94 (0.09)	1.03 (0.10)
			3	2.99 (0.12)	2.38 (0.15)
			5	2.87 (0.11)	2.1 (0.14)
		180	1	1.73 (0.18)	1.02 (0.08)
			2	2.64 (0.08)	2.25 (0.10)
			3	2.67 (0.16)	2.29 (0.12)
			5	2.38 (0.23)	1.32 (0.11)
	1.2	160	3	2.91 (0.08)	1.97 (0.13)
	1.6	160	3	2.74 (0.14)	1.72 (0.56)

* Values in parenthesis are standard deviations.

At the low temperature of 140 °C and short-term pressing time 1–2 min, the flow of HDPE is very poor; HDPE cannot permeate adequately into the vessels and cracks of veneers, resulting in worse strength. In other work [7], it was also found that lower temperatures did not promote an adequate melting of the HDPE. On the other hand, when the plywood samples were pressed at 140 °C for 3 min the result was better. de Barros Lustosa et al. [7] observed a similar trend in their study.

With increasing hot-pressing temperature from 140 to 180 °C, the shear strength of the plywood samples increased by 37.8, 127.3 and 33.9%, respectively, for pressing times of 1 min, 2 min and 3 min. This can be explained by the fact that with increasing hot-pressing temperature, more HDPE can permeate adequately into the vessels and cracks of veneers and help to enhance the bonding strength [19]. It is known that the viscosity of plastic film decreases with increasing temperatures [12,13]. This suggests that high temperatures contribute to the melting of HDPE film, which provides better fluidity, which allows for polyethylene to be more evenly distributed. In turn, this creates better conditions for the penetration of molten polyethylene into the veneer cavities and accordingly creates better conditions for the formation of mechanical locks. Accordingly, this contributes to the increase in the shear strength. Furthermore, a higher hot-pressing temperature can also decrease the proportion of hydrophilic groups in wood, thus enhancing the interfacial compatibility between hydrophilic wood and hydrophobic plastic film [25,27]. This tendency is true for the pressing time of 1–3 min. If the pressing time was increased to 5 min, then a drop in shear strength was observed with an increase in the temperature from 160 to 180 °C. Chang et al. [19] also observed that temperatures higher than 160 °C made the mechanical interlock worse and gave poorer strength to the plywood. It is obvious that at high temperature and prolonged pressing time (Figure 3), the polyethylene film is subjected to decomposition and fracturing, so that the rate of increase in the bond strength decreased.

At both temperatures, 160 and 180 °C, high strength values were obtained that exceed the minimum value according to the European Standard EN 314-2 [38]. Taking into account the energy consumed, the hot-pressing temperature of 160 °C was more economical. The highest shear strength was achieved at 160 °C when the pressing time was increased to 3 min. The smallest shear strength was achieved at 140 °C and the pressing time 1–2 min.

These results are in agreement with a study carried out by Cui et al. [11]. In this cited study, by increasing hot-pressing temperature, the bonding strength of plywood indicated a clear upward trend and then declined.

Hot-pressing time had a significant impact on the shear strength of plywood samples. Too short pressing time is insufficient for good adhesive penetration [16]. The time used should be enough for heating the inner areas of the plywood samples allowing the melting of the HDPE and the drying of the wood veneer at the same time [7]. Increasing the pressing time from 1 to 3 min leads to increased shear strength. The subsequent increase in pressing time to 5 min leads to increasing the time for production and decreasing the shear strength. This can be explained by the fact that after hot pressing for 5 min, the HDPE film was completely melted, and the thickness of the film could be reduced if the pressing time was longer than 3 min because part of the film was melted and flew out from the plywood, resulting in a lack of polyethylene film and, finally, the shear strength of plywood samples decreases. Moreover, with long-term hot pressing, the molten plastic film penetrated into the wood and, consequently, caused a decrease in the shear strength of plywood. Therefore, the hot pressing could not also be so long as to prevent thermal degradation of the wood veneer [7], decomposition and fracturing of the HDPE film. At the various hot-pressing pressures and temperatures, the highest shear strength values are observed for the pressing time of 3 min. A similar trend in the impact of pressing time on the properties of plywood is described in the work [11]. They concluded that the optimal parameters were a plastic use of 100 g/m^2, a hot-pressing temperature of 150 °C, and a hot-pressing time of 6 min.

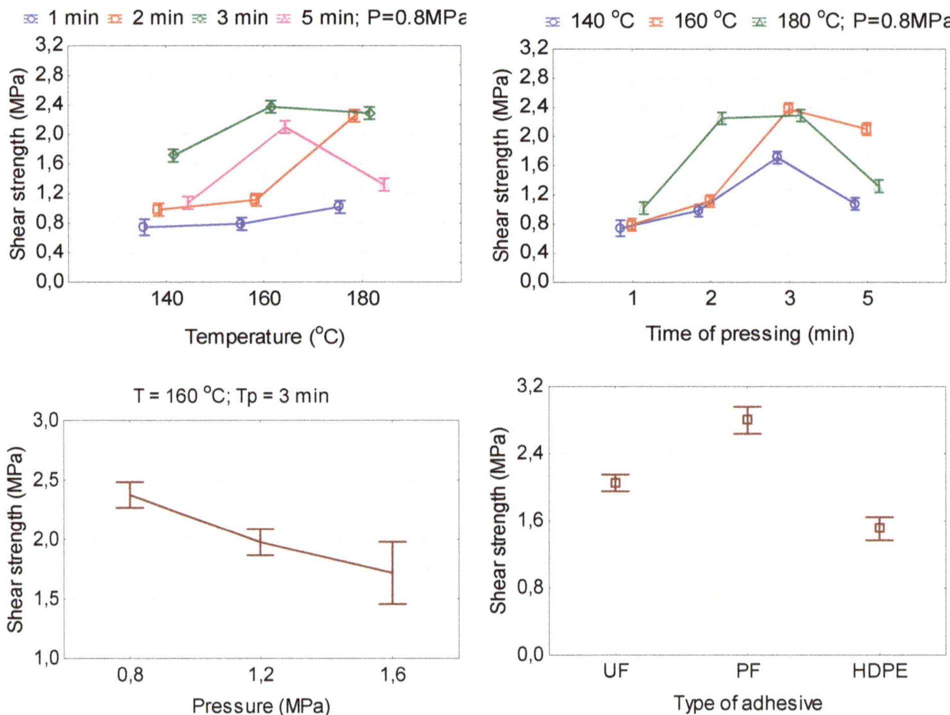

Figure 3. The relationship between the shear strength of plywood and hot-pressing parameters.

The pressure also played an important role because it is responsible for providing close contact between both materials and helping the flow of the HDPE into voids and irregularities of the wood veneer [7]. Therefore, high pressure is recommended for enhancement of adhesion properties with veneer [19]. In our study, with increasing pressing pressure from 0.8 to 1.6 MPa (at 160 °C and time 3 min), the shear strength decreased by 27.7%. This can be explained by the fact that with increasing the hot-pressing pressure, the more molten HDPE film was pressed into the vessels and cracks in the veneer, but less remains between the sheets of the veneer, reducing the bonding strength. Chang et al. [19] also found that when the hot-pressing pressure increased up to 1.3 MPa, the strength decreased. This phenomenon they ascribe to the fact that, with the increase in hot-pressing pressure, the ejection of HDPE resin from vessels and cracks occurs. Thus, unlike liquid adhesives such as UF and PF, the use of which requires high pressing pressure, the use of polyethylene film allows the bonding of plywood panels at significantly lower pressure (by about 50%).

The findings of our work are in good agreement with the results obtained by Smith et al. [41]. They also found that processes that kept the polypropylene molten at the surface of the wood for a longer time, higher temperature, and higher pressure achieved much better interlocking than the short process cycle.

To confirm the above-described processes of the melting and flowing of HDPE film during the hot-pressing operation, measurements of the temperature inside the veneer package for different types of adhesive used and different pressing conditions were made.

Figure 4 shows the temperature distribution inside the veneer package during the hot pressing of plywood samples for different types of adhesive used and different pressing conditions. The most slowly warmed up veneer package was the case of the manufacturing of plywood at a pressing temperature of 105 °C using UF adhesive. The temperature of 100 °C inside the veneer package is reached in 50 s from the start of pressing. The veneer package with the use of PF adhesive in which pressing is carried out already at a higher temperature of 145 °C warmed up faster. The temperature of 100 °C inside the veneer package was reached in 14 s, which is almost 3.5 times faster than in the case of pressing UF plywood. The main factor determining the speed of heating the veneer package with HDPE film is the pressing temperature. The melting temperature of the HDPE film is 135 °C. Such temperature is achieved inside the veneer package in 90, 56, and 34 s at pressing temperatures of 140, 160, and 180 °C, respectively. With the increase in the pressing temperature, the heating rate of the veneer package increases. This has an effect on the bonding strength of plywood using HDPE film. From Figure 3 and Table 5, it can be seen that the temperature of 140 °C and the pressing time of 1 and 2 min are insufficient to warm the package, to melt the plastic film, and to allow it to flow over the surface of the veneer sheet. At a temperature of 140 °C, the film's fluidity is insufficient to move inside the pores of the wood and to form mechanical adhesive bonds, which adversely affects the bonding strength (Figure 3, Table 5). For temperatures of 160 and 180 °C, the pressing time of 1 min is also undesirable because the bonding strength is practically equal to the permissible value of 1.0 MPa in accordance with European Standard EN 314-2 [38].

Figure 5 shows microscopic images of alder veneer bonded with HDPE film. In the plywood samples prepared at 140 °C and a pressing time of 1 min, a bondline was observed between the adjacent two veneers, indicating that the penetration of HDPE films into the veneer surfaces was not desirable, which could easily cause the delamination of HDPE films from the veneer surfaces (Figure 5a). This image clearly indicates the lack of chemical bonds between the plastic film and wood substance because the HDPE film could be easily separated mechanically from the wood after a short immersion in water. This was an indication of poor adhesion of the HDPE film to the wood surface. The lack of chemical bonds between the plastic films and the surface of wood are also indicated by other authors [13,15,19,32,41]. Therefore, the plywood samples prepared using 140 °C and a pressing time of 1 min had the lowest shear strength in Figure 3 and Table 5.

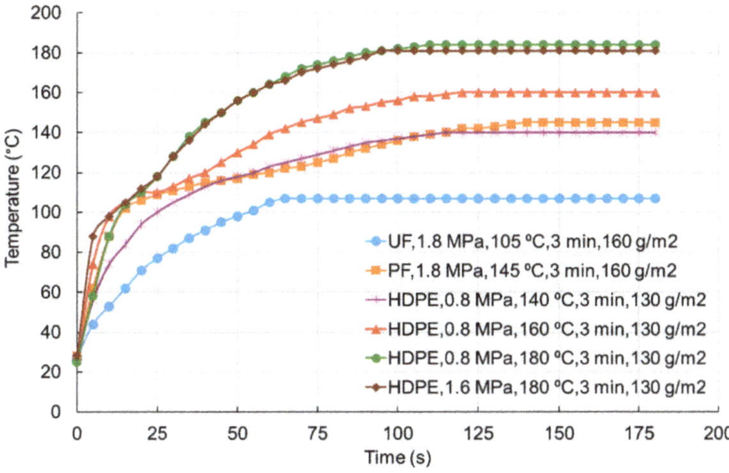

Figure 4. Core-layer temperature curves of three-layer plywood prepared with different adhesives at various hot-pressing conditions.

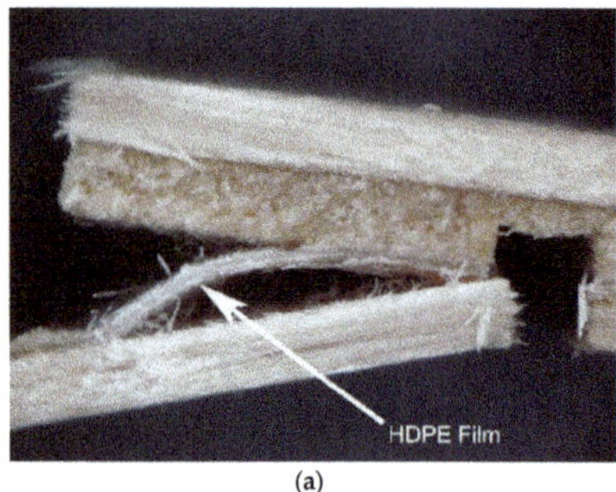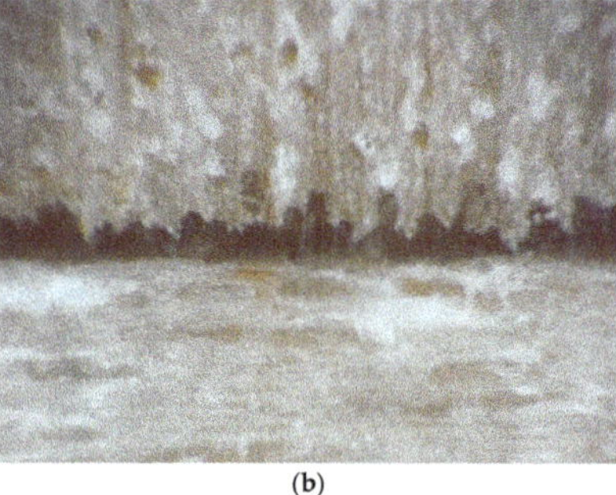

(a)   (b)

Figure 5. Micrograph of the HDPE-bonded plywood samples (**a**) prepared at 140 °C and 1 min after shear strength test; (**b**) prepared at 160 °C and 3 min.

In the plywood samples prepared using higher temperatures of 160 and 180 °C and a longer pressing time of 2–3 min, the penetration of HDPE films into the veneer surfaces was better (Figure 5b) than that in the samples prepared using 140 °C and a pressing time of 1 min. It can be seen that HDPE film flowed and penetrated into the vessels of wood veneer during the hot-pressing stage, forming a continuous bondline and mechanical interlock structure (Figure 5b). The better penetration produces stronger bonded joints in the samples. Therefore, the plywood samples prepared at 160 °C and a pressing time of 3 min had the highest shear strength in Table 5.

In another study [32], it was also shown that the polypropylene melted during hot pressing and made good contact with veneer surfaces penetrating into the lumina of wood cells, lathe checks, and other spaces open on the veneer surface. These authors indicated that the anchoring effect of polypropylene, which had penetrated into various wood elements and spaces in the veneer, contributed dominantly to the gluability [32]. Therefore, because wood is a porous material, a mechanical interlock is the most likely bonding mechanism involved [13,15,19,32,41].

3.4. Bending Strength and Modulus of Elasticity in Bending

The average values of the bending strength and modulus of elasticity (MOE) in the bending of plywood samples are presented in Table 6.

Table 6. MOR and MOE values of plywood panels pressed using different types of adhesive and pressing conditions.

Type of Adhesive	Pressing Conditions			MOR (MPa)		MOE (MPa)	
	Pressure (MPa)	Temperature (°C)	Time (min)	MOR (‖)	MOR (⊥)	MOE (‖)	MOE (⊥)
UF	1.8	105	3	119.8 (4.2) *	22.7 (2.5)	8228.2 (232.2)	775.2 (28.4)
PF	1.8	145	3	134.6 (2.2)	25.9 (1.5)	7698.9 (160.8)	748.6 (30.1)
HDPE	0.8	140	1	26.7 (4.3)	17.5 (0.4)	6996.2 (886.0)	796.8 (37.8)
			2	107.6 (2.8)	21.3 (0.7)	9986.6 (191.5)	891.8 (14.0)
			3	111.6 (2.2)	20.8 (0.2)	9722.7 (150.4)	911.7 (28.3)
		160	1	107.3 (2.6)	20.1 (0.5)	9501.5 (309.9)	816.2 (17.1)
			2	113.7 (3.4)	20.8 (0.2)	10,141.8 (170.8)	891.7 (16.3)
			3	119.3 (5.1)	22.5 (0.6)	9511.0 (214.6)	879.7 (7.3)
		180	1	107.9 (3.3)	20.1 (0.8)	10,136.9 (278.1)	895.7 (17.6)
			2	110.1 (3.1)	19.3 (1.3)	10,307.7 (329.6)	908.4 (37.5)
			3	115.8 (2.6)	20.9 (0.2)	10,119.9 (322.2)	867.9 (15.0)
	1.2	160	3	113.4 (3.9)	22.0 (1.4)	9407.4 (272.2)	826.4 (18.2)
	1.6	160	3	119.9 (4.2)	24.5 (1.0)	10,173.7 (225.5)	923.8 (18.1)

* Values in parenthesis are standard deviations.

The temperature, pressure, and time of hot pressing significantly ($P \leq 0.05$) affect the bending strength in both parallel MOR ($\parallel$) and perpendicular MOR ($\perp$) directions of HDPE-bonded plywood samples (Table 2). With an increase in the pressing temperature from 140 to 180 °C and pressing time from 1 to 3 min, MOR ($\parallel$) and MOR ($\perp$) increased by 304.1 and 3.8% and 14.9 and 0.5%, respectively (Figure 6). For a pressing time of 2 min with an increase in the pressing temperature from 140 to 180 °C, MOR ($\parallel$) increased and MOR ($\perp$) decreased by 2.3 and 9.4%, respectively. As can be seen from Table 6, MOR ($\parallel$) values exceed the MOR ($\perp$) values by a factor of more than five, except the pressing conditions of 140 °C and 1 min. A similar trend can also be observed in the case of increasing the pressing time from 1 to 3 min (Figure 6). In this case, the MOR ($\parallel$) and MOR ($\perp$) increased by 318.0, 11.2, and 7.3% and 18.9, 11.9, and 4.0%, respectively, for pressing temperatures of 140, 160 and 180 °C. The differences in the values of MOR ($\parallel$) and MOR ($\perp$) for the temperatures of 160 and 180 °C are insignificant. Therefore, from an economic point of view, it is more expedient to hot-press at a temperature of 160 °C. The smallest value of MOR ($\parallel$) was obtained at a hot-press pressure of 1.2 MPa (113.4 MPa), and the highest values of MOR ($\parallel$) were obtained at 0.8 MPa (119.3 MPa) and 1.6 MPa (119.9 MPa) (Table 6). The difference in the values of MOR ($\parallel$) for the pressures of 0.8 and 1.6 MPa is insignificant. With increasing pressing pressure from 0.8 to 1.6 MPa, the value of MOR ($\perp$) increased by 8.9%. The difference in the values of MOR ($\perp$) for the pressures of 0.8 and 1.2 MPa is insignificant.

The temperature, pressure, and time of hot pressing also significantly ($P \leq 0.05$) affect the MOE in bending in both parallel MOE ($\parallel$) and perpendicular MOE ($\perp$) directions of HDPE-bonded plywood samples (Table 2). With an increase in the pressing temperature from 140 to 180 °C and pressing time from 1 to 3 min, the values of MOE ($\parallel$) and MOE ($\perp$) increased by 44.9, 3.2, and 4.1% and 12.4, 1.9, and 4.8% (reduction), respectively (Figure 7). With increasing the pressing time from 1 to 2 min, MOE ($\parallel$) values initially increased, and further increase in the pressing time to 3 min led to a decrease in MOE ($\parallel$) (Figure 7). The values of MOE ($\perp$) increased by 6.2–7.8% with an increase in the pressing time from 1 to 3 min. In addition, the difference in the values of MOR ($\perp$) for the pressing times of 2 and 3 min is insignificant.

With increasing hot-pressing pressure from 0.8 to 1.6 MPa, the values of MOE ($\parallel$) increased by 7%. The difference in the values of MOE ($\parallel$) for the pressing pressures 0.8 and 1.2 MPa is insignificant. The lowest values of MOE ($\perp$) were observed at a pressing pressure of 1.2 MPa (826.4 MPa), and the highest at a pressure of 1.6 MPa (923.8 MPa) (Table 6).

It has been established that the type of adhesive also significantly ($P \leq 0.05$) affects the mechanical properties of plywood samples, MOR ($\parallel$) and MOR ($\perp$), MOE ($\parallel$) and MOE ($\perp$) (Table 2). The highest MOR ($\parallel$) and MOR ($\perp$) values were obtained for PF adhesive (134.6 and 25.9 MPa, respectively), followed by UF adhesive (119.8 and 22.7 MPa, respectively) and HDPE film (103.6 and 20.5 MPa, respectively) (Table 6, Figure 8). The lowest MOE ($\parallel$) and MOE ($\perp$) values were recorded for PF adhesive (7698.9 and 748.6 MPa, respectively) and UF adhesive (8228.2 and 775.2 MPa, respectively), and the highest for HDPE films (9702.9 and 873.6 MPa, respectively) (Table 6, Figure 8). The differences in the values of MOE ($\parallel$) and MOE ($\perp$) for the UF and PF adhesives are insignificant.

The effect of adhesive could be explained by the quite different properties between thermoplastic (HDPE) and thermoset (UF and PF) polymers. HDPE is a well-known thermoplastic material, because of the inherent molecular structure the thermoplastic polymers are more plastic and show higher toughness than thermosets [20].

Another study stated that the average mechanical properties of polystyrene-bonded plywood panels tend to increase when the pressing time and temperature are increased during production [16].

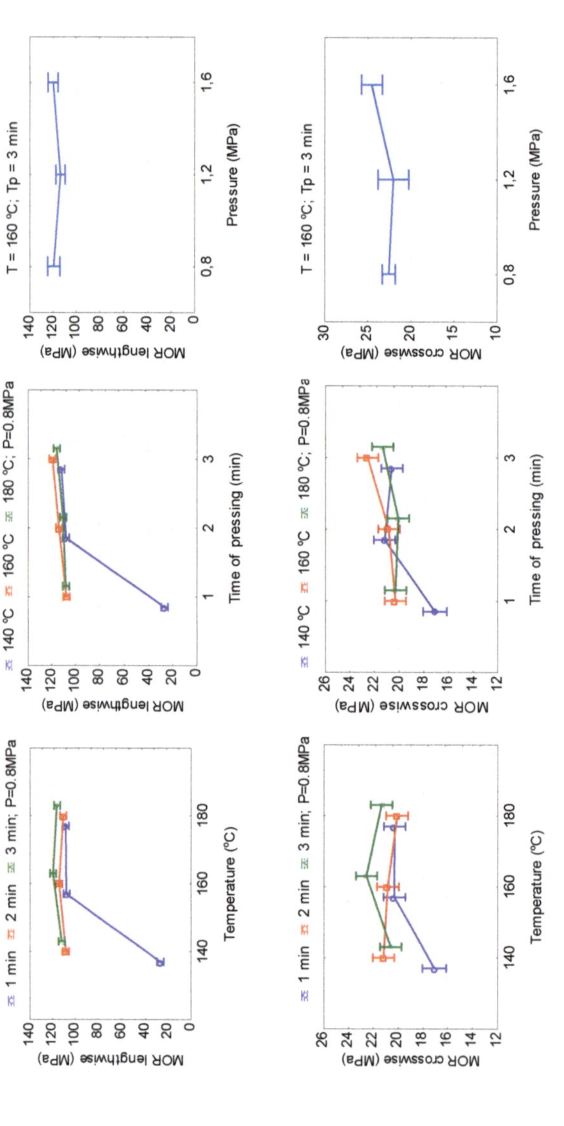

Figure 6. The relationship between bending strength and hot-pressing parameters (lengthwise = in parallel and crosswise = in perpendicular directions).

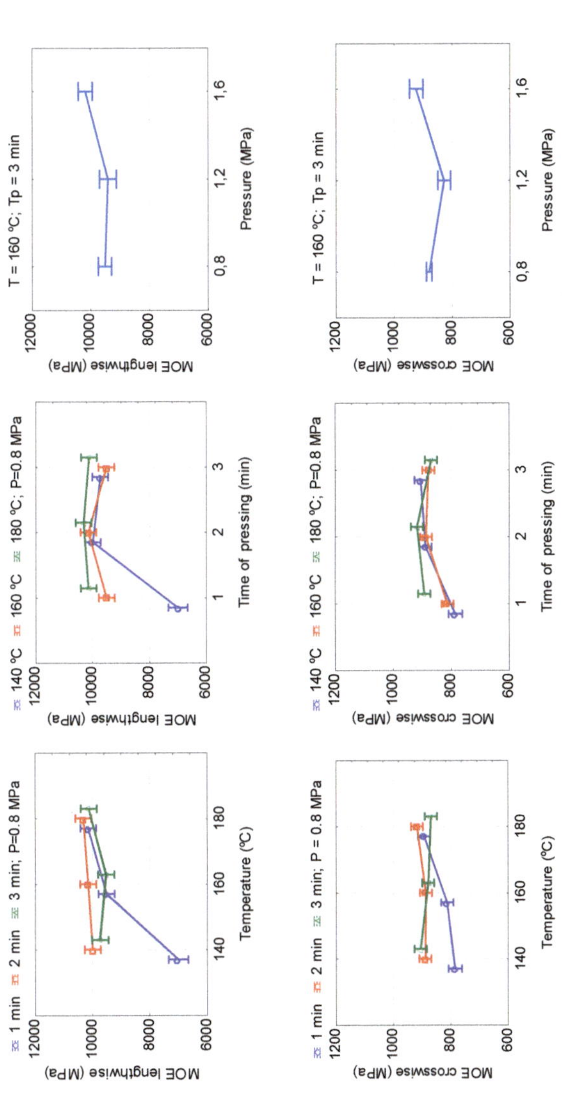

Figure 7. The relationship between the modulus of elasticity in bending and hot-pressing parameters (lengthwise = in parallel and crosswise = in perpendicular directions).

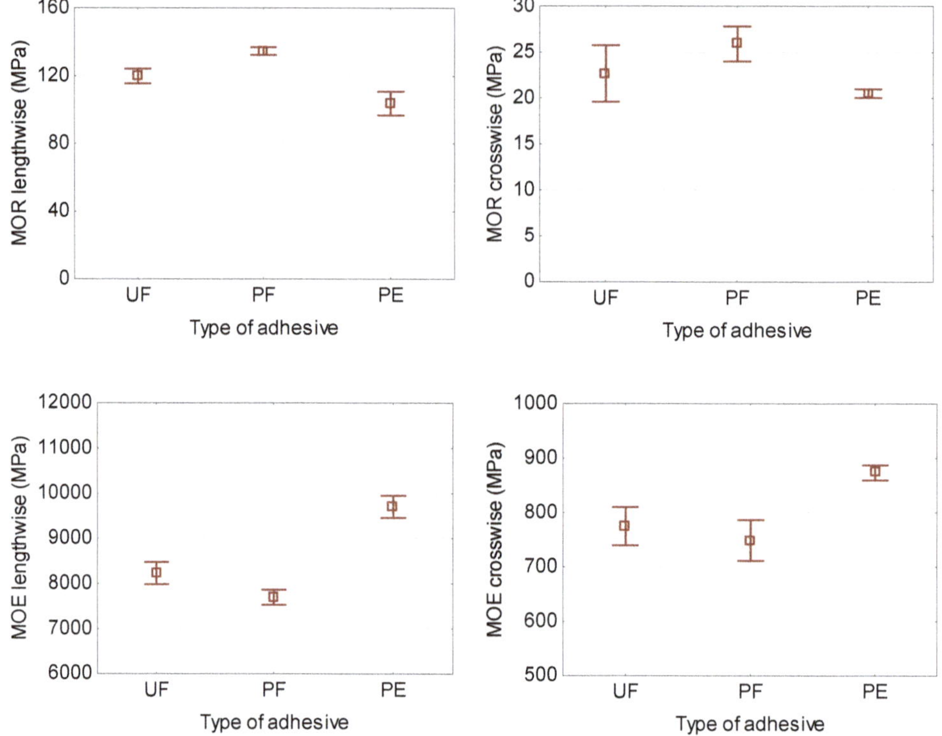

Figure 8. Bending strength and modulus of elasticity in the bending of plywood panels bonded with different types of adhesive (lengthwise = in parallel and crosswise = in perpendicular directions).

Mean values obtained for bending strength and MOE of plywood panels were higher than the limit values for structural purpose solid wood panels (32 MPa for MOR ($\parallel$) and 5 MPa for MOR ($\perp$), 9000 MPa for MOE ($\parallel$) and 600 MPa for MOE ($\perp$)) indicated in European Standard EN 13353 [42] for panels having thicknesses up to 20 mm.

The bending strength and MOE values of HDPE-bonded plywood panels were (Table 6):

- 26.7–119.9 MPa and 6996.2–10,307.7 MPa, respectively, when parallel to the grain direction;
- 17.5–24.5 MPa and 796.8–923.8 MPa, respectively, when perpendicular to the grain direction.

In all specimens, the determined bending strength and MOE values perpendicular to the grain direction were higher than the lower limiting value of 5 and 600 MPa, respectively. The plywood panels produced at 140 °C and 1 min had the smallest bending strength (MOR ($\parallel$) and MOE ($\parallel$)) values in Table 6 and did not meet the European Standard EN 13353 [42]. These panels had the least maximum load to failure, 188.9 N on average, and the least deflection to failure at 1.92 mm. As can be seen from Figure 9, the generated delamination crack longitudinally extended through the sample and this significantly reduced the maximum failure load. This can diminish the bending strength of plywood. As was shown above, the shear strength values of plywood samples bonded at the hot-pressing temperature of 140 °C and pressing time of 1 min were also the smallest.

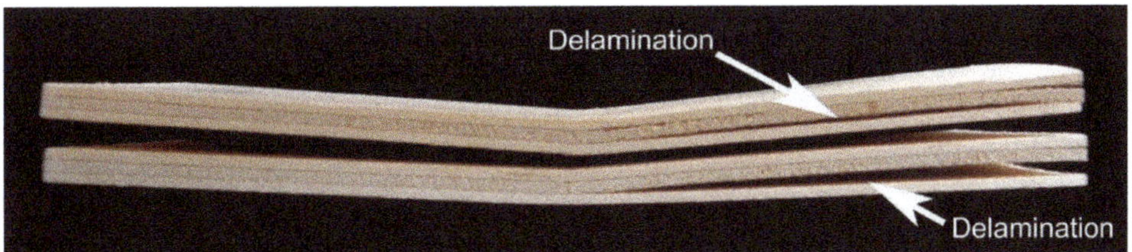

Figure 9. Delamination in the plywood samples prepared at 140 °C and 1 min during the determination of bending strength.

3.5. Thickness Swelling and Water Absorption

The average values of TS and WA of plywood samples are given in Table 7.

The pressing temperature and time insignificantly ($P \leq 0.05$) affect the TS of the HDPE-bonded plywood samples (Table 3). The smallest TS was observed at a pressing time of 2 min, and the difference in the values of TS for the pressing times of 1 and 3 min was insignificant. With increasing time of soaking, the TS increases (Table 7). The smallest TS was observed during soaking for 2 h, and then the TS is increased by 42.7–46.5%. The differences in the values of TS for the time of soaking 24, 48, and 72 h were insignificant.

The pressing pressure significantly ($P \leq 0.05$) affects the TS of HDPE-bonded plywood samples (Table 3). With increasing pressing pressure from 0.8 to 1.6 MPa, the values of TS increased by 17.7%. The difference in the values of TS for the pressing pressures 0.8 and 1.2 MPa was insignificant.

The type of adhesive had a significant ($P \leq 0.05$) effect on TS (Table 3). The smallest value of TS was observed in plywood samples using HDPE film (8.11%), the higher value in samples using UF adhesive (8.70%) and the highest value in samples using PF adhesive (11.09%) (Table 7). This was attributed to the hydrophobic character of the plastic film. This result was also associated with the hot-pressing condition. A higher hot-pressing temperature for wood-based composites can reduce hydrophilic groups in raw materials, which contributes to reducing the water absorption and the TS of plywood panels [25,27]. In this research, the hot-pressing temperature of HDPE-bonded plywood samples was higher than PF-bonded samples, while the hot-pressing temperature of PF-bonded samples was higher than that of UF-bonded samples.

The temperature and time of hot pressing significantly affect the WA of HDPE-bonded plywood samples (Table 3). The lowest values of WA were found at 160 °C (56.03%), and higher values were obtained at temperatures of 140 and 180 °C (58.5% and 57.3%, respectively) (Table 7). The difference in the values of WA for the pressing temperatures of 140 and 180 °C was insignificant. The smaller value of WA was observed at the pressing time of 3 min (56.5%), and higher value at a pressing time of 1 min (58.3%). The difference in the values of WA for the pressing time of 2 and 3 min was insignificant. With increasing time of soaking in water from 2 to 72 h, the WA gradually increased by 68.3%. The largest increase in WA occurs during the first 2 h of soaking in water.

The pressing pressure affects WA insignificantly (Table 3). The type of adhesive had a significant effect on water absorption. The smallest WA was observed in plywood samples using UF adhesive (54.3%), intermediate in samples using HDPE film (57.1%), and the highest in samples using PF adhesive (59.7%) (Table 7).

Table 7. Thickness swelling (TS) and water absorption (WA) of plywood panels pressed using different types of adhesive and pressing conditions.

Type of Adhesive	Pressing Conditions			Physical Properties							
	P (MPa)	T (°C)	Time (min)	TS (2 h)	TS (24 h)	TS (48 h)	TS (72 h)	WA (2 h)	WA (24 h)	WA (48 h)	WA (72 h)
UF	1.8	105	3	6.03	9.38	9.70	9.70	35.84	53.12	60.19	67.94
PF	1.8	145	3	7.94	11.84	12.21	12.38	37.29	61.29	67.67	72.66
HDPE	0.8	160	3	6.15	8.82	8.82	9.00	38.86	51.38	59.51	70.48
	1.2	160	3	5.90	9.22	9.37	9.37	39.08	52.75	61.51	71.76
	1.6	160	3	7.02	10.19	10.61	10.80	40.70	51.83	59.32	66.75
	0.8	140	1	–	–	–	–	–	–	–	–
	0.8	140	2	5.97	8.29	8.54	8.64	44.46	56.14	63.88	74.30
	0.8	140	3	6.53	8.63	8.78	8.85	42.43	53.62	61.22	71.95
	0.8	160	1	6.51	8.90	9.55	9.33	44.58	56.20	63.41	70.71
	0.8	160	2	5.06	8.52	8.82	8.89	38.85	52.10	58.87	67.35
	0.8	160	3	6.15	8.82	8.82	9.00	38.86	51.38	59.51	70.48
	0.8	180	1	5.95	8.35	8.49	8.46	42.52	54.11	61.10	73.60
	0.8	180	2	6.18	8.65	8.86	8.86	42.33	53.51	61.70	71.23
	0.8	180	3	6.08	8.97	9.08	8.94	43.59	54.26	61.50	68.74

– the samples delaminated.

3.6. Effect of Adhesive Types on the Physical and Mechanical Properties of Plywood

Plywood bonded with UF and PF resins were manufactured for comparison with plywood bonded with HDPE film. All plywood contained approximately equal adhesive dosage. Table 8 summarizes the physical and mechanical properties of the three types of plywood.

Table 8. Physical and mechanical properties of plywood panels bonded with different adhesives *.

Adhesive Type	T (mm)	D (kg/m³)	Dry SS (MPa)	Wet SS (MPa)	MOR (∥) (MPa)	MOR (⊥) (MPa)	MOE (∥) (MPa)	MOE (⊥) (MPa)	TS (24 h) (%)	WA (24 h) (%)
UF	4.23	601.7	2.99	2.05	119.8	22.7	8228.2	775.2	9.38	53.12
PF	3.99	619.1	2.95	2.79	134.6	25.9	7698.9	748.6	11.84	61.29
HDPE **	4.63	553.2	2.99	2.38	119.3	22.5	9511.0	879.7	8.82	51.38

* T = thickness; D = density; SS = shear strength; MOR (∥) = bending strength in parallel direction; MOR (⊥) = bending strength in perpendicular direction; MOE (∥) = modulus of elasticity in bending in parallel direction; MOE (⊥) = modulus of elasticity in bending in perpendicular direction; TS = thickness swelling; WA = water absorption.
** The properties values for HDPE film are indicated for HDPE-bonded samples at 160 °C and 3 min.

The mechanical properties of HDPE-, UF-, and PF-bonded plywood panels were comparable despite the fact that these three polymers have considerably different mechanical properties, which affect the final properties of the panels. The lower TS and WA of the plywood could be explained by the fact that HDPE film filled the micropores of the wood veneers and covered a larger surface area of the hygroscopic wood component and thus prevents water penetration into the wood veneers [12,13].

HDPE-bonded plywood samples were made using a lower adhesive dosage (130 g/m^2) and at a lower hot-pressing pressure than UF and PF plywood samples. However, their properties are not inferior to these traditional plywood. In addition, HDPE-bonded plywood can be attributed to environmentally-friendly plywood. The formaldehyde emission of the plywood made from recycled plastics is very low; compared with that of ordinary plywood made with urea–formaldehyde resin, the amount of emission is almost zero [11]. In addition, the use of HDPE film makes plywood more flexible and simplifies the production of different bent constructions from plywood [21].

The values of the mechanical properties of the plywood panels obtained in this work were significantly higher than those obtained in other works [13,16,18,19,24], which mainly used poplar and very rarely eucalyptus [25], as well as different films. It can be assumed that wood species may have an effect on the ability to be bonded with a thermoplastic film.

4. Conclusions

The thermoplastic polymers were successfully used for the bonding of alder plywood. The findings of this work indicate that HDPE film adhesive gave bending strength, MOE in bending, and shear strength values of alder plywood panels that are comparable to those obtained with traditional UF and PF adhesives. Moreover, this is despite the fact that the adhesive dosage and pressing pressure were less than when UF and PF adhesives were used. Among the HDPE-bonded alder plywood panels, all the highest mechanical properties values were obtained for panels produced with high pressing temperature and pressing time. The highest shear strength was achieved at 160 °C when the pressing time was increased to 3 min. The smallest shear strength was achieved at 140 °C and a pressing time of 1–2 min. Environmentally-friendly high-density polyethylene-bonded formaldehyde-free alder plywood panels have been successfully produced using thermoplastic polymers as an adhesive.

Author Contributions: Conceptualization, P.B.; methodology, P.B. and J.S.; investigation, P.B. and J.S.; writing—original draft preparation, P.B.; writing—review and editing, P.B. and J.S.

Funding: The work was supported by the Slovak Academic Information Agency (SAIA) and by the Slovak Research and Development Agency under the contracts No. APVV-14-0506, APVV-16-0177, APVV-17-0583 and APVV-18-0378.

Acknowledgments: The work was supported by the Slovak Academic Information Agency (SAIA) and by the Slovak Research and Development Agency under the contracts No. APVV-14-0506, APVV-16-0177, APVV-17-0583 and APVV-18-0378.

Conflicts of Interest: The authors declare no conflict of interest.

References

1. De Melo, R.R.; Del Menezzi, C.H.S. Influence of veneer thickness on the properties of LVL from Parica (*Schizolobium amazonicum*) plantation trees. *Eur. J. Wood Prod.* **2014**, *72*, 191–198. [CrossRef]
2. FAO. Global Production and Trade of Forest Products in 2017. Available online: http://www.fao.org/forestry/statistics/80938/en/ (accessed on 19 January 2019).
3. Dunky, M. Adhesives in the wood industry. In *Handbook of Adhesive Technology*, 2nd ed.; Pizzi, A., Mittal, K.L., Eds.; Marcel Dekker, Inc.: New York, NY, USA, 2003; Chapter 47; p. 71.
4. Formaldehyde, 2–Butoxyethanol and 1–tert–Butoxypropan–2–ol. In *Monographs on the Evaluation of Carcinogenic Risk to Humans*; World Health Organization—International Agency for Research on Cancer: Lyon, France, 2006; Volume 88.
5. Roffael, E. *Die Formaldehydabgabe von Spanplatten und Anderen Werkstoffen*; DRW–Verlag: Stuttgart, Germany, 1982.
6. Yorur, H. Utilization of waste polyethylene and its effects on physical and mechanical properties of oriented strand board. *BioResources* **2016**, *11*, 2483–2491. [CrossRef]
7. De Barros Lustosa, E.C.; Del Menezzi, C.H.S.; de Melo, R.R. Production and properties of a new wood laminated veneer/high-density polyethylene composite board. *Mater. Res.* **2015**, *18*, 994–999. [CrossRef]
8. Järvelä, P.K.; Tervala, O.; Järvelä, P.A. Coating plywood with a thermoplastic. *Int. J. Adhes. Adhes.* **1999**, *19*, 295–301. [CrossRef]
9. Tervala, O.; Järvelä, P. Thermoplastic coating processes for plywood. *Polym. Plast. Technol. Eng.* **1999**, *38*, 831–848. [CrossRef]
10. Kuusipalo, J. Plastic coating of plywood using extrusion technique. *Silva Fenn.* **2001**, *35*, 103–110. [CrossRef]
11. Cui, T.; Song, K.; Zhang, S. Research on utilizing recycled plastic to make environment-friendly plywood. *For. Stud. China* **2010**, *12*, 218–222. [CrossRef]
12. Fang, L.; Chang, L.; Guo, W.; Chen, Y.; Wang, Z. Manufacture of environmentally friendly plywood bonded with plastic film. *For. Prod. J.* **2013**, *63*, 283–287. [CrossRef]
13. Fang, L.; Chang, L.; Guo, W.; Ren, Y.; Wang, Z. Preparation and characterization of wood-plastic plywood bonded with high density polyethylene film. *Eur. J. Wood Prod.* **2013**, *71*, 739–746. [CrossRef]

14. Fang, L.; Yin, Y.; Han, Y.; Chang, L.; Wu, Z. Effects of a number of film layers on the properties of a thermoplastic bonded plywood. *J. For. Eng.* **2016**, *1*, 45–50. [CrossRef]
15. Song, W.; Wei, W.; Ren, C.; Zhang, S. Developing and evaluating composites based on plantation eucalyptus rotary-cut veneer and high-density polyethylene film as novel building materials. *BioResources* **2016**, *11*, 3318–3331. [CrossRef]
16. Demirkir, C.; Öztürk, H.; Çolakoğlu, G. Effects of press parameters on some technological properties of polystyrene composite plywood. *Kast. Univ. J. For. Fac.* **2017**, *17*, 517–522. [CrossRef]
17. Chang, L.; Guo, W.; Tang, Q. Assessing the tensile shear strength and interfacial bonding mechanism of poplar plywood with high-density polyethylene films as adhesive. *BioResources* **2017**, *12*, 571–585. [CrossRef]
18. Song, W.; Wei, W.; Li, X.; Zhang, S. Utilization of polypropylene film as an adhesive to prepare formaldehyde-free, weather-resistant plywood-like composites: Process optimization, performance evaluation, and interface modification. *BioResources* **2017**, *12*, 228–254. [CrossRef]
19. Chang, L.; Tang, Q.; Gao, L.; Fang, L.; Wang, Z.; Guo, W. Fabrication and characterization of HDPE resins as adhesives in plywood. *Eur. J. Wood Prod.* **2018**, *76*, 325–335. [CrossRef]
20. Zike, S.; Kalnins, K. Enhanced impact absorption properties of plywood. In Proceedings of the 3rd International Conference CIVIL ENGINEERING'11, Jelgava, Latvia, 12–13 May 2011; Volume 3, pp. 125–130.
21. Kajaks, J.; Reihmane, S.; Grinbergs, U.; Kalnins, K. Use of innovative environmentally friendly adhesives for wood veneer bonding. *Proc. Est. Acad. Sci.* **2012**, *61*, 207–211. [CrossRef]
22. Borysiuk, P.; Mamiński, M.Ł.; Parzuchowski, P.; Zado, A. Application of polystyrene as binder for veneers bonding—The effect of pressing parameters. *Eur. J. Wood Prod.* **2010**, *68*, 487–489. [CrossRef]
23. Matuana, L.M.; Balatinecz, J.J.; Park, C.B. Effect of surface properties on the adhesion between PVC and wood veneer laminates. *Polym. Eng. Sci.* **1998**, *38*, 765–773. [CrossRef]
24. Song, W.; Wei, W.; Wang, D.; Zhang, S. Preparation and properties of new plywood composites made from surface modified veneers and polyvinyl chloride films. *BioResources* **2017**, *12*, 8320–8339. [CrossRef]
25. Song, W.; Wei, W.; Ren, C.; Zhang, S. Effect of heat treatment or alkali treatment of veneers on the mechanical properties of eucalyptus veneer/polyethylene film plywood composites. *BioResources* **2017**, *12*, 8683–8703. [CrossRef]
26. Sorensen, R. Dry film gluing in plywood manufacture. In Proceedings of the Semi-Annual Meeting of the American Society of Mechanical Engineers, Chicago, IL, USA, 26 June–1 July 1933; pp. 37–48.
27. Follrich, J.; Muller, U.; Gindl, W. Effects of thermal modification on the adhesion between spruce wood (*Picea abies Karst.*) and a thermoplastic polymer. *Holz Roh Werkst.* **2006**, *64*, 373–376. [CrossRef]
28. Lu, J.; Wu, Q.; McNabb, H. Chemical coupling in wood fiber and polymer composites: A review of coupling agents and treatments. *Wood Fiber Sci.* **2000**, *32*, 88–104.
29. John, R.; Trommler, K.; Schreiter, K.; Siegel, C.; Simon, F.; Wagenführ, A.; Spange, S. Aqueous poly(N-vinylformamide-co-vinylamine) as a suitable adhesion promoter for wood veneer/biopolyethylene composite materials. *BioResources* **2017**, *12*, 8134–8159. [CrossRef]
30. Fang, L.; Chang, L.; Guo, W.; Chen, Y.; Wang, Z. Influence of silane surface modification of veneer on interfacial adhesion of wood–plastic plywood. *Appl. Surf. Sci.* **2014**, *288*, 682–689. [CrossRef]
31. Fang, L.; Xiong, X.; Wang, X.; Chen, H.; Mo, X. Effects of surface modification methods on mechanical and interfacial properties of high-density polyethylene-bonded wood veneer composites. *J. Wood Sci.* **2017**, *63*, 65–73. [CrossRef]
32. Goto, T.; Saiki, H.; Onishi, H. Studies on wood gluing. XIII: Gluability and scanning electron microscopic study of wood-polypropylene bonding. *Wood Sci. Technol.* **1982**, *16*, 293–303. [CrossRef]
33. Tang, L.; Zhang, Z.; Qi, J.; Zhao, J.; Feng, Y. The preparation and application of a new formaldehyde-free adhesive for plywood. *Int. J. Adhes. Adhes.* **2011**, *31*, 507–512. [CrossRef]
34. Tang, L.; Zhang, Z.G.; Qi, J.; Zhao, J.R.; Feng, Y. A new formaldehyde-free adhesive for plywood made by in-situ chlorinating grafting of MAH onto HDPE. *Eur. J. Wood Prod.* **2012**, *70*, 377–379. [CrossRef]
35. *EN 323. Wood-Based Panels—Determination of Density*; European Committee for Standardization: Brussels, Belgium, 1993.
36. *EN 310. Wood-Based Panels—Determination of Modulus of Elasticity in Bending and of Bending Strength*; European Committee for Standardization: Brussels, Belgium, 1993.
37. *EN 314-1. Plywood—Bonding Quality—Part 1: Test Methods*; European Committee for Standardization: Brussels, Belgium, 2004.

38. *EN 314-2. Plywood—Bonding Quality—Part 2: Requirements*; European Committee for Standardization: Brussels, Belgium, 1993.
39. *EN 317. Particleboards and Fibreboards. Determination of Swelling in Thickness after Immersion in Water*; European Committee for Standardization: Brussels, Belgium, 1993.
40. *EN 315. Plywood. Tolerances for Dimensions*; European Committee for Standardization: Brussels, Belgium, 2000.
41. Smith, M.J.; Dai, H.; Ramani, K. Wood-thermoplastic adhesive interface—Method of characterization and results. *Int. J. Adhes. Adhes.* **2002**, *22*, 197–204. [CrossRef]
42. *EN 13353. Solid Wood Panels (SWP)—Requirements*; European Committee for Standardization: Brussels, Belgium, 2008.

Article

New Perspective on Wood Thermal Modification: Relevance between the Evolution of Chemical Structure and Physical-Mechanical Properties, and Online Analysis of Release of VOCs

Jiajia Xu [1,†], Yu Zhang [1,†], Yunfang Shen [2], Cong Li [1], Yanwei Wang [3], Zhongqing Ma [1,*] and Weisheng Sun [1,*]

[1]   School of Engineering, Zhejiang Provincial Collaborative Innovation Center for Bamboo Resources and High-Efficiency Utilization, Zhejiang A & F University, Hangzhou 311300, Zhejiang, China
[2]   Zhejiang Shenghua Yunfeng Greeneo Co. Ltd., Huzhou 313220, Zhejiang, China
[3]   Treessun Flooring Co. Ltd., Huzhou 313009, Zhejiang, China
*   Correspondence: mazq@zafu.edu.cn (Z.M.); sunweisheng@zafu.edu.cn (W.S.);
     Tel.: +86-571-6110-0905 (Z.M. & W.S.)
†   These authors were contributed equally to this manuscript.

Received: 26 May 2019; Accepted: 2 July 2019; Published: 4 July 2019

Abstract: Thermal modification (TM) is an ecological and low-cost pretreated method to improve the dimensional stability and decay resistance of wood. This study systematically investigates the relevance between the evolution of chemical structure and the physical and mechanical properties during wood thermal modification processes. Moreover, the volatility of compounds (VOCs) was analyzed using a thermogravimetric analyzer coupled with Fourier transform infrared spectrometry (TGA-FTIR) and a pyrolizer coupled with gas chromatography/mass spectrometer (Py-GC/MS). With an increase of TM temperature, the anti-shrink efficiency and contact angle increased, while the equilibrium moisture content decreased. This result indicates that the dimensional stability improved markedly due to the reduction of hydrophilic hydroxyl (–OH). However, a slight decrease of the moduli of elasticity and of rupture was observed after TM due to the thermal degradation of hemicellulose and cellulose. Based on a TGA-FTIR analysis, the small molecular gaseous components were composed of H_2O, CH_4, CO_2, and CO, where H_2O was the dominant component with the highest absorbance intensity, i.e., 0.008 at 200 °C. Based on the Py-GC/MS analysis, the VOCs were shown to be mainly composed of acids, aldehydes, ketones, phenols, furans, alcohols, sugars, and esters, where acids were the dominant compounds, with a relative content of 37.05–42.77%.

Keywords: wood; thermal modification; mechanical properties; dimensional stability; color; chemical structure; VOCs

1. Introduction

Regarded as a renewable natural composite material, wood has been widely used to produce construction materials, flooring, furniture, and interior finishing materials because of its versatile properties, e.g., favorable strength-to-weight ratio, ease of shaping with tools, as well as beautiful grain and color [1,2]. However, the outdoor utilization of wood is highly limited by its strong hygroscopicity and low durability. Thermal modification (TM) is an ecological and low-cost pretreatment method to improve the dimensional stability and decay resistance of wood without using any toxic chemicals [3]. It is normally performed at between 160 to 260 °C in a vacuum, nitrogen, air, or oil environments [4–6].

During the last decade, thermal modifications of wood have been extensively studied and applied commercially [7–13]. Bruno et al. found that dimensional stability in the radial and tangential directions increased by 88% and 96% at 200 °C, respectively [14]. Bal et al. studied the effect of temperature (120, 150, and 180 °C) and the duration (4, 6, and 8 h) of thermal modification on the mechanical properties (e.g., MOE, MOR, compression strength, and impact bending) of Eucalyptus grandis; it was found that these properties decreased with an increase of TM temperature and duration. However, the influence of temperature was more remarkable than that of duration [15]. Candelier et al. showed that the bending strength of TM wood decreased by about 45%, and the elastic modulus by about 12% at 230 °C [16]. Lin et al. investigated the variation of chemical structure and composition of wood at different TM temperatures (200, 210, 220 and 230 °C), and found that the hygroscopic hydroxyl and oxygen element was remarkably removed by dehydration reactions, resulting in great improvements in dimensional stability [17]. However, a systematic investigation on the effect of TM temperature and duration on the chemical structure and the physical-mechanical properties (e.g., dimensional stability, color, and surface functional group) of oak (*Quercus alba* L.) has not been reported.

During thermal modification processes, a certain amount of volatile organic compounds (VOCs) will be released from the thermal degradation of wood, such as terpenes, aldehydes, acids, and alcohols. The emission of exhaust organic gas containing high concentrations of VOCs might result in serious atmospheric contamination, which could negatively affect human health. Traditionally, the identification of VOCs comprised two steps: VOCs were collected by condensation or extraction techniques, and then analyzed using a gas chromatography/mass spectrometer (GC/MS) [18–20]. Liu et al. extracted the VOCs adsorbed in activated carbon using methylene chloride and analyzed them using GC/MS. The result showed that butanedioic acid, bis (2-methylpropyl) ester was the dominant component in the volatiles of a pipe thermal modification process, accounting a content of 40.67% [21]. Kačík et al. extracted the terpenes adsorbed in sawdust using hexane and analyzed them using GC/MS; it was found that recent fir wood contained approximately 60 times more terpenes than older wood (186 vs. 3.1 mg/kg) [22]. Manninena et al. compared the content and components of VOCs in air-dried and heat-treated pine wood, and found that the former released about 8 times more VOCs than the latter [23]. Hyttinen et al. also found that the levels of VOCs released from heat treated wood were much lower than those from air-dried wood [24]. However, according to the literature, the traditional detection of VOCs is a complex process, requiring a long-period of experimentation and showing poor repeatability. Therefore, developing a quick and simple online detection method is essential for gaining a better understanding of the properties of VOCs.

A thermogravimetric analyzer coupled with Fourier transform infrared spectrometry (TGA-FTIR) and a pyrolizer coupled with a gas chromatography/mass spectrometer (Py-GC/MS) were traditionally used to online investigate the components of pyrolysis volatiles of lignocellulosic biomasses [25–27]. TGA-FTIR analysis makes it possible to investigate weight loss characteristics during biomass thermal degradation processes, as well as to identify the evolved gas components in real time, especially for the small molecular weight gas components (H_2O, CO_2, CO, and CH_4) [28,29]. Py-GC/MS was developed for the further qualitative and quantitative real-time analysis of each organic component in the volatiles, providing the advantages of rapid analyses, high sensitivity, and effective identification of complex organic compounds released from the wood thermal modification processes [30,31]. Until now, these two instruments have been extensively employed to analyze the components of pyrolysis volatiles of different lignocellulosic biomasses [29,32,33], or their three pseudo components, cellulose [34], hemicellulose [35,36], and lignin [26,31,37]. However, the application of these two instruments in the research field of wood TM has not been reported.

In this study, the thermal modification of white oak (*Quercus alba* L.) was carried out at different temperatures (160, 180 and 200 °C) and holding times (3, 6, 9 h). Then, the relevance between the evolution of chemical structure (e.g., elementary composition, surface functional group, and crystallinity) and the physical-mechanical properties (e.g., dimensional stability, MOE, MOR, color, and contact angle) were systematically investigated using a Universal Testing Machine, Elementary Analyzer, Colorimeter, FTIR, and XRD. Furthermore, the release characteristics of VOCs was online detected by TGA-FTIR and Py-GC/MS.

2. Materials and Methods

2.1. Materials

White Oak (*Quercus alba* L.) was used for the thermal modification experiment; samples were purchased from Treessun Flooring Co. Ltd., Huzhou City, Zhejiang Province, China. The Oak was first cut into different sizes, depending on the different test methods, such as a dimension of 300 mm × 20 mm × 20 mm (*L* × *W* × *H*) for the test of bending strength, 30 mm × 20 mm × 20 mm (*L* × *W* × *H*) for the test of shrinkage, and powder with particle sizes between 220 to 280 meshes for the test of the release of VOCs. Before the thermal modification experiment, the sample was packaged inside sealed plastic bags and stored in a dryer at room temperature. The flow diagram of the wood thermal modification experiment is shown in Figure 1. The statistical analysis (standard deviation and *p*-value) of all physical and mechanical properties of TM wood are listed in Tables S1–S7 (Supplementary Materials).

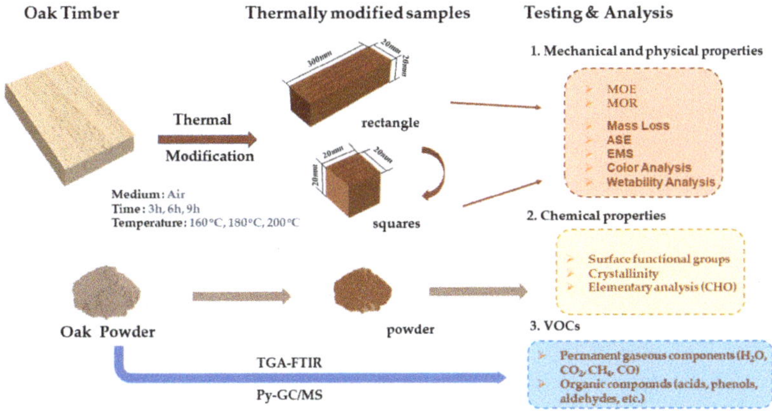

Figure 1. Flow diagram of wood thermal modification experiment.

2.2. Thermal Modification Experiment

The thermal modification experiment of oak was carried out in an oven (WFO-710, Shanghai Ailang instrument Co., LTD, Shanghai, China) with an air atmosphere. The designed temperatures for heat treatment were 160, 180, and 200 °C and the durations were 3, 6, and 9 h. The thermally-modified (TM) samples at different temperatures and durations were labeled as TM-xxx-y, where "xxx" represented the temperature of the heat treatment and "y" the duration. Fox example, TM-180-3 represented a sample which was treated at 180 °C for 3 h. All thermal modification experiments were repeated at least 3 times.

2.2.1. Mechanical and Physical Properties

The mass loss (ML) of wood after thermal modification was determined by Equation (1), where m_0 is the initial mass of the untreated sample, and m_1 is the mass after thermal modification.

$$\text{ML (\%)} = 100 \times (m_0 - m_1)/m_0 \tag{1}$$

Prior to the test of mechanism properties, the raw and thermally-modified samples were conditioned at 20 °C and 65 % relative humidity for the necessary time to stabilize the mass of the samples. The equilibrium moisture content (EMC) and anti-swelling efficiency (ASE) was tested according to the national standard of GB/T 1931-2009 and GB/T 1934.2-2009, respectively. The modulus of elasticity (MOE) and modulus of rupture (MOR) were tested according to the national standards, i.e., GB/T 1936.2-1991 and GB/T 1936.1-2009, respectively.

2.2.2. Color Analysis

The variations of color of the TM samples were measured by a colorimeter (DC–P3, Beijing Xingguang Color Measuring Instrument Co., Ltd., Beijing, China). In this instrument, a D65 light source, a 10° visual field, and a sensor head with 6 mm diameter were employed. Then, the color parameters of control and TM samples, namely L^* (lightness coordinate), a^* (red and green coordinates), and b^* (yellow and blue coordinates) were recorded. In order to ensure the accuracy of the results, the color was measured on three specific places on each sample. Finally, the total color differences (ΔE^*) were calculated according to Equation (2):

$$\Delta E^* = \sqrt{\Delta L^{*2} + \Delta a^{*2} + \Delta b^{*2}}, \tag{2}$$

where ΔL^*, Δa^*, and Δb^* are the differences of parameters before and after TM.

2.2.3. Chemical Properties

The ultimate analysis (C, H, and O) of the raw and TM samples was performed using an elemental analyzer (Vario EL III, Elementary, Germany). The chemical functional groups of the raw and TM samples were tested by Fourier transform infrared spectrometry (Nicolet 6700, Thermo Fisher Scientific, Massachusetts, USA). The mass ratio of KBr to bio-char was 100. The resolution and spectral region of the recorded FTIR spectra were 4 cm^{-1} and 4000–400 cm^{-1}, respectively, and the spectrum scan time was set at 8 s intervals.

The crystallographic structure of the raw and TM samples was tested using an X-ray diffractometer (XRD 6000, Shimadzu, Kyoto, Japan) with Cu radiation at 40 kV and 30 mA. Scans were performed at a speed of 0.5° min^{-1} over an angle (θ) range of 5° to 40°. The crystallinity index (CrI) was calculated using Equation (3) according to Segal et al. [38], where C_rI is the crystallinity index, I_{002} represents the intensity of the 200 crystalline peaks, and I_{am} represents the intensity of the diffraction of the amorphous part.

$$\text{CrI (\%)} = 100 \times (I_{002} - I_{am})/ I_{002} \tag{3}$$

The contact angle was measured using optical contact angle measuring and contour analysis systems (OCA 200, Data Physics Instruments GmbH, Filderstadt, Germany), and through the disposition of a distilled water droplet (5 μL) in three distinct points of a tangential section of the wood samples. The data was recorded after 20 s of the droplet contacting the sample surface.

2.3. Online Analysis of VOCs by Using TGA-FTIR and Py-GC/MS

2.3.1. TGA-FTIR Analysis

A thermogravimetric analyzer coupled with Fourier transform infrared spectrometry (TGA-FTIR) allowed us to investigate the weight loss characteristics during wood thermal modification processes, as well as to identify the evolution of the gas components in real time, especially for small molecular weight gaseous components (H_2O, CO_2, CO, and CH_4). The instrument models of the TGA and FTIR were TGA-8000 and Frontier, respectively, both of which were made by PerkinElmer Co., Ltd, Waltham, MA, USA. The settled thermal modification temperatures were 160, 180, and 200 °C, with a fixed heating rate of 10 °C min^{-1} and a holding time of 30 min. In order to enhance the intensity of the infrared characteristic absorption peaks of the permanent gas components, 35 mg wood powder was used in each experiment. The carried gas was high-purity nitrogen (99.999%) with a flow rate of 40 mL min^{-1}. The resolution and spectral region of the FTIR were 4 cm^{-1} and 4000–400 cm^{-1} respectively, and the spectrum scan time was set at 8 s intervals. More detailed information on the experiment may be found in our previous publications [27,39].

According to the Lambert-Beer law, the intensity of absorbance of a characteristic infrared absorbance band is linearly dependent on the concentration of the evolved gas component. In order to normalize FTIR data for comparison, first, the same initial mass (35 mg) of samples was used for TGA-FTIR analysis. Then, the experimental parameters in the TGA-FTIR analysis were fixed, e.g., heating rate (10 °C min^{-1}), holding time (30 min), flow rate of carrier gas (40 mL min^{-1}), spectrum scan time (8 s). The only variable was the thermal modification temperature (160, 180, and 200 °C), to investigate the effect of TM temperature on the properties of VOCs.

2.3.2. Py–GC/MS Analysis

Py–GC/MS was used for the qualitative and semi-quantitative analyses of the organic compounds during wood thermal modification processes. The organic compounds released from such processes were on-line analyzed using a pyrolizer (5200, Chemical Data Systems Analytical, Oxford, Pennsylvania, USA) coupled with a gas chromatography/mass spectrometer (7890B-5977B, Agilent Technology, Palo Alto, California, USA). First, 0.5 mg of wood powder was put into a quartz filler tube and then heated to the target torrefaction temperatures of 160, 180, and 200 °C at a heating rate of 10 °C ms^{-1}, and maintained for 20 s. The GC oven was first heated to 40 °C for 3 min, then raised to 290 °C (10 °C min^{-1}) and maintained at that temperature for 3 min. The organic components were identified according to the NIST library and the literature. Other experimental information may be found in references [30,32,40].

3. Results and Discussion

3.1. Mass Loss

The effect of TM on mass loss (ML) is shown in Figure 2. The results showed that higher temperatures and longer durations led to an increase of ML, ranging between 10.78% and 19.10%. In addition, temperature had a more remarkable influence on ML than duration. This result was confirmed by other researchers [7]. Srinivas et al. found that ML gradually increased from 3% to 18% with an increase of temperature and duration of TM from 210 °C and 2 h to 240 °C and 8 h, respectively [41]. Wang et al. also reported that ML reached its maximum value of 18.1% under the severest TM conditions, i.e., 190 °C and 6 h [42]. In the low temperature range (<160 °C) of TM, the mass loss was a consequence of the evaporation of free and hydroscopic water [17]. However, at higher temperatures (>160 °C), the mass loss was mainly due to the thermal degradation of hemicellulose [3]. This was caused by the fact that hemicellulose showed the lowest thermal stability among the three biomass pseudo components (cellulose, hemicellulose, and lignin) within a temperature range of 100–365 °C [28]. The mass loss in wood transfers into other small molecular weight components, such as CO, CO_2, H_2O, CH_4, and VOCs, etc.

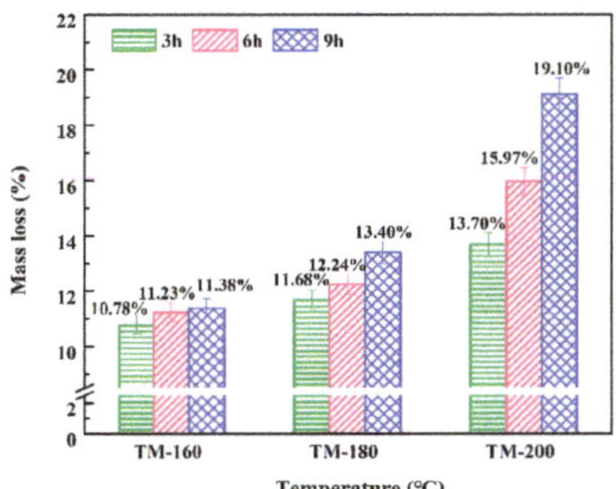

Figure 2. Effects of thermal treatment on the mass loss of wood.

3.2. XRD Analysis

The XRD spectra of control and thermally-modified samples are shown in Figure 3a. Four characteristic diffraction peaks at 2θ of 15.2°, 16.5°, 22.2° and 34.6° were clearly observed, corresponding to the triclinic cellulose I_α ($1\overline{1}0$ and 110), monoclinic cellulose I_β (200), and the glucan chains of cellulose (400) [43]. The most remarkable variation was observed in the peak of monoclinic cellulose I_β (200); the intensity of this peak slightly increased with the increase of temperature and duration, indicating an increase in the crystallinity index (CrI) of cellulose. This result was caused by the thermal degradation of part of the hemicellulose and the rearrangement of cellulose molecules in the amorphous region [42,44]. As shown in Figure 3b, the value of CrI increased from 41.81% in the control sample to 44.38% of TM-200-9. Other references also reported the similar conclusions [5,45]. Okon et al. reported that CrI increased from 38.83% to 63.78% of the control sample from an oil heat treatment at 210 °C and 8 h [5]. Wang et al. also found that the CrI increased from 55.55% of the control sample to 63.33% after TM treatment at 190 °C and 6 h [42]. The decrease of the amorphous region in cellulose would result in a decrease of hydroxyl (–OH), leading to a significant reduction in the hygroscopicity of wood after TM [46].

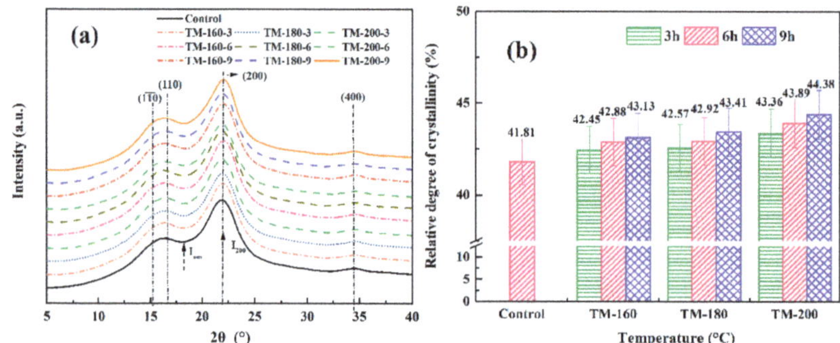

Figure 3. XRD analysis of the control and thermally-modified wood: (**a**) the XRD spectra; (**b**) the crystallinity index (CrI).

3.3. FTIR Analysis

The evolution of the surface functional groups of wood at different TM temperatures and durations is exhibited in Figure 4. Five characteristic absorbance bands may be clearly observed in the IR spectra of control and TM wood. The first band is the stretching vibration of hydroxyl (−OH) at the wavenumber of 3460 cm^{-1} [47,48]. The band at 1706 cm^{-1} is ascribed to the stretching vibration of C=O, which is mainly derived from the carbonyl (−C=O) and carboxyl functional groups (−COOH) [49]. The characteristic absorbance peak between 1680−1440 cm^{-1} is the stretching vibration of a benzene ring skeleton (C=C) from lignin [47]. The absorbance peak at 1190−950 cm^{-1} is attributed to the stretching vibration of C−O and C−H derived from aliphatic −CH$_3$ or phenolic−OH bonds [17,50]. The band at 592 cm^{-1} is mainly due to aliphatic −CH$_2$ and alkanes −CH$_3$ [39,51].

Figure 4a shows the effect of TM temperature on the surface functional groups of wood. Overall, the absorbance intensity of the five characteristic adsorption bands decreased as the torrefaction temperature increased from 160 to 200 °C. The decrease of the intensity of −OH indicated that a series of deacetylation reactions had occurred; this resulted in the formation of H$_2$O during TM process, and in a remarkable increase in hydrophobicity [3,50]. The decrease of the intensity of C=O was likely due to decarboxylation and decarbonylation reactions within the structures of cellulose and hemicellulose. The slight decrease of the intensity of C=C or the benzene ring skeleton indicated the thermal degradation of lignin. The decrease of C−O and C−H indicated the degradation of methyl and hydroxyl groups. The band of the aliphatic −CH$_2$ and alkanes −CH$_3$ exhibited a reduced absorbance intensity, indicating that the aliphatic regions of cellulose and hemicellulose were degraded.

Figure 4b shows the effect of TM duration on the surface functional groups of wood. Overall, longer TM durations also resulted in the decrease of the absorbance intensity of the five adsorption bands. FTIR was also employed by Cademartori et al. and Sikora et al. to investigate the effect of TM on the content of surface functional groups of wood. Their results also confirmed that the intensities of several characteristic absorbance peaks decreased with the increase of TM temperature and duration [4,52].

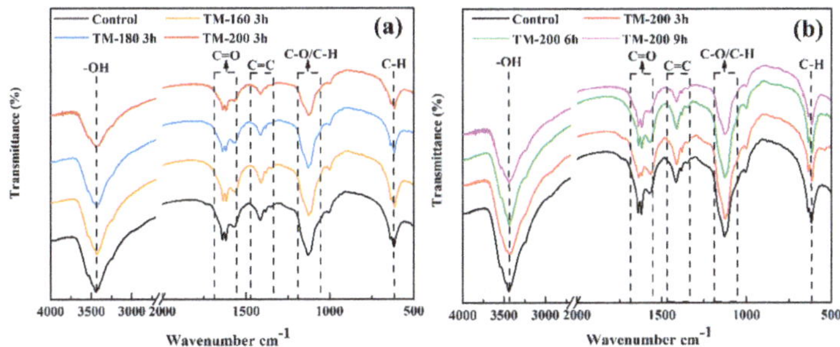

Figure 4. FTIR analysis of the control and thermally-modified wood.

3.4. Ultimate Analysis

The ultimate analysis (C, H, and O element) of the control and TM wood is shown in Table 1. The content of H and O was slightly decreased with an increase of the severity of TM. Fox example, the content of H and O decreased from 6.01% and 47.26% of TM-160-3 to 5.92% and 45.84% of TM-200-9, respectively. The reduction of H and O strongly supports the conclusion that a series of dihydroxylation (−OH) reactions occurred during the TM process, resulting in an increase in dimensional stability [3,53]. However, the content of C increased from 46.74% of TM-160-3 to 48.14% of TM-200-9 [54]. Boonstra et al. also found that the content of C slightly increased from 49.6%, in non-treated samples, to 50.6% in TM samples at 180 °C [3].

Table 1. Ultimate analysis of the control and thermally-modified wood.

Element Content/ (wt.%)	Control and Thermally-Modified Wood									
	Control	TM-160-3	TM-160-6	TM-160-9	TM-180-3	TM-180-6	TM-180-9	TM-200-3	TM-200-6	TM-200-9
C	45.83	46.74	46.89	46.95	47.07	47.68	47.62	47.62	47.93	48.14
H	6.46	6.01	5.99	5.93	6.04	6.04	5.95	6.00	5.97	5.92
O	47.60	47.26	47.10	47.00	46.99	46.16	46.34	46.26	45.95	45.84

3.5. Contact Angles

The wettability of TM samples was evaluated by the contact angle, where a large contact angle corresponded to greater hydrophobicity and better dimensional stability [55]. Figure 5a shows the pictures of the contact angle testing process. Higher TM temperatures and longer durations would result in round-shaped water droplets on the surface; otherwise, flatter shapes were observed, since the liquid was rapidly absorbed, leading to small contact angles [56].

Figure 5b shows the effect of the severity of thermal modification on the contact angle in the tangential section of samples. For the control sample, the contact angle was only 67.63°. However, with the increase of the severity of TM, the contact angles gradually increased to 132.7° for the sample of TM-200-9, indicating that TM was an effective method to improve the hydrophobicity of wood. This result was caused by the dehydration reaction of carbohydrates during thermal modification, reducing the number of hydrophilic groups and restraining the accessibility of free hydroxyl groups to water [51,53]. A similar trend was confirmed by other researchers. Bakar et al. found that the the contact angle of red oak increased from 68° to 143° of the control sample under TM conditions of

190 °C and 8 h [56]. Lee et al. reported that the contact angle of bamboo increased from 53.2° to 116.8° of control sample at TM conditions of 210 °C and 4 h [57]. After treatment at higher temperatures, the contact angles were all higher than 90°, suggesting that TM pretreatment is highly beneficial for surface hydrophobicity.

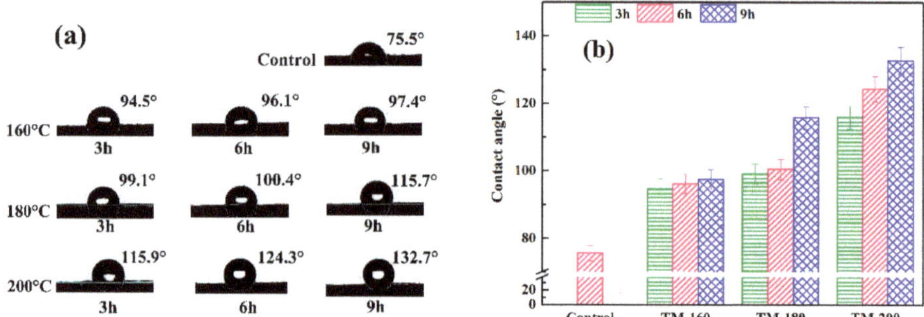

Figure 5. Contact angle in the tangential section of the control and thermally-modified woods.

3.6. EMC and ASE

Figure 6a shows the effect of TM on the equilibrium moisture content (EMC). Compared to the control sample, the EMC of thermally-modified samples was markedly decreased. Meanwhile, increasing the severity of TM would result in a lower value of EMC, decreasing from 7.39% of TM-160-3 to 5.59% of TM-200-9. The decrease of wettability was caused by the elimination of hydroxyl (–OH) linked on hemicellulose and the reduction of the amorphous region of cellulose, which reduces hydrogen bond interactions between the hemicellulose/cellulose and the water from the humid atmosphere [2,3,58].

As shown in Figure 6b, the anti-shrink efficiency (ASE) was also improved with an increase in the severity of TM. The ASE was increased from 23.56% of TM-160-3 to 36.24% of TM-200-9. This result was also confirmed by Ayrilmis et al. and Gonzálezpeña et al [48,59]. Wang et al. found that the EMC of the control sample decreased from 11.3% to 6.6% in TM-190-6 [42]. The ASE of TM samples is positively correlated with TM severity, and a maximum value of 56% was observed in TM-190-6. Although the EMC and ASE of wood are dominantly affected by TM temperature and duration, the influence factor of wood species cannot be ignored [59]. For example, eucalyptus has higher swelling and lower dimensional stability in nature [59]. In conclusion, TM is an effective pretreatment method to improve the dimensional stability of wood.

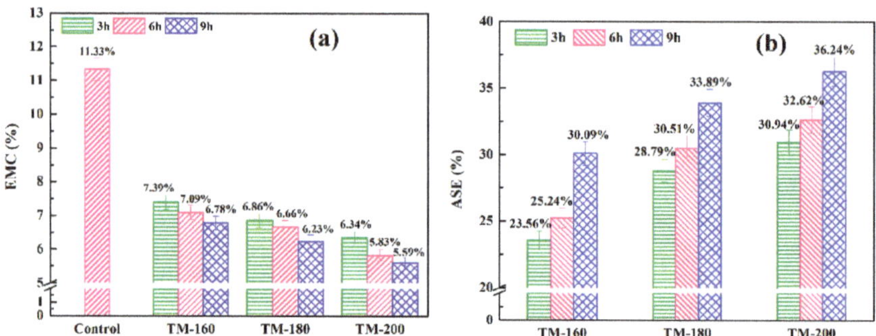

Figure 6. Effect of thermal modification on the EMC (**a**) and ASE (**b**).

3.7. MOR and MOE

The Modulus of Elasticity (MOE) and Modulus of Rupture (MOR) are two important mechanical properties of wood. Figure 7 shows the effect of TM temperature and duration on MOE and MOR. As show in Figure 7b, the MOR was gradually decreased with an increase of temperature and duration. Fox example, the MOR was gradually decreased from 203.85 MPa of the control sample to 169.28 MPa of TM-200-9. The decrease of MOR was mainly due to the acceleration of the thermal degradation of hemicellulose at higher temperatures and longer durations [41,60,61]. This result was confirmed by other researches [59,60]. Ayrilmis et al. reported that after TM at 180 °C, the MOR and MOE of eucalyptus wood fibers decreased by 5–19% and 7–22%, respectively [59]. The decrease of MOR was highly related to the thermal degradation of cellulose and hemicellulose.

However, the MOE firstly increased from 9.23 GPa of the control sample to 10.84 GPa of TM-160-9, and then gradually decreased to 7.64 GPa of TM-200-9 (shown in Figure 7a). Similar results were obtained in studies conducted by Guo et al. and Yildiz et al. [62,63]. Guo et al. studied the effect of TM on the mechanical properties of white poplar (Populous tomentosa.). The results showed that the MOE firstly increased by 13% after TM at 200 °C for 1 h, before decreasing by 9% after TM at 250 °C for 5 h [62]. The increase of MOE at 160 °C and 180 °C was mainly due to the increase of the crystallization of cellulose and condensation of lignin via cross-linking reactions with furfural produced from the thermal degradation of hemicellulose [15,42,64].

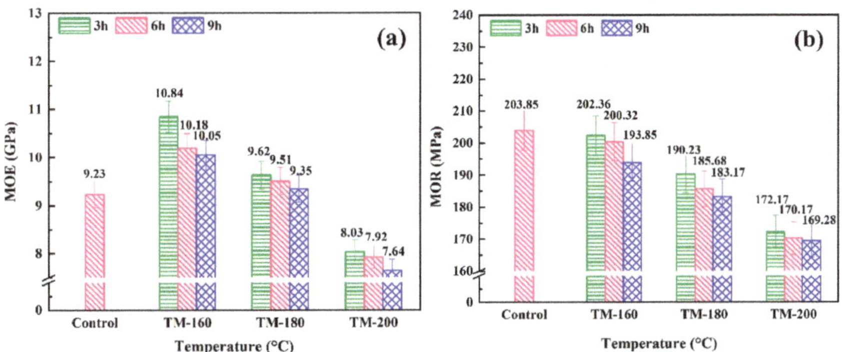

Figure 7. Effect of thermal modification on MOE (**a**) and MOR (**b**).

3.8. Color Analysis

Figure 8a shows the surface color of the control and TM woods. The results showed that TM could induce remarkable color variations on the wood surface; higher temperatures and durations were associated with darker color. The variations of color parameters (ΔL^*, Δa^*, Δb^*, ΔE^*) before and after thermal modification are shown in Figure 8b. The ΔL^* (lightness coordinate), Δa^* (red/green coordinate), and Δb^* (yellow /blue coordinate) were negative values derived from the difference values before and after thermal modification. Researchers reported that higher difference values indicated large color variation [4,57]. The absolute values of ΔL^*, Δa^*, Δb^*, and ΔE^* all increased with an increase of the severity of TM. The increase of the absolute value of ΔL^* from −32.20 of TM-160-3 to −107.50 of TM-200-9 indicated a reduction of lightness. A similar trend was observed on Δa^*, which increased from −7.23 of TM-160-3 to −23.25 of TM-200-9, resulting in the surface taking on a reddish color. The increase of the absolute value of Δb^* from −1.85 of TM-160-3 to −6.86 of TM-200-9 resulted in the surface taking on a yellow color. The increase of total color differences (ΔE^*) indicated the variation trend of surface color towards darker tones, which was in accordance with the color variation shown in Figure 8a.

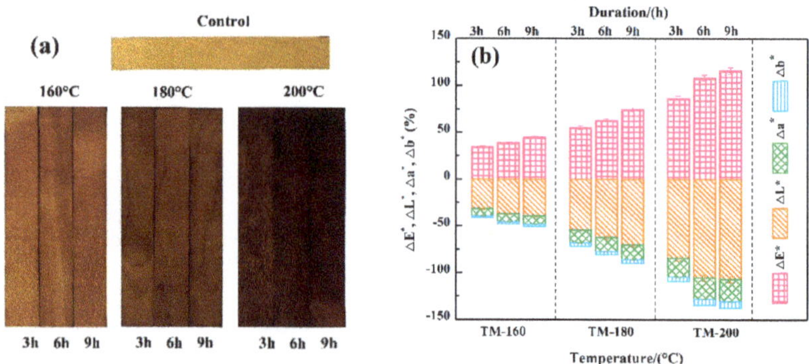

Figure 8. Surface color (**a**) and color parameters (**b**) of the control and thermally-modified woods.

The variation of surface color was caused by the increase of chromophores formed in the TM process. Firstly, the acetic acid formed by the deacetylation of hemicellulose will act as a catalyst to promote the oxidation and dehydration reactions of lignin or carbohydrates to form new chromophores, particularly, carbonyl and carboxyl groups [41,51]. Then, the acetic acid will also promote the substitution reaction of free hydroxyl groups to form ether bonds, and polycondensation of phenolic hydroxyl groups to form oxidation products such as conjugated aromatic ketone and quinones, resulting in a progressively darker wood color [2,60,65]. Bekhta et al. reported that changes in the color of spruce wood, i.e., becoming darker and redder, were due to the enrichment of phenolics on the wood surface. The Δb^* of wood after TM at 150 °C was 6–7 times that of the control sample [66]. In addition, besides the TM temperature and duration, Sundqvist et al. reported that the content of extractives (e.g., phenols, ketones and quinones) also has a strong influence on the color [41,65].

3.9. TG-FTIR Analysis

TGA-FTIR analysis makes it possible to investigate weight loss characteristics during biomass thermal degradation processes, as well as to identify the evolution of gas components in real time, especially for the small molecular weight bio-gas components (H_2O, CO_2, CO, and CH_4). Figure 9 shows the thermogravimetry (TG) and derivative thermogravimetry (DTG) curves of a wood thermal modification process with a heating rate of 10 °C min^{-1}. Based on the TG curves shown in Figure 9a, the residual mass decreased from 94.91% to 94.30% as the thermal modification temperature increased from 160 to 200 °C. This result indicates that higher thermal modification temperatures lead to higher mass loss in wood.

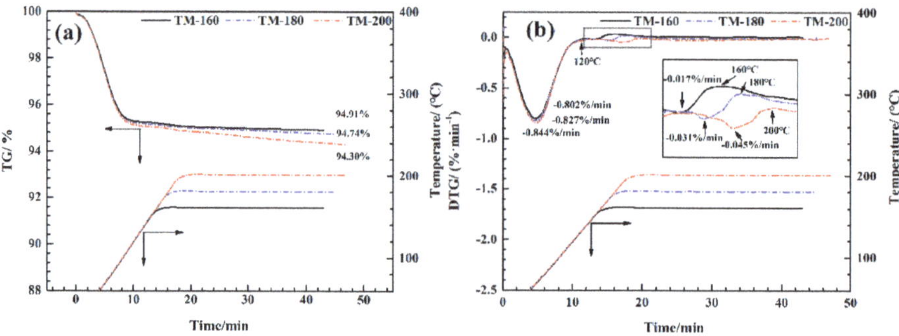

Figure 9. TG (**a**) and DTG (**b**) curves of thermal modification process the wood.

As shown in Figure 9b, two distinct mass loss peaks were observed. The first indicates the dehydration stage (30–120 °C), resulting from the evaluation of free and bound water in wood [17,39]. With the increase of TM temperature from 160 to 200 °C, the mass loss rate at the first peak was slightly increased, i.e., 0.802% to 0.844%/min. The second peak was formed by the thermal degradation of hemicellulose and cellulose in the wood [50,67]. Higher thermal modification temperatures resulted in a wider temperature range of mass loss and in higher mass loss rates. For example, the temperature range of this peak increased from 120–160 °C to 120–200 °C, and the mass loss rate increased from 0.017% to 0.045%/min. The mass lost in this stage mainly transferred into small molecular weight gaseous components (CO, CO_2, CH_4 and H_2O) and VOCs [3,50,52].

3.10. 3D–FTIR Analysis

Figure 10a–c shows the 3D-FTIR spectra of wood thermal degradation at three different temperatures (160, 180, and 200 °C). The intensity of absorbance significantly increased as the thermal modification temperature increased from 160 to 200 °C, indicating that higher temperatures promote the formation of evolved gas components. In order to identify the components of the evolved gas, a 2D-FTIR diagram with the wavenumber as the x-axis and the adsorbance intensity as the y-axis is presented (Figure 10d). Some permanent gaseous components could be easily identified according to their characteristic infrared absorbance bands, such as those of H_2O at 3735 cm^{-1}, CH_4 at 2938 cm^{-1}, CO_2 at 2358 cm^{-1}, and CO at 2181 cm^{-1} [49,50].

After the identification of components at characteristic infrared absorbance bands, the evolution of each detected gas component (H_2O, CH_4, CO_2, and CO) was determined; see Figure 11. According to the Lambert-Beer law, the intensity of the absorbance of a characteristic infrared absorbance band is linearly dependent on the concentration of the evolved gas components [68,69]. As shown in Figure 10, the intensities of the characteristic infrared absorbance bands of four gas components gradually increased with the increase of thermal modification temperature. Among the four gas components, H_2O had the highest absorbance intensity, i.e., 0.008, at a thermal modification temperature of 200 °C, followed by CH_4, i.e., 0.006, CO, 0.0058, and CO_2, about 0.0038.

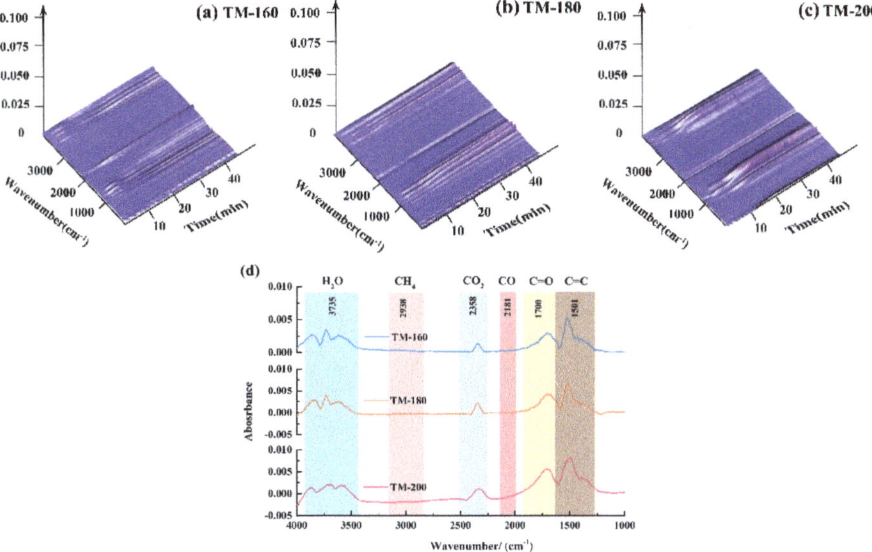

Figure 10. 3D-FTIR analysis of wood thermal modification at different temperatures: (**a**) 3D-FTIR of TM-160; (**b**) 3D-FTIR of TM-180; (**c**) 3D-FTIR of TM-200; (**d**) 2D-FTIR at the points of maximum weight loss from the three thermal modification samples.

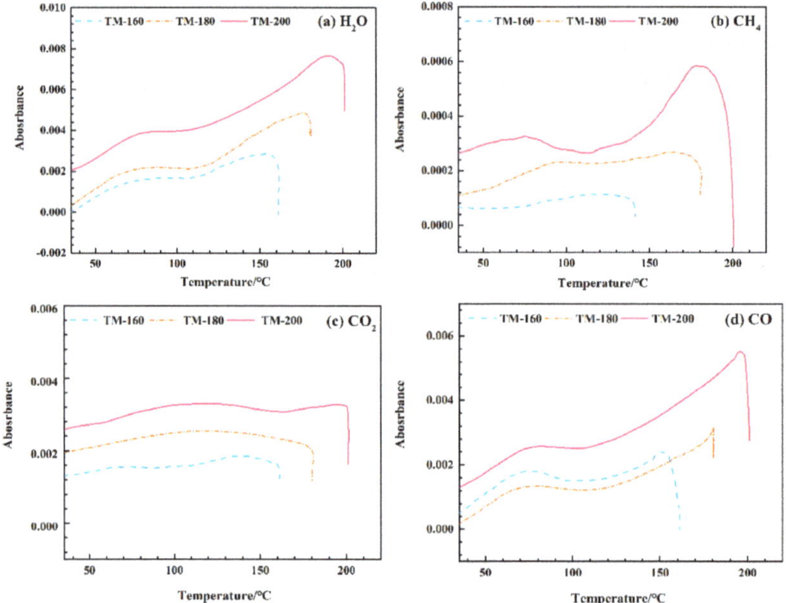

Figure 11. Evolution of evolved gas components during wood thermal modification at different temperatures: (**a**) H_2O; (**b**) CH_4; (**c**) CO_2; and (**d**) CO.

The release of H_2O with increasing TM temperature can be divided into two stages. At temperatures lower than 100 °C, the release is due to the evaporation of free water [17], while at higher temperatures, it is due to the breakage of hydroxyl groups linked to the glycosyl in hemicellulose [25]. The release of CO_2 can be mainly attributed by the decarbonylation and decarboxylation reactions of C=O and –COOH groups linked to the glucuronic acid units of hemicellulose [39,49,50]. The formation of CO was found to be mainly due to the cracking of carboxyl groups during ring-opening reactions of the glycosyl unit in hemicellulose [4,48,49]. The formation of CH_4 was also mainly due to breakages of methyl (–CH_3) and methylene (–CH_2–) linked to the lateral chain of the glycosyl unit in hemicellulose [39,63]. The acceleration of the thermal degradation of hemicellulose resulted in an increased release of these four gas components at higher thermal modification temperatures.

3.11. VOCs Analysis by PY-GC/MS

The VOCs released from wood TM were online detected by Py-GC/MS. The compounds and their relative contents are listed in Table 2. According to the characteristic functional groups, the VOCs may be divided into eight groups, namely acids (37.05–42.77%), aldehydes (11.67–18.99%), ketones (11.49–18.94%), phenols (9.6–15.56%), furans (11.54–16.67%), alcohols (3.09–5.2%), sugars (1.53–3.22%), and esters (1.25–2.16%). The effect of thermal modification temperature on the relative contents of these groups is shown in Figure 12. Several publications have reported that monoterpenes (e.g., α-pinene, β-pinene, camphene, limonene, and β-phellandrene) are common compounds in VOCs derived from the TM of softwoods, such as Pine and Fir [20,22,23,70]. However, these compounds were not detected in the TM process of hardwood (white oak). The difference might be caused by the different chemical compounds in the extracts of soft- and hard- wood.

Among the eight groups' chemicals, acids presented the highest relative content, i.e., 37.05–42.77%. The formation of acetic acid was mainly due to the thermal degradation of hemicellulose, which can occur via two pathways [35]. Firstly, acetic acid can be formed by the elimination of O–acetyl groups linked to xylan side chains at the C_2 position; secondly, it can occur by the ring-opening reaction of the 4–O–methylglucuronic acid unit after the breakage of carbonyl and O–methyl groups. Sundman et al. and Hyttinen et al. also found that acids were the dominant compounds in VOCs. Sundman et al. also found that the maximum emission of acids was 2800 $\mu g/(m^2 \cdot h)$ [19]. Hyttinen et al. reported that acetic acid reached its maximum emission rate, i.e., 170 $\mu g/(m^2 \cdot h)$ on the 28th day over a five-week testing period [24].

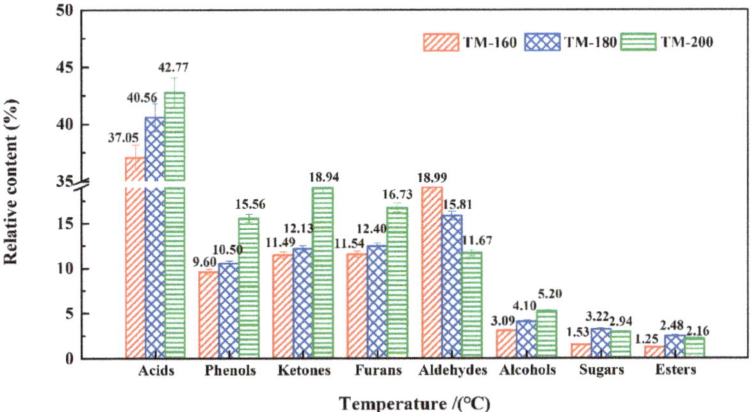

Figure 12. The relative contents of components in VOCs released from different thermal modification temperatures.

The relative content of acids gradually increased with an increase of thermal modification temperature. This result indicates that more hemicellulose was degraded at higher temperatures. It is worth noting that about 50% of the acids was composed of long-chain fatty acids, such as heptadecanoic acid (10.2–12.43%) and eicosanoic acid (15.08–15.7%). This result indicates that lower thermal degradation temperature promotes the formation of long-chain fatty acids.

The relative contents of phenols increased from 9.6% to 15.56% at higher TM temperatures. Phenols were produced by the thermal degradation of lignin [26,31]. As shown in Table 2, trans-isoeugenol and (E)-2,6-dimethoxy-4-(prop-1-en-1-yl) phenol are the two dominant components in phenols, with relative abundances of 2.23–2.26% and 3.74–5.77%, respectively. These two components could be directly obtained from the cleavage of β–O–4 linkages in the lignin [25].

With the increase of the thermal modification temperature, the relative contents of ketones and furans increased from 11.49% and 11.54 % to 15.94% and 16.73%, respectively, while the content of aldehydes decreased from 18.99% to 11.67%. The increase in furans and ketones was mainly attributed to ring-opening and depolymerization reactions of the basic structural units of glucan in cellulose, and glycosyl in hemicellulose [25,71]. Under lower TM temperatures, hemicellulose thermal degradation was the dominant process. However, as the TM temperature increased to over 200 °C, cellulose started to degrade, and therefore, higher TM temperatures promoted the formation of furans and ketones. Hyttinen et al. investigated the effect of heat treatment on the release content of VOCs, and found that furans were the major degradation products of hemicellulose, and that the content of furans gradually increased from 7 $\mu g/(m^2 \cdot h)$ on the second day to 37 $\mu g/(m^2 \cdot h)$ on the 28th day [24].

Table 2. Compounds and their relative contents in VOCs released from wood thermal modification at different temperatures.

Category	RT (min)	Compounds	Relative Content (Peak Area%)		
			TM-160	TM-180	TM-200
Acids	4.44	Acetic acid	1.53	3.32	5.51
	7.91	2-Hydroxy-6-methyl-3-cyclohexen-1-carboxylic acid	1.47	1.39	1.29
	9.09	Z-3-Methyl-2-hexenoic acid	0.91	0.56	0.78
	9.43	1,2-Dimethylcyclopropanecarboxylic acid	0.23	0.20	0.17
	9.79	(E)-3-Methyl-4-decenoic acid	0.31	0.21	0.16
	11.87	2-Hydroxy-6-methyl-3-cyclohexen-1-carboxylic acid	0.21	0.32	0.47
	12.50	(E)-3-Hexenoic acid	0.28	0.75	1.02
	12.92	3-Ethyl-3-methyl-pentanedioic acid	0.16	0.25	0.37
	16.25	Dodecanoic acid	1.27	1.10	/
	17.67	7-Methoxybenzofuran-2-carboxylic acid	0.88	0.71	0.79
	17.97	3,5-Dimethoxy-4-hydroxyphenylacetic acid	0.76	0.33	0.70
	18.49	Tetradecanoic acid	1.06	1.23	0.39
	19.53	3,5-Dimethoxy-4-hydroxyphenylacetic acid	0.00	0.44	0.31
	20.35	n-Hexadecanoic acid	0.00	0.30	0.33
	20.60	Heptadecanoic acid	10.2	11.32	12.43
	21.50	(Z,Z)-9,12-Octadecadienoic acid	0.16	0.26	/
	22.19	Oleic Acid	0.69	1.36	1.41
	22.23	Octadecanoic acid	1.23	1.43	1.32
	22.50	Eicosanoic acid	15.70	15.08	15.32
		Total	37.05	40.56	42.77
Phenols	13.11	2-Methoxy-4-vinylphenol	1.33	0.90	1.07
	13.62	2,6-Dimethoxy-phenol	0.29	0.33	0.37
	13.70	Eugenol	0.40	0.35	0.28
	14.91	trans-Isoeugenol	2.26	2.23	2.24
	16.78	2,6-Dimethoxy-4-(2-propenyl)-phenol	0.47	0.43	0.63
	17.90	(E)-2,6-Dimethoxy-4-(prop-1-en-1-yl) phenol	3.74	4.18	5.77
	18.70	Desaspidinol	0.61	0.71	1.34
	19.98	5-(3-Hydroxypropyl)-2,3-dimethoxyphenol	0.34	0.31	0.50
	27.14	3,5-bis(1,1-Dimethylethyl)-1,2-benzenediol	0.00	0.00	0.31
	31.97	2,6-bis(1,1-Dimethylethyl)-1,4-benzenediol	0.16	1.06	3.05
		Total	9.60	10.50	15.56
Aldehydes	14.28	Vanillin	3.13	2.33	0.59
	15.28	4-(t-Butyl)benzaldehyde	0.57	0.38	0.24
	17.48	4-Hydroxy-3,5-dimethoxy-benzaldehyde	3.71	2.97	1.83
	18.31	4-Hydroxy-2-methoxycinnamaldehyde	5.8	5.26	4.02
	20.86	3,5-Dimethoxy-4-hydroxycinnamaldehyde	5.78	4.87	4.99
		Total	18.99	15.81	11.67
Ketones	8.38	2,4-Hexanedione	7.50	8.18	15.71
	15.96	1-(4-Hydroxy-3-methoxyphenyl)-2-propanone	0.38	0.36	0.38
	16.33	3′,5′-Dimethoxyacetophenone	1.27	1.73	1.88
	16.63	1-(2-Hydroxy-4-methoxyphenyl)propan-1-one	1.98	1.56	0.34
	19.34	1-[2-(5-hydroxy-1,1-dimethylhexyl)-3-methyl-2-cyclopropen-1-yl]-ethanone	0.36	0.30	0.28
	24.41	3-Tridecanoyl-3-cyclohexen-4-ol-1-one	0.00	0.00	0.35
		Total	11.49	12.13	18.94
Furans	5.39	2(5H)-Furanone	0.00	0.00	0.40
	6.18	Furfural	8.45	9.34	13.41
	11.99	5,6-Dihydro-6-pentyl-2H-pyran-2-one,	0.68	0.76	0.89
	12.23	6-Ethoxy-3,6-dihydro-3-hydroxy-2H-pyran-2-methanol	0.28	0.34	0.45
	13.24	5-Butyldihydro-4-methyl-2(3H)-furanone	0.38	0.25	0.14
	17.76	5-(1-Hexynyl)-furan-2-carboxylic acid	0.00	0.20	0.22
	19.26	5-tert-Butyl-2-(4-tert-butylphenoxymethyl)-furane-3-carboxylic acid	1.75	1.51	1.22

Table 2. Cont.

Category	RT (min)	Compounds	Relative Content (Peak Area%)		
			TM-160	TM-180	TM-200
		Total	11.54	12.40	16.73
Alcohols	15.06	α-Ethyl-4-methoxy-benzenemethanol	0.76	0.22	0.29
	16.33	(2α,3α,4α)-2-(3,4-dimethoxyphenyl)-3,4-dihydro-6-methyl-2H-1-benzopyran-3,4-diol	1.27	0.73	1.88
	17.35	4-Hydroxy-3-methoxy-benzenepropanol	0.67	0.64	0.62
	27.786	(all-E)-(±)-2,6,10,15,19,23-hexamethyl-1,6,10,14,18,22-tetracosahexaen-3-ol	0.39	2.51	2.41
		Total	3.09	4.10	5.20
Sugars	11.16	2-Deoxy-D-galactose	0.91	0.66	0.44
	13.98	D-Allose	0.62	0.7	0.75
	15.45	1,6-Anhydro-β-D-glucopyranose	0.00	1.86	1.75
		Total	1.53	3.22	2.94
Esters	10.14	Heptamethylene diacetate	0.38	0.42	0.55
	12.23	Carbonic acid-but-2-yn-1-yl undecyl ester	0.28	0.38	0.45
	22.11	Heptadecanoic acid-16-methyl-methyl ester	0.38	0.29	0.11
	23.73	Octadecanoic acid-2-hydroxy-1,3-propanediyl ester	0.00	0.74	0.21
	25.68	Diisooctyl phthalate	0.00	0.30	0.51
	25.96	Cholesteryl formate	0.21	0.35	0.33
		Total	1.25	2.48	2.16

4. Conclusions

The connection between the evolution of both chemical structure and physical-mechanical properties during wood TM process, as well as the release characteristics of VOCs, were systematically investigated. The results indicated that the dimensional stability (e.g., anti-shrink efficiency, contact angle, equilibrium moisture content) improved markedly due to the reduction of hydrophilic hydroxyl (–OH). However, the mechanical properties (MOE and MOR) decreased after thermal modification due to the thermal degradation of hemicellulose and cellulose. Based on the TGA-FTIR analysis, the small molecular gaseous components were composed of H_2O, CH_4, CO_2, and CO, where H_2O was the dominant component with the highest absorbance intensity, i.e., 0.008 at 200 °C. Based on a Py-GC/MS analysis, the VOCs were mainly composed of acids, aldehydes, ketones, phenols, furans, alcohols, sugars, and esters, where acids were the dominant compounds, with relative contents of 37.05−42.77%.

Supplementary Materials: The following are available online at http://www.mdpi.com/2073-4360/11/7/1145/s1, Table S1: The P-level analysis of data from the physical-mechanical properties, Table S2: Mean and standard deviation values of mass loss of thermal modified wood, Table S3: Mean and standard deviation values of crystallinity index (CrI) of the control and thermal modified wood, Table S4: Mean and standard deviation values of ultimate analysis of the control and thermal modified wood, Table S5: Mean and standard deviation values of the EMC and ASE of the control and thermal modified wood, Table S6: Mean and standard deviation values for the MOE and MOR of the control and thermal modified wood, Table S7: Mean and standard deviation values of the surface color of the control and thermal modified wood.

Author Contributions: Conceptualization and experiment design, Z.M. and W.S.; material preparation, characterization, and performances test, J.X., Y.Z., Y.W., Y.S. and C.L.; original draft preparation, J.X. and Y.Z., review and editing, Z.M. and W.S.; data analysis and scientific discussion, all authors. J.X. and Y.Z. contributed equally to this work.

Funding: This research was funded by National Key R & D Plan of the "13th Five-Year" (2017YFD0601105), Natural Science Foundation of China (51706207), Natural Science Foundation of Zhejiang Province (LQ19E060009), the Young Elite Scientists Sponsorship Program by CAST (2018QNRC001), the Fund of Zhejiang Provincial Collaborative Innovation Center for Bamboo Resources and High-Efficiency Utilization (2017ZZY2-02), Undergraduate Research Training Program of Zhejiang A & F University (KX20180111), and National Undergraduate Innovation and Entrepreneurship Training Program.

Conflicts of Interest: The authors declare no conflict of interest.

References

1. Mahlberg, R.; Paajanen, L.; Nurmi, A.; Kivistö, A.; Koskela, K.; Rowell, R.M. Effect of chemical modification of wood on the mechanical and adhesion properties of wood fiber/polypropylene fiber and polypropylene/veneer composites. *Holz. Roh. Werkst.* **2001**, *59*, 319–326. [CrossRef]

2. Tomak, E.D.; Ustaomer, D.; Yildiz, S.; Pesman, E. Changes in surface and mechanical properties of heat treated wood during natural weathering. *Measurement* **2014**, *53*, 30–39. [CrossRef]

3. Boonstra, M.J.; Tjeerdsma, B. Chemical analysis of heat treated softwoods. *Holz. Roh. Werkst.* **2006**, *64*, 204–211. [CrossRef]

4. Sikora, A.; Kačík, F.; Gaff, M.; Vondrová, V.; Bubeníková, T.; Kubovský, I. Impact of thermal modification on color and chemical changes of spruce and oak wood. *J. Wood Sci.* **2018**, *64*, 406–416. [CrossRef]

5. Okon, K.E.; Lin, F.; Chen, Y.; Huang, B. Effect of silicone oil heat treatment on the chemical composition, cellulose crystalline structure and contact angle of chinese parasol wood. *Carbohydr. Polym.* **2017**, *164*, 179–185. [CrossRef] [PubMed]

6. Hill, C.A.S.; Abdul Khalil, H.P.S. Effect of fiber treatments on mechanical properties of coir or oil palm fiber reinforced polyester composites. *J. Appl. Polym. Sci.* **2000**, *78*, 1685–1697. [CrossRef]

7. Hakkou, M.; Pétrissans, M.; Gérardin, P.; Zoulalian, A. Investigations of the reasons for fungal durability of heat-treated beech wood. *Polym. Degrad. Stabil.* **2006**, *91*, 393–397. [CrossRef]

8. Hamada, J.; Pétrissans, A.; Mothe, F.; Ruelle, J.; Pétrissans, M.; Gérardin, P. Intraspecific variation of european oak wood thermal stability according to radial position. *Wood Sci. Technol.* **2017**, *51*, 785–794. [CrossRef]

9. Tjeerdsma, B.F.; Militz, H. Chemical changes in hydrothermal treated wood: Ftir analysis of combined hydrothermal and dry heat-treated wood. *Holz. Roh. Werkst.* **2005**, *63*, 102–111. [CrossRef]

10. Hill, C.A.S.; Ramsay, J.; Gardiner, B. Variability in water vapour sorption isotherm in japanese larch (larix kaempferi lamb.) – earlywood and latewood influences. *Inter. Wood Prod. J.* **2015**, *6*, 53–59. [CrossRef]

11. Rautkari, L.; Honkanen, J.; Hill, C.A.S.; Ridley-Ellis, D.; Hughes, M. Mechanical and physical properties of thermally modified scots pine wood in high pressure reactor under saturated steam at 120, 150 and 180 °C. *Eur. J. Wood Prod.* **2014**, *72*, 33–41. [CrossRef]

12. Rowell, R.M.; Ibach, R.E.; McSweeny, J.; Nilsson, T. Understanding decay resistance, dimensional stability and strength changes in heat-treated and acetylated wood. *Wood Mater. Sci. Eng.* **2009**, *4*, 14–22. [CrossRef]

13. Hill, C.A.S. Wood modification: Chemical, thermal and other processes. In *Wood modification*; Stevens, C.V., Ed.; John Wiley & Sons: Hoboken, NJ, USA, 2006; Volume 5, pp. 99–127.

14. Esteves, B.; Domingos, I.; Pereira, H. Improvement of technological quality of eucalypt wood by heat treatment in air at 170–200 °C. *For. Prod. J.* **2007**, *57*, 47–52.

15. Bal, B.C.; Bektaş, İ. The effects of heat treatment on some mechanical properties of juvenile wood and mature wood of eucalyptus grandis. *Dry Technol.* **2013**, *31*, 479–485. [CrossRef]

16. Candelier, K.; Dumarçay, S.; Pétrissans, A.; Gérardin, P.; Pétrissans, M. Comparison of mechanical properties of heat treated beech wood cured under nitrogen or vacuum. *Polym. Degrad. Stabil.* **2013**, *98*, 1762–1765. [CrossRef]

17. Lin, B.-J.; Colin, B.; Chen, W.-H.; Pétrissans, A.; Rousset, P.; Pétrissans, M. Thermal degradation and compositional changes of wood treated in a semi-industrial scale reactor in vacuum. *J. Anal. Appl. Pyrol.* **2018**, *130*, 8–18. [CrossRef]

18. Jensen, L.K.; Larsen, A.; Mølhave, L.; Hansen, M.K.; Knudsen, B. Health evaluation of volatile organic compound (voc) emissions from wood and wood-based materials. *Arch. Environ. Health* **2001**, *56*, 419–432. [CrossRef] [PubMed]

19. Risholmsundman, M.; Lundgren, M.; Vestin, E.; Herder, P. Emissions of acetic acid and other volatile organic compounds from different species of solid wood. *Holz. Roh. Werkst.* **1998**, *56*, 125–129. [CrossRef]

20. Makowski, M.; Ohlmeyer, M.; Meier, D. Long-term development of voc emissions from osb after hot-pressing. *Holzforschung* **2005**, *59*, 519. [CrossRef]

21. Liu, Y.; Shen, J.; Zhu, X.-D. Influence of processing parameters on voc emission from particleboards. *Environ. Monit. Assess.* **2010**, *171*, 249–254. [CrossRef]

22. Kačík, F.; Veľková, V.; Šmíra, P.; Nasswettrová, A.; Kačíková, D.; Reinprecht, L. Release of terpenes from fir wood during its long-term use and in thermal treatment. *Molecules* **2012**, *17*, 9990–9999. [CrossRef] [PubMed]

23. Manninen, A.M.; Pasanen, P.; Holopainen, J.K. Comparing the voc emissions between air-dried and heat-treated scots pine wood. *Atmos. Environ.* **2002**, *36*, 1763–1768. [CrossRef]

24. Hyttinen, M.; Masalinweijo, M.; Kalliokoski, P.; Pasanen, P. Comparison of voc emissions between air-dried and heat-treated norway spruce (picea abies), scots pine (pinus sylvesteris) and european aspen (populus tremula) wood. *Atmos. Environ.* **2010**, *44*, 5028–5033. [CrossRef]

25. Chen, D.; Gao, A.; Cen, K.; Zhang, J.; Cao, X.; Ma, Z. Investigation of biomass torrefaction based on three major components: Hemicellulose, cellulose, and lignin. *Energ. Convers. Manage.* **2018**, *169*, 228–237. [CrossRef]

26. Ma, Z.; Sun, Q.; Ye, J.; Yao, Q.; Chao, Z. Study on the thermal degradation behaviors and kinetics of alkali lignin for production of phenolic-rich bio-oil using tga–ftir and py–gc/ms. *J. Anal. Appl. Pyrol.* **2016**, *117*, 116–124. [CrossRef]

27. Zhang, Y.; Ma, Z.; Zhang, Q.; Wang, J.; Ma, Q.; Yang, Y.; Luo, X.; Zhang, W. Comparison of the physicochemical characteristics of bio-char pyrolyzed from moso bamboo and rice husk with different pyrolysis temperatures. *BioResources* **2017**, *12*, 4652–4669. [CrossRef]

28. Ma, Z.; Chen, D.; Jie, G.; Bao, B.; Zhang, Q.; Ma, Z.; Chen, D.; Jie, G.; Bao, B.; Zhang, Q. Determination of pyrolysis characteristics and kinetics of palm kernel shell using tga–ftir and model-free integral methods. *Energ. Convers. Manage.* **2015**, *89*, 251–259. [CrossRef]

29. Chen, D.; Gao, A.; Ma, Z.; Fei, D.; Chang, Y.; Shen, C. In-depth study of rice husk torrefaction: Characterization of solid, liquid and gaseous products, oxygen migration and energy yield. *Bioresour. Technol.* **2018**, *253*, 148–153. [CrossRef]

30. Chen, D.; Mei, J.; Li, H.; Li, Y.; Lu, M.; Ma, T.; Ma, Z. Combined pretreatment with torrefaction and washing using torrefaction liquid products to yield upgraded biomass and pyrolysis products. *Bioresour. Technol.* **2017**, *228*, 62–68. [CrossRef]

31. Ma, Z.; Wang, J.; Zhou, H.; Zhang, Y.; Yang, Y.; Liu, X.; Ye, J.; Chen, D.; Wang, S. Relationship of thermal degradation behavior and chemical structure of lignin isolated from palm kernel shell under different process severities. *Fuel Process Technol.* **2018**, *181*, 142–156. [CrossRef]

32. Chen, D.; Cen, K.; Jing, X.; Gao, J.; Li, C.; Ma, Z. An approach for upgrading biomass and pyrolysis product quality using a combination of aqueous phase bio-oil washing and torrefaction pretreatment. *Bioresour. Technol.* **2017**, *233*, 150–158. [CrossRef] [PubMed]

33. Ma, Z.; Yang, Y.; Ma, Q.; Zhou, H.; Luo, X.; Liu, X.; Wang, S. Evolution of the chemical composition, functional group, pore structure and crystallographic structure of bio-char from palm kernel shell pyrolysis under different temperatures. *J. Anal. Appl. Pyrol.* **2017**, *127*, 350–359. [CrossRef]

34. Qiang, L.; Yang, X.C.; Dong, C.Q.; Zhang, Z.F.; Zhang, X.M.; Zhu, X.F. Influence of pyrolysis temperature and time on the cellulose fast pyrolysis products: Analytical py-gc/ms study. *J. Anal. Appl. Pyrol.* **2011**, *92*, 430–438.

35. Shen, D.K.; Gu, S.; Bridgwater, A.V. Study on the pyrolytic behaviour of xylan-based hemicellulose using tg–ftir and py–gc–ftir. *J. Anal. Appl. Pyrol.* **2010**, *87*, 199–206. [CrossRef]

36. Wang, S.; Ru, B.; Dai, G.; Sun, W.; Qiu, K.; Zhou, J. Pyrolysis mechanism study of minimally damaged hemicellulose polymers isolated from agricultural waste straw samples. *Bioresour. Technol.* **2015**, *190*, 211–218. [CrossRef] [PubMed]

37. Ma, Z.; Wang, J.; Li, C.; Yang, Y.; Liu, X.; Zhao, C.; Chen, D. New sight on the lignin torrefaction pretreatment: Relevance between the evolution of chemical structure and the properties of torrefied gaseous, liquid, and solid products. *Bioresour. Technol.* **2019**, *288*, 121528. [CrossRef] [PubMed]

38. Segal, L.; Creely, J., Jr.; Martin, A.E.J.; Conrad, C. An empirical method for estimating the degree of crystallinity of native cellulose using the x-ray diffractometer. *Text Res. J.* **1959**, *29*, 786–794. [CrossRef]

39. Ma, Z.; Wang, J.; Yang, Y.; Zhang, Y.; Zhao, C.; Yu, Y.; Wang, S. Comparison of the thermal degradation behaviors and kinetics of palm oil waste under nitrogen and air atmosphere in tga-ftir with a complementary use of model-free and model-fitting approaches. *J. Anal. Appl. Pyrol.* **2018**, *134*, 12–24. [CrossRef]

40. Cen, K.; Zhang, J.; Ma, Z.; Chen, D.; Zhou, J.; Ma, H. Investigation of the relevance between biomass pyrolysis polygeneration and washing pretreatment under different severities: Water, dilute acid solution and aqueous phase bio-oil. *Bioresour. Technol.* **2019**, *278*, 26–33. [CrossRef]

41. Srinivas, K.; Pandey, K.K. Effect of heat treatment on color changes, dimensional stability, and mechanical properties of wood. *J. Wood Chem. Technol.* **2012**, *32*, 304–316. [CrossRef]

42. Wang, X.; Chen, X.; Xie, X.; Wu, Y.; Zhao, L.; Li, Y.; Wang, S. Effects of thermal modification on the physical, chemical and micromechanical properties of masson pine wood (pinus massoniana lamb.). *Holzforschung* **2018**, *72*, 1063. [CrossRef]

43. Ma, Z.; Yang, Y.; Wu, Y.; Xu, J.; Peng, H.; Liu, X.; Zhang, W.; Wang, S. In-depth comparison of the physicochemical characteristics of bio-char derived from biomass pseudo components: Hemicellulose, cellulose, and lignin. *J. Anal. Appl. Pyrol.* **2019**, *140*, 195–204. [CrossRef]

44. Yin, J.; Yuan, T.; Lu, Y.; Song, K.; Li, H.; Zhao, G.; Yin, Y. Effect of compression combined with steam treatment on the porosity, chemical compositon and cellulose crystalline structure of wood cell walls. *Carbohyd. Polym.* **2017**, *155*, 163–172. [CrossRef] [PubMed]

45. Akgül, M.; Gümüşkaya, E.; Korkut, S. Crystalline structure of heat-treated scots pine (pinus sylvestris l.) and uludag fir [abies nordmanniana (stev.) subsp. Bornmuelleriana (mattf.)] wood. *Wood Sci. Technol.* **2007**, *41*, 281–289. [CrossRef]

46. Bhuiyan, M.T.R.; Hirai, N.; Sobue, N. Changes of crystallinity in wood cellulose by heat treatment under dried and moist conditions. *J. Wood Sci.* **2000**, *46*, 431–436. [CrossRef]

47. Esteves, B.; Marques, A.V.; Domingos, I.; Pereira, H. Chemical changes of heat treated pine and eucalypt wood monitored by ftir. *Maderas. Cienc. y Tecnol.* **2013**, *15*, 245–258. [CrossRef]

48. González-Peña, M.M.; Curling, S.F.; Hale, M.D.C. On the effect of heat on the chemical composition and dimensions of thermally-modified wood. *Polym. Degrad. Stabil.* **2009**, *94*, 2184–2193. [CrossRef]

49. Popescu, M.-C.; Froidevaux, J.; Navi, P.; Popescu, C.-M. Structural modifications of tilia cordata wood during heat treatment investigated by ft-ir and 2d ir correlation spectroscopy. *J. Mol. Struct.* **2013**, *1033*, 176–186. [CrossRef]

50. Yang, H.; Rong, Y.; Chen, H.; Dong, H.L.; Zheng, C. Characteristics of hemicellulose, cellulose and lignin pyrolysis. *Fuel* **2007**, *86*, 1781–1788. [CrossRef]

51. Tjeerdsma, B.F.; Boonstra, M.; Pizzi, A.; Tekely, P.; Militz, H. Characterisation of thermally modified wood: Molecular reasons for wood performance improvement. *Holz. Roh. Werkst.* **1998**, *56*, 149–153. [CrossRef]

52. Cademartori, P.H.G.; dos Santos, P.S.B.; Serrano, L.; Labidi, J.; Gatto, D.A. Effect of thermal treatment on physicochemical properties of gympie messmate wood. *Ind. Crop Prod.* **2013**, *45*, 360–366. [CrossRef]

53. Gérardin, P.; Petrič, M.; Petrissans, M.; Lambert, J.; Ehrhrardt, J.J. Evolution of wood surface free energy after heat treatment. *Polym. Degrad. Stabil.* **2007**, *92*, 653–657. [CrossRef]

54. Chaouch, M.; Dumarçay, S.; Pétrissans, A.; Pétrissans, M.; Gérardin, P. Effect of heat treatment intensity on some conferred properties of different european softwood and hardwood species. *Wood Sci. Technol.* **2013**, *47*, 663–673. [CrossRef]

55. Huang, S.; Ma, Z.; Nie, Y.; Lu, F.; Ma, L. Comparative study of the performance of acetylated bamboo with different catalysts. *Bioresources* **2018**, *14*, 44–57.

56. Bakar, B.F.; Hiziroglu, S.; Md Tahir, P. Properties of some thermally modified wood species. *Mater. Des.* **2013**, *43*, 348–355. [CrossRef]

57. Lee, C.-H.; Yang, T.-H.; Cheng, Y.-W.; Lee, C.-J. Effects of thermal modification on the surface and chemical properties of moso bamboo. *Constr. Build. Mater.* **2018**, *178*, 59–71. [CrossRef]

58. Kamdem, D.P.; Pizzi, A.; Jermannaud, A. Durability of heat-treated wood. *Holz Roh Werkst* **2002**, *60*, 1–6. [CrossRef]

59. Ayrilmis, N.; Jarusombuti, S.; Fueangvivat, V.; Bauchongkol, P. Effect of thermal-treatment of wood fibres on properties of flat-pressed wood plastic composites. *Polym. Degrad. Stabil.* **2011**, *96*, 818–822. [CrossRef]

60. Kačíková, D.; Kačík, F.; Čabalová, I.; Ďurkovič, J. Effects of thermal treatment on chemical, mechanical and colour traits in norway spruce wood. *Bioresour. Technol.* **2013**, *144*, 669–674. [CrossRef]

61. Boonstra, M.J.; Acker, J.V.; Tjeerdsma, B.F.; Kegel, E.V. Strength properties of thermally modified softwoods and its relation to polymeric structural wood constituents. *Ann. For. Sci.* **2007**, *64*, 679–690. [CrossRef]

62. Guo, F.; Huang, R.; Lu, J.; Chen, Z.; Cao, Y. Evaluating the effect of heat treating temperature and duration on selected wood properties using comprehensive cluster analysis. *J. Wood Sci.* **2014**, *60*, 255–262. [CrossRef]

63. Yildiz, S.; Tomak, E.D.; Yildiz, U.C.; Ustaomer, D. Effect of artificial weathering on the properties of heat treated wood. *Polym. Degrad. Stabil.* **2013**, *98*, 1419–1427. [CrossRef]

64. Gunduz, G.; Aydemir, D.; Karakas, G. The effects of thermal treatment on the mechanical properties of wild pear (pyrus elaeagnifolia pall.) wood and changes in physical properties. *Mater. Des.* **2009**, *30*, 4391–4395. [CrossRef]

65. Sundqvist, B.; Karlsson, O.; Westermark, U. Determination of formic-acid and acetic acid concentrations formed during hydrothermal treatment of birch wood and its relation to colour, strength and hardness. *Wood Sci. Technol.* **2006**, *40*, 549–561. [CrossRef]

66. Bekhta, P.; Niemz, P. Effect of high temperature on the change in color, dimensional stability and mechanical properties of spruce wood. *Holzforschung* **2003**, *57*, 539–546. [CrossRef]

67. Chen, W.H.; Peng, J.; Bi, X.T. A state-of-the-art review of biomass torrefaction, densification and applications. *Renew. Sustain. Energy Rev.* **2015**, *44*, 847–866. [CrossRef]

68. Wang, S.; Guo, X.; Wang, K.; Luo, Z. Influence of the interaction of components on the pyrolysis behavior of biomass. *J. Anal. Appl. Pyrol.* **2011**, *91*, 183–189. [CrossRef]

69. Weiland, J.J.; Guyonnet, R. Study of chemical modifications and fungi degradation of thermally modified wood using drift spectroscopy. *Holz. Roh. Werkst.* **2003**, *61*, 216–220. [CrossRef]

70. Banerjee, S.; Su, W.; Wild, M.P.; Otwell, L.P.; Hittmeier, M.E.; Nichols, K.M. Wet line extension reduces vocs from softwood drying. *Environ. Sci. Technol.* **1998**, *32*, 1303–1307. [CrossRef]

71. Werner, K.; Pommer, L.; Broström, M. Thermal decomposition of hemicelluloses. *J. Anal. Appl. Pyrol.* **2014**, *110*, 130–137. [CrossRef]

MDPI

St. Alban-Anlage 66

4052 Basel

Switzerland

Tel. +41 61 683 77 34

Fax +41 61 302 89 18

www.mdpi.com

Polymers Editorial Office

E-mail: polymers@mdpi.com

www.mdpi.com/journal/polymers